On the track of Ice Age mammals

ON THE TRACK OF

Antony J Sutcliffe

Harvard University Press
Cambridge, Massachusetts
1985

ICE AGE MAMMALS

Title page: *The edge of the Greenland ice sheet, Kronprins Christians Land; outwash area with braided streams and patterned ground in the foreground. During the Ice Age similar conditions prevailed far beyond their present limits, for example across the United States of America and Europe, including England (see map, fig. 2.6, p. 19). (Photo: Geodaetisk Institut, Copenhagen).*

10 9 8 7 6 5 4 3 2 1

Printed in Great Britain

Library of Congress Cataloging-in-Publication Data
Sutcliffe, Antony John
On the track of Ice Age Mammals.

Bibliography: p. 212
Includes index.
1. Mammals, Fossil. 2. Paleontology—Quaternary.
3. Glacial epoch. I. Title.
QE881.S96 1985 569 85–24805
ISBN 0–674–63777–1

Contents

Preface

The Ice Age has long held a fascination for man and how fortunate it is for us that most of its ice has subsequently melted, allowing habitation and agriculture over much of the area previously covered. But, what if the ice should return and the zones of agriculture and habitation be correspondingly reduced? The human population has increased by millions since the waning of the ice sheets that introduced the present phase of genial climate and a relatively sudden contraction of habitable land could lead to the most catastrophic economic and political consequences. It is therefore very much in man's interest to be able to predict and be prepared for whatever the climate may hold for him in the future.

Although it has long been recognized that the Ice Age was a complex event, with the ice sheets advancing and shrinking more than once, only more recently have scientific studies demonstrated what a large number of glacial cycles actually occurred, how short usually were the warm episodes between them and how suddenly the cooling and warming often took place. The warm phase in which we are living now has already lasted 10 000 years. How much longer do we have?

It is not surprising therefore that Ice Age related studies have taken on a new significance and the need, in particular, for a multi-disciplinary approach has received increasingly wide recognition. In this book, which I have written as a mammalian palaeontologist, I have attempted to show how a specialist in one field must try to follow the researches of specialists working in other disciplines if he is to get the most from his own line of study. In return he may hopefully contribute to the general story of the Ice Age with evidence which is unique to his own.

Although discussions relate to fossil mammalian discoveries of world-wide distribution, the book does not set out to be a comprehensive treatise on the Ice Age mammalian faunas of all the continents of the world. It is written primarily for the serious reader with little previous knowledge of the subject, but who would like to be able to follow the way in which investigations are undertaken. It is hoped that there may be something in it for the specialist also. For those seeking more detailed information a selected list of further reading follows chapter 14 (mostly the publications are from the last 15 years although the earliest goes back to 1697), with additional lists of references available from the bibliographies of the books and papers cited.

The British Isles, situated on the eastern shore of the North Atlantic, lie within the zone of climatic influence of the constantly migrating 'polar front', an east–west line of abrupt sea temperature change where warm currents from the south meet polar currents. They have been chosen for slightly more detailed examination than the other geographic areas, because they show such extraordinary contrasts in their mammalian faunal sequence.

baby mammoth (fig. 1.1) was a sensation in both popular and scientific circles throughout the world.

So far we have considered such mammals mainly for their popular appeal. But they also have an important role in the more practical aspect of the study of the Ice Age. It has long been recognized that this event was not an isolated incident but that the great ice sheets and glaciers of the high latitudes and high altitudes advanced far beyond their present limits and shrank again repeatedly. Between the glacial advances were temperate phases broadly known in mid latitudes as

Fig. 1.1 *Carcass of a baby woolly mammoth, found on 23 June, 1977, in frozen ground near the River Kirgilyakh, Siberia. The baby was 115 cm long, 104 cm high, had a trunk 57 cm long and was six or seven months old when it died. In life it had a hairy coat, most of which has fallen out since death, though the part covering the feet is still intact. The carcass, preserved by freezing for over 40 000 years, is now displayed in Leningrad Museum. (Photo: USSR Academy of Sciences).*

interglacials, as warm as or warmer than at the present day. Thirty years ago, it was generally believed that there had been only four or five major glacial advances during the Quaternary; though recent studies (especially the evidence provided by cores of sediment obtained from the sea bottom) have now shown this to be an underestimate and that there may have been as many as seventeen cold – warm cycles, with an average duration of only 100 000 years. Superimposed on these major climatic fluctuations were many minor oscillations of shorter duration.

We are in an interglacial, which began about 10 000 years ago, now. Often this period of time has been called the 'Post-Glacial', implying that the Ice Age had ended and that we have now entered a stable period of more congenial climate. But this is clearly not the case. The climate is still changing all the time. A human lifetime is too short for anyone to witness a complete change from an interglacial to a glacial event, or vice versa, but the evidence for a minor climatic change is plain to see. Existing glaciers still wane or increase in size whilst elsewhere fluctuations of rainfall, causing changes in the extent of desert areas, are all part of the same process of climatic change.

Fig. 1.2 *The woolly mammoth still holds a special place in the popular imagination. (With acknowledgements to* Punch*).*

Before the advent of the deep sea studies it had generally been believed that the few recognized interglacials had been of very long duration, some of them perhaps as much as 200 000 years. There was little need for man to worry about the present interglacial coming to an end. The deep sea evidence, however, showed not only that there had been many more glacial episodes than had previously been accepted, but that by geological standards climatic changes had sometimes been extremely rapid; furthermore that glacial or cold conditions had prevailed during a much greater part of Quaternary time than interglacial conditions, which were relatively abnormal; individual interglacials seldom having a duration of more than about 10–15 000 years. We will examine these events in greater detail later.

What does the climate hold for us in the future, and what will be its effect upon the world's most numerous mammal – man? Will we soon enter another glacial episode; how would man re-organize himself in face of such an event? What would happen to the inhabitants of such low-lying cities as London if the remaining water locked up in the world's ice sheets were to return in consequence of further melting to the sea and submerge their homes? Even quite small fluctuations of climate can have far-reaching effects on man, especially on his food supply (fig. 1.3). Within historic times the minor cold period that has become known as the 'Little Ice Age', which lasted from about AD 1500–1850, had disastrous effects on some human communities, especially in Europe, Iceland and Greenland, as increased frost caused crop failures, starvation and emigration.

Large areas of the world at high latitudes, including nearly half the USSR, have since the Ice Age been underlain by deeply and permanently frozen ground known as permafrost, which creates special engineering problems when buildings, roads and bridges are constructed upon it. A rise of temperature sufficient to cause even local melting would have far reaching economic repercussions in those countries.

Droughts can be quite as disastrous for man as freezing or thawing. The great drought that turned parts of the United States and Canadian prairies into a dust bowl between 1928 and 1937 caused the abandonment of farms which had taken years to establish. As recently as the beginning of the last decade, 100 000 people and millions of animals died as a result of loss of pastures along the southern margin of the Sahara Desert. Unhappily the drought was renewed in 1983, and, at the time that this book goes to press in 1985, there is widespread abandonment of agricultural land resulting from successive crop failures, especially in northern Ethiopia, with thousands of destitute farmers and their families walking long distances to

The Big Snow grips America from the great plains to the Gulf of Mexico

Express Jan 11. 1982

It's colder than the South Pole!

By JOHN CHRISTOPHER and NORMAN LUCK

BRITAIN was colder yesterday than the South Pole.

While Antarctic penguins basked in −21 Centigrade, British upper lips were stiffened by several degrees below even that.

Crash boy's five days in ice tomb

BLIZZARDS PARALYSE BRITAIN

Brazil fears drought will kill millions

From Alan McGregor, Geneva

America's coldest winter to date claims 32 lives

By IAN BALL in New York

Drought brings famine threat to one million Ethiopians

From Charles Harrison, Nairobi

Death amid the cold hits the old

AN exceptionally high number of old people died over Christmas—many, it is thought, because of the cold weather.

D. TELEGRAPH 5.10.84

Devastating damage feared as the Earth gets hotter

By Pearce Wright Science Editor

TIMES 22.10.83

Second catastrophe threatens Sahel

By Tony Samstag

Worst winter for 50 years in US

From Trevor Fishlock, New York

Science report

Four more years of drought predicted

By the Staff of Nature

Food shortage in Ethiopia may cost 500,000 lives

By DAVID ADAMSON Diplomatic Correspondent

Fig. 1.3 *Some newspaper headlines, 1980–1985, reporting episodes of extreme weather in various parts of the world and forecasts of climatic change. If a short term cold spell or drought can cause so much distress and financial loss, what would be consequences of a more deep-seated change of climate?*

reach refugee camps (some of them in neighbouring Sudan), where they are totally dependent on relief aid. On the way and after arrival, deaths caused by starvation and famine-related diseases reach new peaks. Against a background of world-wide appeals for famine relief the drought was described by the United Nations Emergency Aid Chief as 'potentially the biggest catastrophe humanity has ever known' (fig. 1.4). Many other aspects of man's economy, including fuel consumption, are also climate-related.

Man's continued orderly existence depends, in no small measure, on his ability to understand the factors controlling the world's climate and to be able to anticipate what the future holds for him. Climatic change has no respect for political boundaries and, if the food-growing area on which a population has been dependent shifts from one country to another, those concerned could find themselves unable to follow and without adequate water and food. Man's already determined attempts to maintain the quality of life for the earth's inhabitants by slowing the population explosion can be of only limited value if based on the assumption that future generations will find the world's climate and geographical zonation the same as they are today.

So important has man's ability to predict future changes of climate become that all aspects of climatology – present, past and future – are now subjects of intense study by scientists in a great diversity of disciplines throughout the world. It is in this context that the events of the Ice Age take on special importance, since the most detailed record available of the changes of climate that have occurred in the past is a fundamental prerequisite to an understanding of what may occur in the future. The climatologist cannot simulate long periods of time in his experiments, but he can refer to the historical and geological record instead. He can obtain information from various sources. Meteorological records contain precise details of climatic events during the latest period; historical accounts of crop failures, of the state of glaciers, of rivers being frozen in winter, of droughts and famines and of dates when great civilizations flourished in now desert areas take us far back into historical time. But for earlier information we have only the geological record to which to look for evidence. Here a vast fund of information is steadily being assembled, and some of this has been provided by the mammalian palaeontologist. From his understanding of the changes of mammalian faunas in the various parts of the world, in response to climatic change, he contributes to the data required by the climatologist.

There are other aspects of man's survival where the study of Ice Age mammals can be of direct application. Can anything be learned from palaeoecological studies of the mammals that populated the various vegetational zones of the world during the Ice Age that would help improve man's future food supply? How, for example, did some of the arctic regions support such a vast mammalian biomass (including woolly mammoth, woolly rhinoceros, horse, bison and lion) at the end of Ice Age, whereas an equivalent weight of mammals could not find sustenance there at the present day?

In such studies the Ice Age mammalogist cannot work in isolation but must also be aware of what is going on in other fields. Without the palaeobotanist to look at the contents of mammoths' stomachs, how would he know what these animals ate or what climate prevailed when the food plants were growing; without the geochemist, how old such remains are; or without the archaeologist, whether man was involved in the mammoth's extinction?

In this book we will look at some examples of Ice Age mammalian studies, chosen to illustrate man's changing attitude to this subject over the centuries and how the mammalogist works today, in collaboration with other specialists, in the interpretation of his evidence.

Fig. 1.4 *The tragedy of the great mid-twentieth-century drought in the Sahel. Above, carcass of a camel, Niger. (Photo: Oxfam) and below, human distress in Ethiopia. (Photo: Mike Wells, Save the Children).*

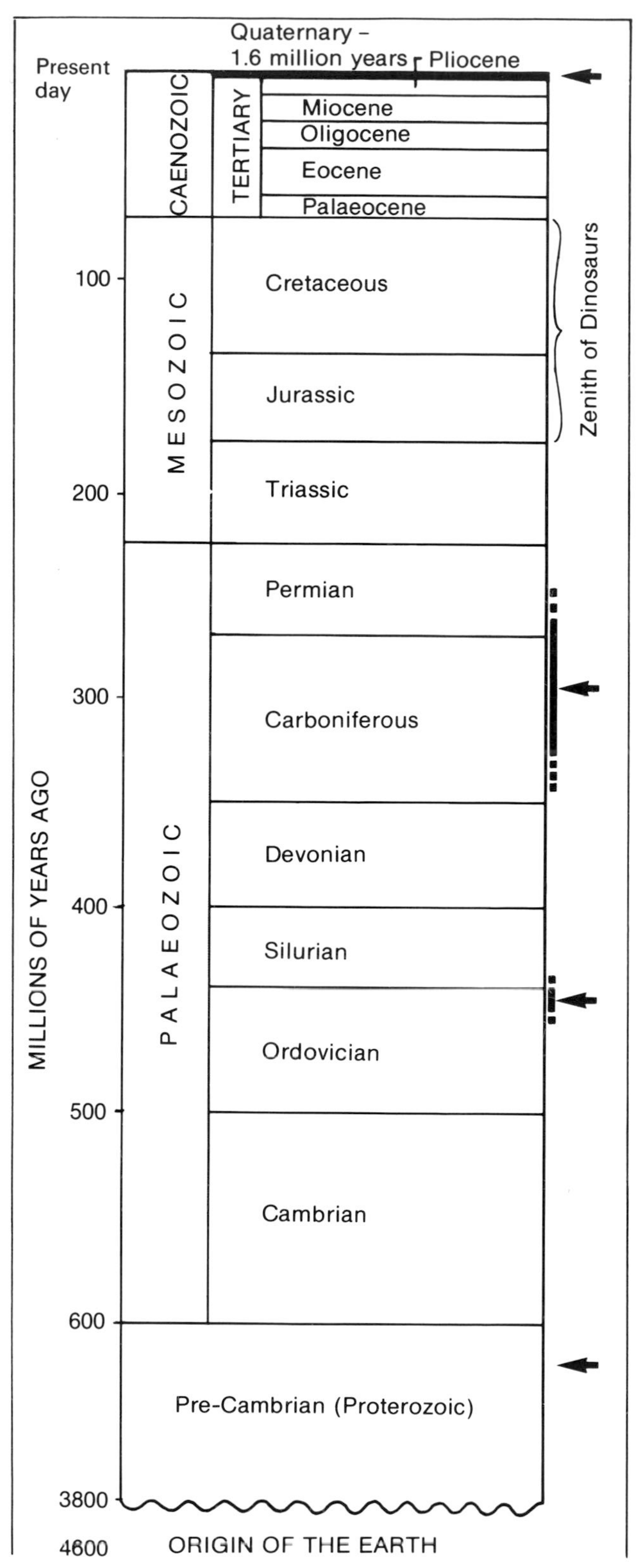

Fig. 2.1 *Geological time scale, showing major periods of glaciation (arrowed) since the late Pre-Cambrian.*

Chapter 2 **What was the Ice Age?**

As we have already seen, any portrayal of the Ice Age as a simple event would be very misleading. Let us look at some of the evidence more closely, in order to establish a broader picture. Firstly, how many glacial episodes have occurred; and what was their position in geological time? If more than one, then what is implied by the term '*The Ice Age*'? It has long been recognized that deposits that can only be interpreted as being of glacial origin occur in sedimentary strata of diverse ages intermittently throughout the geological column, from early times up to the present day. Ice ages are apparently not a unique phenomenon, but occur repeatedly (fig. 2.1).

Of the various glacial episodes the most recent are the most easy to interpret, since the evidence is most perfect and the extent of the ice covered areas can be compared with that of the surviving ice centres of the present day. The Quaternary can readily be described as an ice age, or series of ice ages, since there was an expansion of the world's ice sheets to many times their present size during this period.

The interpretation of glacial evidence from the earlier geological periods is less simple. The occurrence, in the Sahara Desert, of glacial deposits of Ordovician age – 450–500 million years old – might seem surprising, yet there can be no doubt about their origin. Striated rock pavements, and many other readily recognizable glacial features are constantly being exposed by denudation in an area, not far from the present equator, where summer temperatures today sometimes exceed 70°C! If such a glaciation were the result of increase in the size of the polar ice caps, as we know them today, then it would be necessary to postulate that, in Ordovician times, the world had been ice covered from pole to pole, which seems very unlikely. There are, however, other features of the evolution of the earth's geography that must also be taken into consideration: changes in the position of the earth's axis of rotation, and consequently of its poles (fig. 2.2); and changes in the position of the continents relative to one another (continental drift, see chapter 13). Where glacial deposits occur at low latitudes consideration must be given to the effects of both actual expansion of the world's ice cover, as occurred, for example during the late Ordovician and Permo-Carboniferous, and to the wandering of continents by continental drift. The relative northward movement of Africa in relation to the south pole, for example, can be demonstrated by the migration of glacial conditions from North Africa in the Upper Ordovician to South Africa during the Carboniferous and Australia during the Permian. Allowing for the known magnitude of such changes since Ordovician times, Saharan glaciation becomes entirely feasible.

Having observed that there were also earlier glacial episodes, let us now devote all our attention to the last of them, that which began to build up at the end of the Tertiary (during the later part of the Pliocene, about three million years ago; in South America there is also evidence of a glacial episode some time between 7 and 4·6 million years ago) and which continued throughout the Quaternary. It is this series of glacial and interglacial fluctuations that is most usually referred to as '*The Ice Age*'. Since it all happened so recently the evidence is still very fresh. So recent is it, indeed, by geological standards that many geologists regard its study as barely a geological concern at all. Broadly speaking we are still in the Ice Age today.

For convenience the Quaternary (also known in the USSR as Anthropogene, implying the period of time that man has been present on earth) is sub-divided into two parts, the Pleistocene (from about 1·6 million to 10 000 years ago); and the Holocene, Post-Glacial or Recent, continuing to the present day. Special names are given to the various human cultures of the Quaternary. In Europe, Palaeolithic or Old Stone Age man lived during the Pleistocene; with Mesolithic (Middle Stone Age), Neolithic (New Stone Age) and later cultures following during the Holocene. Pleisto-

Fig. 2.2 *Map showing the apparent migration of the south pole and polar ice cap from the Ordovician, about 500–450 million years ago, to late Carboniferous, about 275 million years ago, on the super-continent of Gondwanaland (see also chapter 13).*

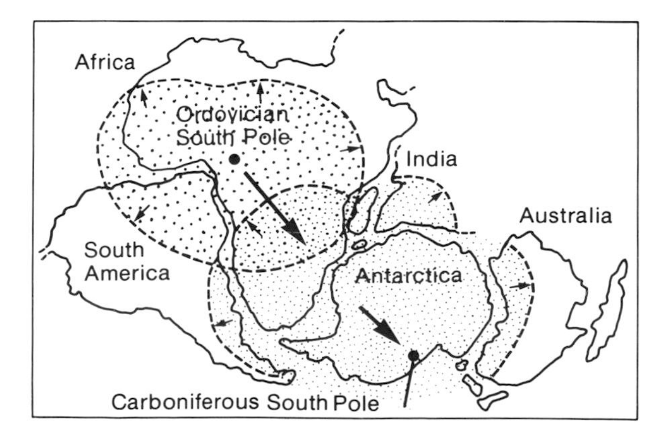

Present day			Industries (Europe)
	QUATERNARY	HOLOCENE POST-GLACIAL or RECENT	BRONZE AGE and later industries
			NEOLITHIC
10 000 years ago →			MESOLITHIC
about 1·6 million years ago →		PLEISTOCENE	PALAEOLITHIC (beginning in the Pliocene in Africa)
	TERTIARY	PLIOCENE	

Table 1 *Relationship between archaeological industries of Europe and the Quaternary time scale*

cene and Palaeolithic are not wholly synonymous words, however, since the various human industries developed at different rates in different parts of the world. In Africa, the earliest man-made tools are apparently of Pliocene age.

Studies of the Quaternary can be followed through a series of stages related to the philosophical attitudes and available scientific techniques of the times. In the western world the earliest workers, influenced by the Biblical account of the Creation in the *Book of Genesis* and with only primitive scientific methods available to them, sought only a few thousands or tens of thousands of years to cover all the events back to the creation of the earth; although, as scientific methods improved, so did arguments develop for a longer span of time. Fossil mammalian remains were commonly attributed to animals that had been drowned by the Biblical Deluge.

From the 1830s the theory of a world-wide deluge was rapidly superceded by the acceptance of a single ice age and, by 1860, excavations in France and England had demonstrated man's coexistence with mammals previously regarded as ante-diluvian. By the end of the nineteenth century it had become apparent, from studies in Europe, America and Asia that there had been more than one glacial episode during the Pleistocene; and a chronology of four glaciations and three interglacials, which was widely accepted until well into the 1950s, became firmly entrenched in the literature. The value of these climatic fluctuations for dating purposes was quickly realized, and this glacial – interglacial sequence became the yardstick against which Pleistocene events were most commonly measured. Such dates were nevertheless only relative, and there was still no means of determining the actual number of years involved.

It was not until the 1950s, with the advent of carbon^{14} and other methods of radiometric dating, that it began to be possible to obtain ages for the divisions of the Pleistocene in actual number of years before the present. Previously attempts to establish a detailed Pleistocene chronology had been based mainly on the study of terrestrial deposits and the fossils in them. An inherent problem of working with such evidence has always been the fragmentary nature of the deposits available for study. The period of time represented at any one locality is usually quite short, the time relationship between the deposits of different localities is often obscure, and it is difficult to recognize periods of time unrepresented by any deposits. The development of methods of recovering cores of sediment, representing continuous accumulation over long periods of time, from the sea bottom, and of studies of oxygen isotope ratios and of geomagnetic reversals, opened a new field of research from which the history of expansion and melting of the continental ice sheets has now been established, with an increasingly accurate time scale, back to before the beginning of the Pleistocene. Today, the sequence of seventeen cold – warm cycles, demonstrated within the Pleistocene by sea bottom evidence, provides the most complete key available to what is also likely to have been happening on the land, and is becoming more and more widely accepted as a chronological framework to which terrestrial deposits (which include most of the mammal-bearing deposits) can be related. Although good correlations have already been demonstrated for the upper part of the Pleistocene, the equivalence of climatic events becomes increasingly more obscure as we pass back into Quaternary time. We will consider the problems of working with a glacial – interglacial time scale at greater length in chapter 6.

Interpreting the Quaternary – comparing past and present

All studies related to Quaternary climatic fluctuations are based on a very simple principle. At the present day the world's surface is divided into a variety of inter-related geographical regions. For our present purpose we need to consider only four such types of region – the glaciated areas; the areas around the ice sheets; the temperate and tropical areas; and the oceans (and their shores). Basically the divisions within the continents (tundra, forest, deserts etc.) correspond to zones of latitude; but altitude, the effects of warm and cold ocean currents and whether potentially rain-supplying oceans are frozen or not are additional controlling factors. The result is a much more irregular pattern of geographical regions than a simple latitudinal one.

These various types of region are not stable but change their position or extent, or may even disappear, in response to climatic change. Often the sediments that are being laid down now show characters that are absolutely diagnostic of the climatic conditions locally prevailing. Contained organic remains may have similar significance, since the animals and plants concerned change their distribution in response to climatic change. Determine, by comparison with present-day processes, the nature of the local climate when a deposit of Quaternary age was laid down; and then also determine its age by one of the many methods of dating now available, and a record has been obtained of climatic conditions prevailing at one place on the earth's surface, at one moment in Quaternary time. At stratified sites more than one climatic event may be apparent in a single sequence of deposits. Obtain information of this nature from many thousands of sites of a wide range of ages, from all parts of the world, and it becomes possible to reconstruct Quaternary climatic changes in very great detail.

In this chapter we will consider the sort of evidence that is available for palaeoclimatic studies in each of the types of geographical region outlined above and, as an example, see what equivalent evidence from late Pleistocene deposits tells us about conditions in the world at the time of the last major advance of the world's ice sheets, which spread to their greatest extent about 18 000 years ago in the northern hemisphere, slightly earlier in the southern hemisphere. Later, in chapter 6, we will take a general look at more of the Quaternary.

Glacial regions

At the present day these are restricted to high latitudes and high altitudes. Very large continental ice sheets still exist on Greenland (frontispiece) and Antarctica; the Arctic Ocean is permanently frozen; and there are extensive glaciers on such mountains as the Himalayas, the Alps, the Rockies and the Southern Andes; with the snow line ranging in altitude from sea-level at the poles to over 5000 metres on such equatorial mountains as Ruwenzori, Mount Kenya and Kilimanjaro in Africa.

Ice sheets and glaciers leave very characteristic evidence of their passing (fig. 2.4). The underlying basement rock is scoured and polished, valleys are eroded until they become U-shaped in section and morainic deposits are laid down that may contain far-travelled blocks of foreign rocks ('erratics', fig. 2.5). Beyond the ice-front escaping melt water leaves behind sheets of glaci-fluvial gravels and sands. Eskers (sinuous ridges of sand and gravel, often of great length, laid down in drainage channels beneath glaciers) are further typical features of a recently glaciated landscape.

From a study of such lines of evidence a very detailed picture of the state of the world's ice at the time of the last major glacial advance has now been established. All the ice sheets and mountainous regions experienced massive increase in ice cover at this time. Huge ice sheets covered northern Europe (including all of Scandinavia and most of the British Isles) and North America, as far south as New York, building up to depths of over 3 km in some places. Remarkably Alaska (where the climate was more continental than further east) remained largely unglaciated, except for the Brook's Mountains and the tongue of ice extending along the Aleutian chain (fig. 2.6).

There were similar smaller increases in the extent of the ice cover in the southern hemisphere, especially the Andes (fig. 2.7) and New Zealand. Tasmania carried a small ice cap where there is none today and glaciers extended to a lower altitude on the high equatorial mountains of Africa. There was a great increase in the amount of sea ice around Antarctica.

The regions around the ice sheets

Around the present margins of the earth's ice sheets (especially in the northern hemisphere) are vast tracts of polar desert and tundra which, although unglaciated today, nevertheless have a very severe climate. These areas are described as periglacial. Often they are snow-covered in winter, although this snow provides insufficient insulation to prevent the ground below becoming deeply frozen – the phenomenon of permafrost. Locally, for example in northern Canada and Siberia, permafrost extends to depths of 4–500 metres, and there is a unique record of 1450 metres at the Shalagontsi Settlement in N.W. Yakutia, Siberia.

The types of deposit that accumulate under periglacial conditions are of very distinctive character. Especially important is the process known as solifluction or mass downhill movement of water-saturated sediment on slopes (fig. 2.8). This process is not restricted to periglacial areas, though it reaches its greatest development there, where it is known as gelifluction. In areas of permafrost, the surface of the ground thaws in summer to a depth from a few centimetres up to a metre or two (the active layer) and, since there can be no drainage into the underlying frozen ground, the resulting sludge moves gradually downhill into the valley bottoms, this process halting each winter as the ground becomes frozen again.

Fig. 2.3 *Air view of a high latitude low altitude currently glaciated area; Inglefield Mountains, Ellesmere Island, Canada, 78°N. Two distinct types of ice body are visible–valley glaciers and blanket cover on the higher ground. (Photo: A. J. Sutcliffe).*

Fig. 2.4 *A formerly glaciated valley, Wasdale Head, Lake District, Cumbria, England. The U-shaped valley in the middle distance, with heaps of morainic debris in its bottom, is a typical glacial feature; likewise the valley in the foreground, though this has subsequently also been filled with alluvial sediments. At the time of glaciation the scene probably closely resembled that shown in fig. 2.3. (Photo: Cambridge University Collection of Air Photos).*

Fig. 2.5 *Evidence of past glaciation. Transported erratic block of Silurian grit abandoned in late Pleistocene times by a melting glacier on a pavement of Carboniferous limestone at Norber, Austwick, Yorkshire. (Photo: British Geological Survey).*

Fig. 2.6 *Nearly equal area projection map of the world, showing the distribution of ice sheets at the present day (black) and at the time of the glacial expansion centred upon about 18 000 years ago (white). The boundary shown represents the furthest limits of this expansion, not attained precisely synchronously everywhere. (Ice sheet data from Denton & Hughes, 1981).*

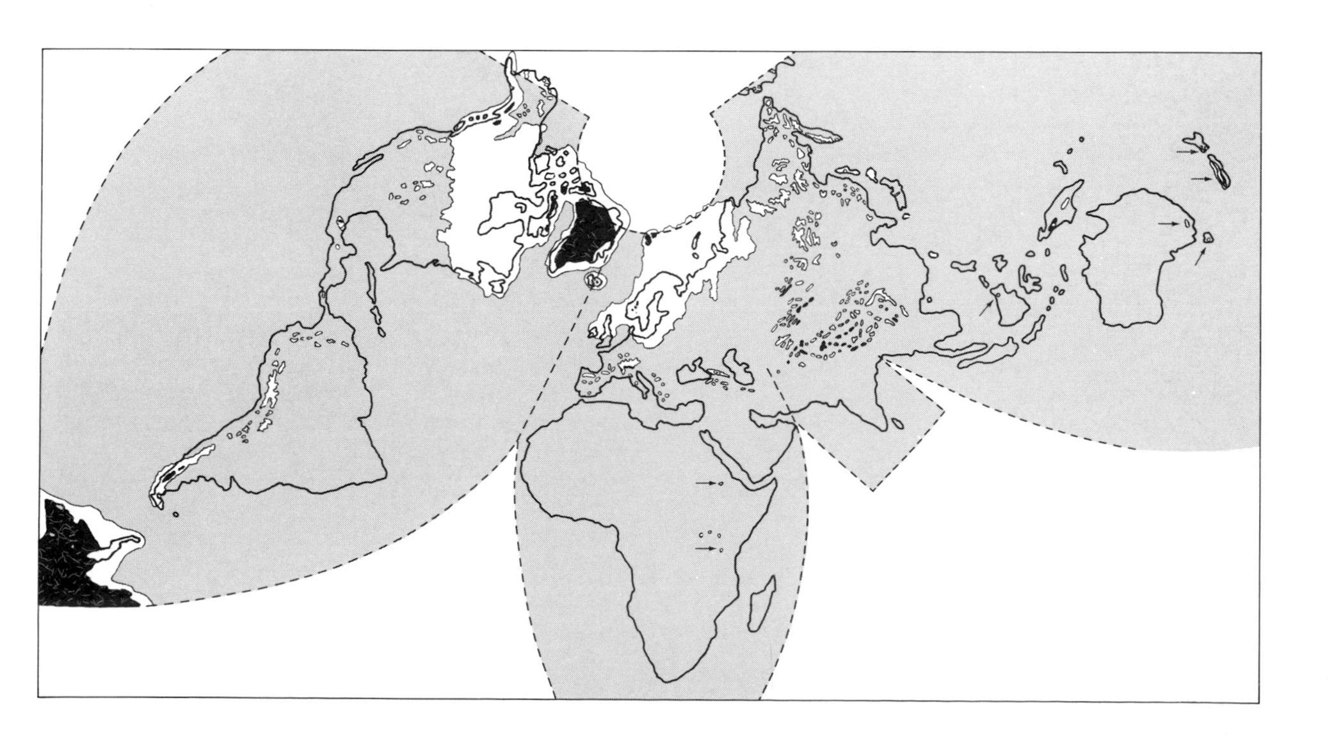

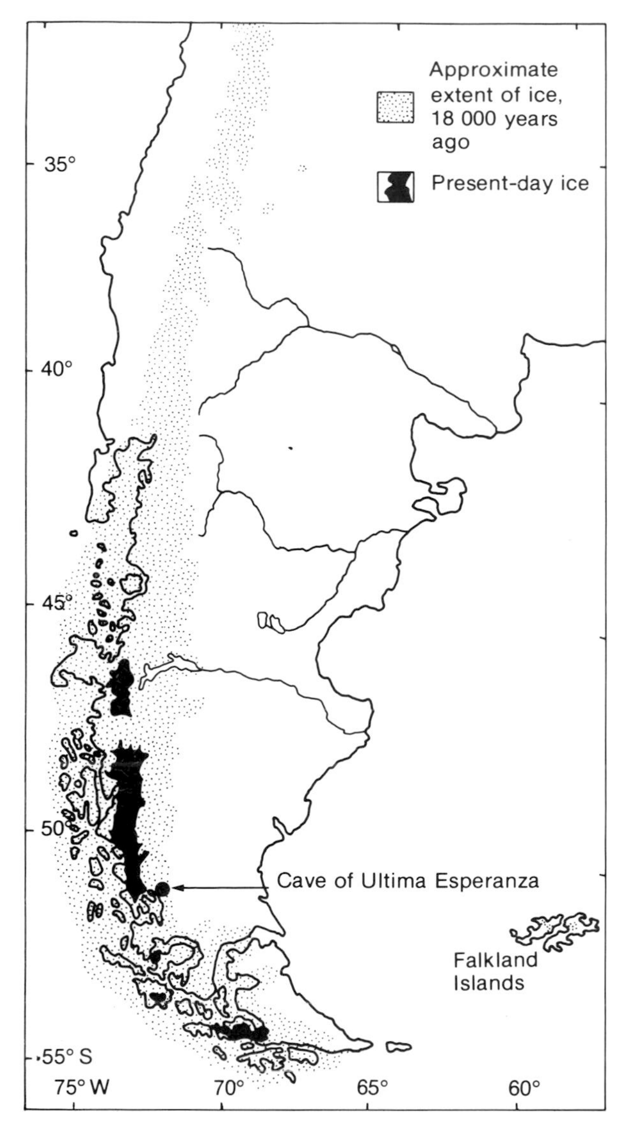

Fig. 2.7 Extent of ice cover in the southernmost Andes, South America, today and at the maximum advance of the last glaciation, about 18 000 years ago (after CLIMAP). The location of the Cave of Ultima Esperanza, discussed in chapter 12, is also shown.

Old gelifluction deposits of Pleistocene age are widely distributed outside the present-day periglacial areas and give valuable evidence of former colder conditions.

A further important periglacial feature is that wide variety of phenomena known as patterned ground. Frost heaving and the annual contraction and expansion of the ground may lead to the most remarkable sorting of rock fragments; often arranged on the surface as polygonal structures. When viewed in section these polygons are seen as a series of vertical wedges, sometimes ice filled (figs 2.9 and 2.10). Collapsed ice wedges are a common feature in many former periglacial areas.

In craggy terrain frost shattering of the bed rock may give rise to massive screes and other deposits of angular rock fragments.

A further very important sediment type characteristic of the regions around the ice sheets is the wind blown dust known as loess, readily identifiable by geologists because of the uniformity of size of the winnowed silt particles of which it is composed. Deposition was greatest at the time of retreat of the Pleistocene ice sheets, when winds blowing away from the remaining ice centres passed over wide expanses of newly deglaciated and still unvegetated cold desert, picking up the smallest rock particles, which were carried along in suspension until caught up in the sparse vegetation beyond.

Pleistocene loess deposits are widespread on many continents, especially in Asia (where they reach their greatest thickness in China), Europe and North

Fig. 2.8 Hillside gelifluction flows (periglacial mudflows), Seward Peninsula, Alaska. (Photo: US Geological Survey).

Fig. 2.9 *Natural exposure of an ice wedge in the permafrost of Wilbur Creek, Alaska. Note V-shaped cross section and the way in which the strata on either side have been pushed up by ice pressure. In plan such wedges form a polygonal network. (Photo: T. L. Péwé).*

Fig. 2.10 *Artificially prepared section through a fossil ice wedge of Pleistocene age in a gravel pit near Isleworth, London. Note how the upper gravel layers have collapsed into the wedge; and the graben-like small scale faulting of the strata on either side. This section provides a wealth of palaeoclimatic information. In the lower deposits remains of bison and reindeer occur in association with and above a layer of temperate plant remains dated by the radiocarbon method as 43 000 years old (a minor warm interval during the last glacial stage). The ice wedge cuts these deposits and must therefore be later. It can be inferred that at some time after 43 000 years ago, not yet more accurately dated, there was a cooling of climate, with permafrost to a depth of at least 5 metres below present ground level. The climate subsequently ameliorated, with no permafrost remaining in the British Isles at the present day. (Photo: A. J. Sutcliffe).*

America. Frequently these deposits can be interpreted as evidence that periglacial conditions formerly extended far beyond their present margins, but caution must be exercised in the drawing of such conclusions. Loess-like deposits are also known from some other environments for example the 'dust loams' of the hot Negev Desert, Israel; and deposits secondarily spread by wind from glacially derived river silt in the Mississippi basin, USA.

The temperate and tropical regions

Although, at low latitudes, glaciation can occur only at high altitudes and is therefore a relatively unimportant process, the fluctuations of rainfall are barely less dramatic. Studies of the past water levels of African lakes, for example, show that these have fluctuated repeatedly. Lake Chad was formerly nearly 1000 km long, requiring a water intake 16 times greater than at present, in an area that is now mostly desert; while giraffes, elephants and hippos found sufficient vegetation to enable them to survive over a large part of the Sahara Desert. Such episodes have become known as 'pluvials'. In contrast, dead sand dunes in the forest areas south of the Sahara show that the desert area has been even more extensive at other times during the past than it is today, with the great mass of the Zaire rain forest greatly reduced in extent.

At low and middle, just as at high latitudes, there are many sediment types that can be recognized by the geologist as climatically diagnostic. We have already mentioned the spread of continental sand dunes as an evidence of former aridity. Also important are the soil horizons formed by weathering, which reach their greatest development in damp temperate regions. The soil specialist can often provide an accurate reconstruction of the climatic conditions under which a fossil soil formed from a study of its composition and texture. Many of the Pleistocene loess deposits alternate repeatedly with soil horizons that are interpreted as evidence of minor ameliorations of climate when loess deposition ceased and vegetation could become established.

Although no ice is directly involved in the above mentioned processes, such changes are nevertheless all related to the climatic fluctuations that caused the variations in the size of the ice sheets and must all be considered together as parts of the same story. For a long time it was uncertain whether the glacial episodes of the high latitudes and high altitudes corresponded in time with the pluvials of the lower latitudes or with the more arid (interpluvial) periods between them. Twenty years ago a correlation with the pluvials was generally considered most probable. With the advent of better dating methods, however, it was shown that this was not correct, the interpluvials corresponding broadly with the glacial maxima, the pluvials with the temperate stages between them.

Writing in 1978, Sarnthein reviewed the dates when desert dune fields had been most active. He found that, whereas at the present day about 10 per cent of the land area between 30°N and 30°S is so covered, at the height of the last glaciation, 18 000 years ago, the figure was nearly 50 per cent; with the tropical rain forest and adjacent savannahs reduced to a narrow corridor, in some places only a few degrees of latitude wide. Especially extensive development of desert occurred in the Sahara and Kalahari–Congo basin areas of Africa, Arabia, NW India, Australia and parts of North and South America. There were nevertheless local exceptions to this general rule in some desert areas, for example in western North America, where there is evidence of quite high rainfall at this time. Not only did Sarnthein compare the distribution of desert areas of 18 000 years ago with that of the present day but also with that of the Holocene climatic optimum, 6000 years ago, when greater humidity prevailed and there were especially high water levels in many desert areas.

Other studies have shown that the monsoons which today bring rain to the Indian subcontinent are not a permanent feature of the earth's climate, but have fluctuated greatly in intensity, virtually ceasing at the time of the glaciation of 18 000 years ago.

At the present day, the deserts of the world occupy areas intermediate in extent between those of the 18 000-year-ago arid stage and those of the 6000-year-ago humid stage.

The oceans and their shores

The continents are not the only parts of the world that provide us with evidence about the Ice Age. Land makes up only about 28 per cent of the earth's surface. The other 72 per cent is covered by the oceans, the bottoms of which are unending traps for sediment containing evidence of past sea temperatures and, incidentally, of past climatic conditions on the surface. In recent years techniques have been perfected for obtaining cores of sediment from the sea bottom; and a treasure house of palaeoclimatic information has been brought to light. The value of such sediments is two-fold. Not only can past climatic conditions be reconstructed from palaeontological studies of the contained small marine organisms known as foraminifera and by chemical studies (notably on the relative proportions of the isotopes 16 and 18 of oxygen), but the record is longer and more detailed than that yet obtained by any other means. From a single core it is possible to obtain information about not just one cold and warm oscillation, but (since sedimentation on the ocean floor goes on continuously with the oldest strata gradually being buried by the later ones) about all the climatic events represented down to the ocean's bed rock or the limit to which the equipment can reach. Many thousand cores have now been obtained from all parts of the world and the record of climatic change that they reveal, at site after site, is so consistent as to leave no doubt about the supremacy of this method in reconstructing the story of Quaternary climatic change. The reality of the glacial episode of 18 000 years ago, first demonstrated from continental evidence, is confirmed and the details are amplified by the deep sea record from all the oceans of the world. Formerly such records were available only from relatively shallow depths beneath the sea bed, representing but the later part of Quaternary time. Recently, however, a 21-metre hole in the Pacific has penetrated the entire Pleistocene and upper part of the Pliocene, an estimated time-span of 2·1 million years. We will examine the ocean bottom record more closely in chapters 5 and 6.

A further aspect of the study of the oceans that is of great importance in the interpretation of past changes of climate is the way in which sea-level varies in response to the advance and melting of continental ice sheets. Interglacials are characterized by high sea-levels, glacials by falls of sea-level, which accompanied the transfer of water into the great ice sheets. There is widespread evidence for such changes having occurred in the past in the form of submerged land surfaces and 'elevated' beach deposits along most of the world's shorelines.

As during the other glacial episodes, the glacial advance of 18 000 years ago was also a time of massive fall of sea-level, believed to have amounted to more than 100 metres. Great areas of the present-day sea floor became dry land. The coastline of Siberia retreated along part of its length 400 km north of its present position, and a land bridge more than 1000 km wide connected Alaska and Siberia where the Bering Straits exist today. Other areas of dry land included the George's Bank off the east coast of North America, most of the English Channel and the southern part of the North Sea. In the southern hemisphere, Australia was connected to New Guinea (fig. 13.6, p. 192).

The temperature of the sea was also greatly affected. More of the surface was frozen, especially around Antarctica, with pack ice and polar water extending further from the poles in both hemispheres, for example along the west coast of Portugal.

The effects of climatic change on flora and fauna

Studies of sediments and geomorphological studies, such as those described above, are not the only means of reconstructing Quaternary climatic events. Of equal importance is the palaeontological evidence. The geographical distribution of plants and animals alters

Fig. 2.11 *Map showing the vegetational zones of eastern North America; left, about 18 000 years ago, at the height of the Wisconsinan glaciation; right, 200 years ago, before large scale human interference. Although the zones, which migrated northwards as the climate ameliorated, have an approximately E–W disposition, other factors besides latitude (such as mountains and rivers) also influenced their distribution. In the western part of the area, in the rain shadow of the Rocky mountains, there was development of prairie glasslands. (After Delcourt & Delcourt, 1981).*

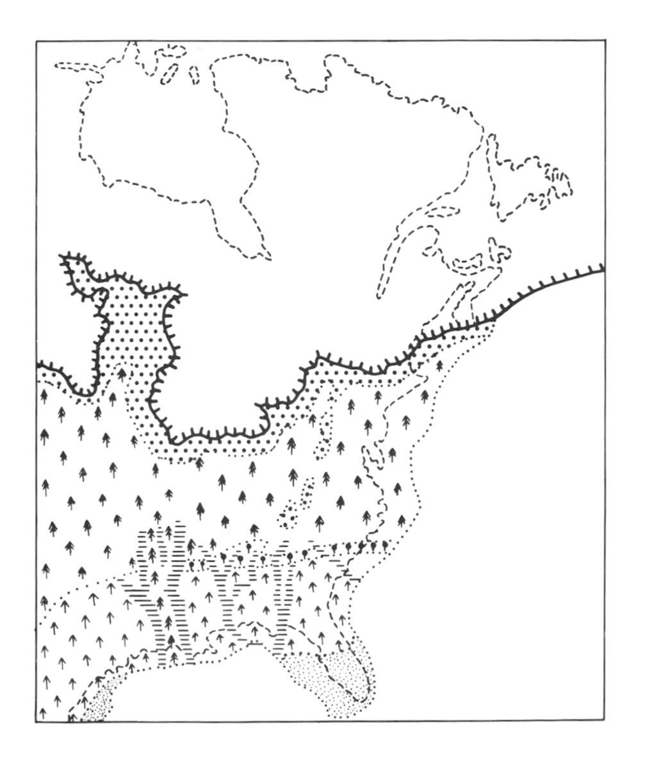

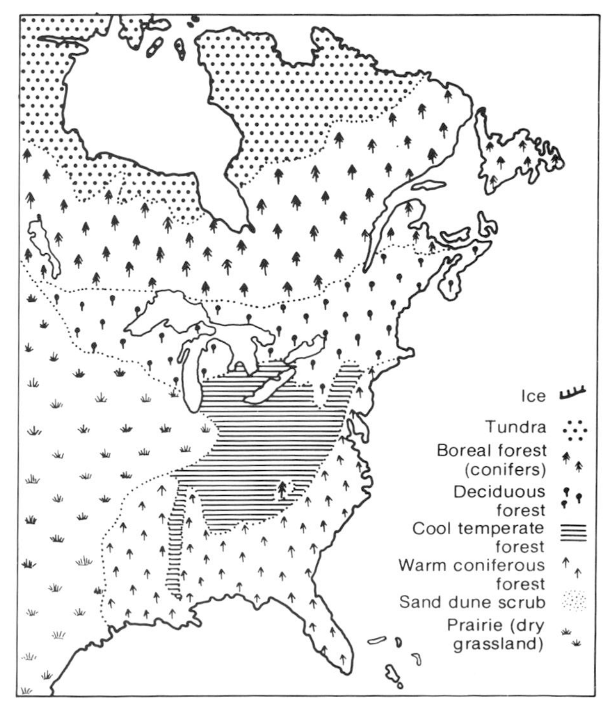

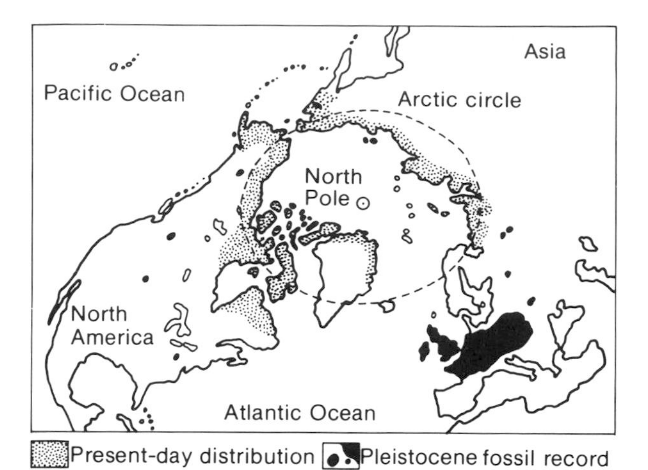

Fig. 2.12*a) Map showing the present-day distribution of the collared lemming and localities where fossil remains have been found in North America and Europe (Asian fossil distribution not shown).*

Fig. 2.12*b) The collared lemming in its summer coat. Note the reduced ears, almost obscured by fur, and its fur-covered feet. Bathurst Island, Canada, 75°N. (Photo: A. J. Sutcliffe).*

in response to changing conditions. Most of the Pleistocene species of plants and many of the animals still survive in some parts of the world at the present day, allowing accurate comparisons. Where the fossil remains of any of these occur outside their present range, or where there is evidence of past changes of distribution, climatic change can often be inferred.

The expansion of the ice sheets during each of the glacial episodes was accompanied by displacement of the world's vegetational zones away from the centres of refrigeration. Desert regions, steppes, grasslands and outwash plains expanded at the expense of the forests. This was not a simple shifting of all the vegetational belts (southwards in the northern hemisphere, northwards in the southern hemisphere) but was a complex process dependent also on many other factors, including the local continentality of the climate. This in turn was partly dependent on the distribution of unfrozen oceans from which evaporation could occur, the areas of which had been reduced by fall of sea-level and increase in the extent of sea ice.

The overall result was that the distribution of the world's vegetational zones was very different from that of the present day (fig. 2.11). Permafrost occurred far beyond its present range. In northern Asia there existed a belt of polar desert where there is tundra today. The tundra belt of the northern hemisphere was situated where there is now forest, including northern USA and Europe as far south as France. Steppe vegetation was widespread in the more continental areas, especially in Asia and parts of North America. The interglacials were times of equivalent changes in the opposite direction.

These displacements of the world's vegetational zones also caused displacement of its mammalian and other faunas. The occurrence, for example, in Europe of the fossil remains of the collared lemming (fig. 2.12 a,b), and musk ox (now inhabitants of the Arctic tundra) may be interpreted as evidence of the former southward displacement of the tundra during a glacial phase; the occurrence of hippopotamus in northern England as evidence of previously warmer conditions. The two groups of mammals never co-existed at the same place, however, because of their different climatic tolerance. Where remains of both occur in different strata at the same site, a cooling or warming of climate can be inferred.

The interpretation of such biological evidence is nevertheless not always straightforward. Some mammals, such as the wolf, have wide climatic tolerance, and do not provide such information. The rate at which plants and animals can alter their distribution in response to climatic change varies from one group to another. Insects can move more quickly than plants and there are instances where the palaeoclimatic evidence which they provide appears conflicting. It is explained by differential rates of movement. Later the plants catch up with the insects and, at times of climatic stability, the climatic evidence provided by each is likely to be the same.

It has also been shown that in the not very distant past there may sometimes have existed ecological assemblages of plants and animals that no longer occur in the same combination anywhere at the present day – for instance, the so-called 'mammoth steppe' or 'arctic steppe' of the late Pleistocene of northern Siberia, Alaska and Canada. We will consider this phenomenon of 'arctic steppe' at greater length in our discussion of woolly mammoths in chapter 9.

For reasons such as these, too close a climatic analogy must not be claimed from any single line of biological evidence. As many different fields of study as possible (both biological and physical) must be considered together. In this book we are concerned primarily with the mammals themselves, and with what these other lines of evidence tell us about them. The mammals are likely to respond especially to changes of flora, for instance; so that an understanding of the botanical history of an area may help us to understand the reasons for migrations or extinctions. The disappearance of the 'arctic steppe' has been widely acclaimed as a likely cause for the final extinction of the woolly mammoth. Continental drift and fluctuations of sea-level, related to climatic change, are other important related topics. Although continental drift was not a factor that directly affected Quaternary biotic changes, since the period of time involved was too short, consideration of what occurred in pre-Pleistocene times does nevertheless help us to understand in our Pleistocene studies the restricted geographical distribution of certain mammalian faunas, such as the marsupials of Australia and South America; and likewise the occurrence of faunal breaks, such as that between Asia and Australia.

Changes of sea-level resulting from climatic change were of great importance in allowing or preventing faunal movements. During the low sea-level of the last glacial advance, as we have already observed, the southern part of the North Sea was dry land, allowing free movements of mammals between Britain and the continent of Europe. Woolly mammoths roamed over a vast plain, now similarly covered by the sea, connecting the New Siberian Islands (75°N) to Siberia; and this plain also connected Siberia and Alaska, allowing faunal movements across the Bering Straits. But the change that provided extra territory for the mammoths also closed a seaway for the marine mammals. Whales and walruses, which today migrate seasonally between the Pacific and Arctic oceans through the Bering Straits, were unable to do so.

Changes in the extent of the world's ice sheets also affected mammalian movements on land. At the time of lowest sea-level mammals migrating east across the Bering Straits soon found their further movement blocked by a continuous or almost continuous ice sheet extending from the Aleutian Islands across Canada to Greenland. Faunally, at this time, unglaciated central Alaska was a continuation of Asia, with some Asian species present, such as the yak and saiga antelope. Geographers have named this formerly united land area 'Beringia' (fig. 2.13).

Fig. 2.13 *Generalized map of Beringia during the peak of the last (Wisconsinan) glaciation, showing approximate shoreline position and glacial cover. Also shown are localities where remains of saiga antelope have been found and possible lowland migration routes from Siberia to Alaska. Not all the remains are of the same age but represent migrations on more than one occasion during the upper Pleistocene. (After Harington, 1981).*

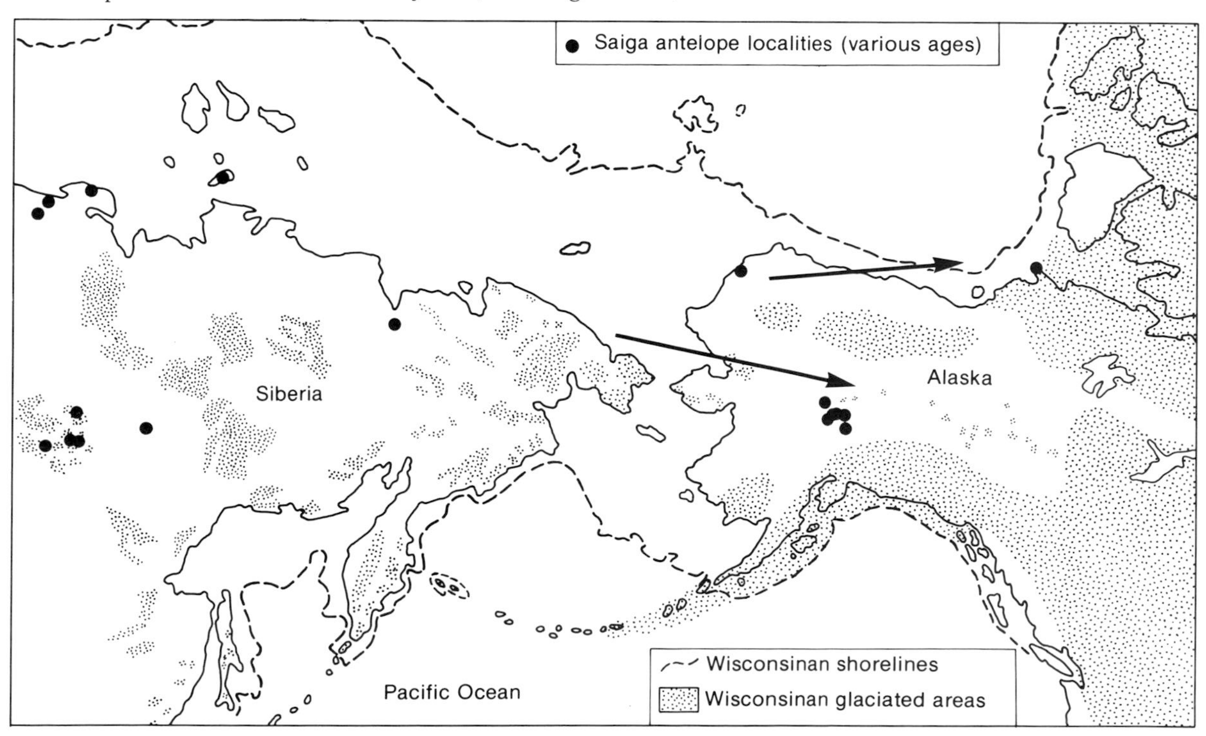

Sometimes the rising sea-level after the glacial stages caused mammals to become shut off from the continent where they had originated. Marooning by this cause is nevertheless sometimes difficult to distinguish from immigration by swimmers. Remains of elephants, hippos and deer occur especially abundantly on islands of the Mediterranean, suggesting that these mammals were better able to make the crossing from North Africa or Europe than some others that might be expected to have accompanied them, but which failed to do so. The subsequent history of such island mammals, after isolation, is often remarkable. On several of the Mediterranean islands the elephants, hippos and deer became independently dwarfed; and dwarf elephants are also known from islands off the Californian coast. Sometimes islands were inhabited by herbivores with no accompanying carnivores, with consequent problems of overcrowding and sickness.

At times when physical barriers (such as ice sheets and seaways) disappeared, the mixing of mammalian faunas that had previously been isolated from one another sometimes took place. Where, in consequence, two groups of mammals occupying the same ecological niche found themselves on the same terrain, biological competition followed that was a potential cause of extinction.

Lastly, this chapter would be incomplete without consideration of some of the effects of the Quaternary climatic change on man himself. Certainly there are many instances during historical times where man has been geographically displaced by climatic change; for example by reduced rainfall during the last 5000 years in the Middle East, where ruined cities now stand in the desert; and in Greenland and Northern Europe during the 'Little Ice Age'. It is happening today in the Sahel (southern Sahara).

The detailed picture that is now available of the Ice Age world 18 000 years ago – of approximately synchronous expansion of both the ice sheets and deserts – is not a comforting one for man. It is true that at the same time much new land appeared as a result of the fall of sea-level, but a large part of this was tundra which, if similar conditions returned today, could not compensate in terms of agricultural productiveness, for the terrain lost.

Chapter 3 **Dragons, unicorns, giants and saints**

The twentieth-century palaeontologist, confronted with the problem of identifying fossil mammalian remains, can usually name the animals concerned with a high degree of precision. He has at his disposal a vast accumulation of zoological and palaeontological knowledge built up over several centuries. Sometimes he can show that fossil remains (especially those of late Pleistocene age) are of mammalian species which, although extinct in the neighbourhood of the discovery, still survive today in other parts of the world, allowing present-day comparisons. Even if the animals concerned are long extinct, their approximate relationship to living and other extinct species can usually be determined. With such ground work now at an advanced stage he is mainly concerned with studying the details of such relationships.

Those who found fossil mammalian remains in the past, even as recently as the eighteenth century, had few such advantages; and some of their attempts to explain what they had discovered, if considered out of the context of their time, might seem bizarre and even ridiculous today.

One of the first problems that had to be considered was whether the objects concerned were really remains of long dead animals; or whether they were concretions or 'sports of nature', the resemblance being entirely accidental; a complex problem that sometimes confuses palaeontologists even today.

As late as 1696, a mammoth skeleton found near Gotha in Germany was classified at the Gotha College of Medicine as being of inorganic origin, although its true origin was correctly determined by Tentzel shortly afterwards.

When it had to be accepted that some of the objects (some of which we now know to be mammalian remains) were indeed of organic origin, it became necessary to identify the animals concerned. Whilst some observers regarded their discoveries as parts of 'sea animals' (for instance a mastodon tooth figured in Grew's *Catalogue of the Royal Society's Repository*, 1681) or noted the similarity to recognizable animals such as elephants (sometimes and depending on locality, those of Claudius or Hannibal or Alexander), others attributed them to giants, dragons or unicorns; frequently to those which had been drowned by the Biblical Deluge. Often the remains of these animals were regarded as of great medicinal value and were traded at very high prices.

As late as 1841 a mastodon skeleton unearthed by Albert Koch in Hickory County, Missouri, USA, was identified as the Leviathan described in the 41st Chapter of the *Book of Job*.

As the true identity of the various remains became known such fanciful identifications became less frequent although, even today, they persist in some regions. 'Dragons' teeth' continue to be an important item of medical commerce in China.

Let us look more closely at the identity of the remains supposed to represent some of these legendary animals.

Giants

The discovery of large fossil bones and teeth has repeatedly led to claims for the former existence of giants. The bones found in 1519 near Tlascala, Mexico; those described by Plot, 1676, from London; and the skeleton seen by Boccaccio in a cave at Trapani in Sicily are examples of such instances.

The Trapani giant is especially fascinating since Boccaccio, writing in 1472, apparently believed that the skeleton was that of the carnivorous one-eyed cyclops Polyphemus, of whom Homer spoke in *The Odyssey*. The story relates that Polyphemus imprisoned Odysseus and some of his men in a cave, from which those who had not already been eaten escaped only after they had blinded him and made their way out with his sheep and goats when he opened the gate in the morning.

A possible explanation for the cyclops myth has been put forward by the German palaeontologist, Othenio Abel. Writing in 1939, he pointed out that the nasal openings of the skull of an elephant, when seen by someone with no anatomical knowledge, could readily be mistaken for two eye holes fused together. If wandering sea travellers of the Homeric age had found skulls of fossil elephants in sea caves on the coast of Sicily (where twentieth-century excavations show that they do indeed occur) these could readily have been mistaken for the skulls of giant ogres with a single large eye on the forehead.

In the USSR, bones of mammoths found at Kostienki ('Bone Village') in the Don Valley were attributed in Russian folklore to the giant Inder, who lived underground and died there.

In all parts of the world from which remains of giants have been claimed, the story can usually be followed back to the finding of very large bones.

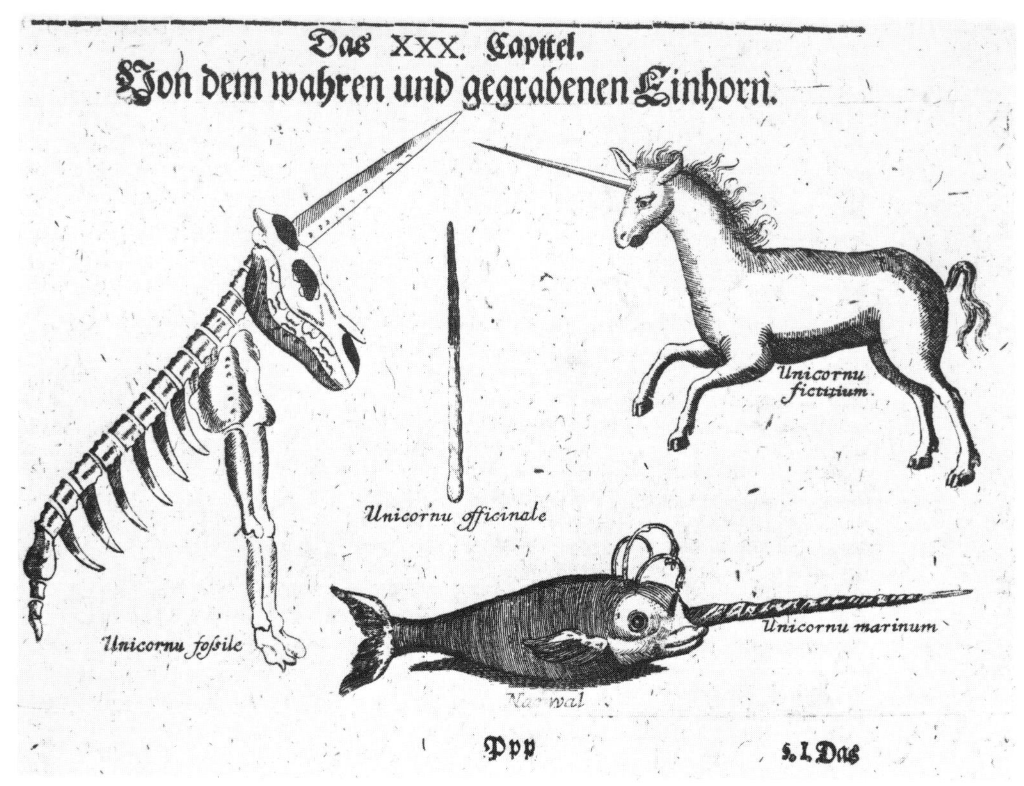

Fig. 3.1 *The various types of unicorn distinguished by Valentini in* Museum Museorum, *1704.*

Unicorns

Although everyone now agrees that the unicorn never existed, such unanimity is quite recent. Belief in this animal reached its climax during the Middle Ages, when its supposed remains fetched high prices as a cure for illness and as an antidote against poisoning (especially by arsenic).

But even then the unicorn story was one of long standing. Ctesias, a Greek physician in the Persian court at the end of the fifth century BC, wrote of the existence of unicorns in India. Unicorns are also mentioned in the Old Testament of the *Bible:*

> *'Will the unicorn be willing to serve thee, or abide by thy crib?*
> *Canst thou bind the unicorn with his band in the furrow? or will he harrow the valleys after thee?*
> *Wilt thou trust him, because his strength is great? or wilt thou leave thy labour to him?*
> *Wilt thou believe him, that he will bring home thy seed, and gather it into thy barn?'*
>
> (Job, XXXIX, 9–13)

Probably both unicorns were based on real animals of the time. Ctesias may have heard of the Indian rhinoceros, which has only one horn; and Job's unicorn was apparently a wild ox, the name unicorn having been substituted when the original Hebrew was translated into Greek. Another possible origin of the unicorn legend is the oryx, an antelope which, when seen in profile, appears to carry but a single almost straight horn on its forehead.

The unicorns of the various accounts differ in appearance and fierceness according to the date and country of their origin. Whilst most were believed to be very fierce, the Chinese unicorn is said to have been so gentle that when it walked it was careful not to tread on the tiniest living creature and it would not even eat live grass, but only that which was dead.

By the Middle Ages the unicorn had become a superbeast in Europe, where it was one of the favourite subjects for artists. Usually it is shown to resemble a horse, except that it had cloven hoofs, the tail of a lion, and, on its forehead, a straight, spirally twisted, tapering horn.

Such were the reputed medicinal properties of unicorn horn (especially during the seventeenth century, when no other medicine was considered so effective), that any object that could be so identified, besides being used as evidence for the reality of the unicorn, was very highly prized. Unicorn horn was even included in the British *Pharmacopoeia.* As it was in short supply it fetched very high prices, sometimes greater than its own weight in gold. Not all such horn, however, was of the same origin. Valentini, writing in 1704, distinguished four types of unicorn, only some of which he accepted as real (fig. 3.1). He was already familiar with the whale known as the narwhal, 'horns' of which had been brought back from Greenland by sailors. The male of this animal has a straight, forward pointing, spirally-twisted, tapering tusk which clearly provides the basis for many medieval reconstructions of the unicorn. He called this *Unicornu marinum.*

Valentini also recognized unicorn remains that had been dug out of the ground; and these he called *Unicornu fossile* (fig. 3.2). It is these that mainly concern us here. He recorded that the fossil remains

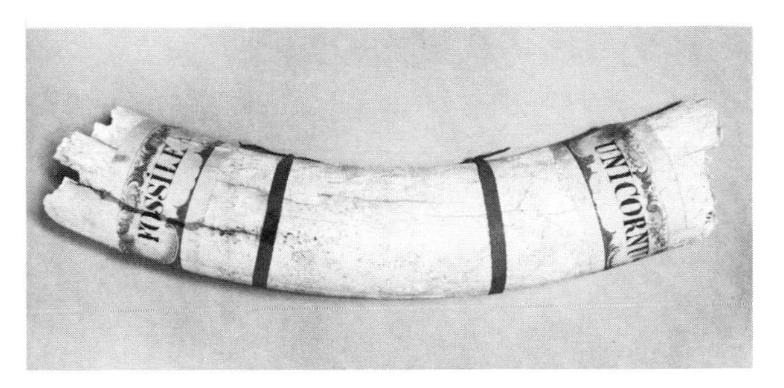

Fig. 3.2 *Fragment of mammoth tusk, labelled* Unicornu fossile, *which was given as a gift by the chemist Strahling to Princess Caroline Luise of Baden about 1760. Preserved in Karlsruhe Museum. (Photo: H. Heckel).*

had an earthy taste and would stick to the tongue; and his illustration shows an imaginative reconstruction of the skeleton of a unicorn based on some bones of mammoth and possibly woolly rhinoceros found in a gypsum sink hole at Quedlinburg, Germany, in 1663. He also recorded that similar remains could be found in Baumann's Cave in the Harz Mountains (well known to present-day palaeontologists as a finding place for remains of woolly mammoth) and at other localities.

One of the greatest accumulations of supposed unicorn remains was found in a clay pit on the bank of the River Neckar near Cannstatt in April, 1700. Duke Eberhard Ludwig ordered systematic digging of the area and, within six months, over sixty mammoth tusks had been recovered together with many bones and teeth of animals of various sizes, including lion. Several cart loads of fossils were driven to the residence of the Duke who kindly thought it proper to present some of his surplus to the City of Zürich. He was duly thanked by the Mayor for these remains of the Schwabian unicorn. The greatest part of the find was transferred to the court apothecary and sold to the faithful sick for six batzen for half an ounce.

It needed a further century to get rid of the belief in Europe of the reality of the unicorn and to bring final proof that the fossil remains were those of elephants and other mammals. At the same time, the growth of the whaling industry made narwhal tusks familiar objects and so unicorn horn ceased to be sought as a medicine and its value rapidly declined.

But, even today, the unicorn is not dead. The horn is still used as a symbol by German drug stores, some of which carry the sign 'Einhorn Apotheke'; and the unicorn is a frequent feature of heraldry, including the British Royal coat of arms, where it represents Scotland. It appears at the head of the notepaper of the British Museum (Natural History).

Saints

For many centuries bones of saints and other religious figures have been treasured as holy relics in diverse parts of the world. The examination of such relics, by present-day zoologists and palaeontologists, has sometimes led to unexpected results.

At least one saint, St Christopher, the former patron saint of ferrymen, was by tradition a giant. There are many examples on record, in Western Europe, of bones and teeth of mammoths being venerated as his remains; for example a thigh bone of a mammoth that formerly hung near the Great Gate of St Stephen's Cathedral, Vienna.

Places of veneration, where it is believed that the bodies of saints may have been buried, occur at several places in Cyprus, for example near Cytheria and in the cave of St Elias above Pascalis Chiftlik.

Of special interest is the rock-hewn chapel of Ayios Phanourios on Cape Pyla in the southeast part of the island. The chapel opens directly onto a bone-bearing deposit that apparently accumulated in the bed of a small water course. Gunnis, writing in 1936, gave the following description of the site:

'Near the sea the tiny rock-cut Chapel of St Phanourios; beneath this chapel the rocks are full of fossil bones, called by the villagers the bones of St Phanourios, but in reality remains of pygmy hippopotamus. Villagers dig out from the rock the fossil bones and, powdering them, mix them with a drink of water; a sovereign cure for nearly every known disease. St Phanourios was a youth who lived in Asia Minor and heard the call of Christ, and came across in a small open boat with only his faithful horse as his companion, and landing, tried to ride up the steep cliff, but his horse slipped and fell and he and his steed were killed, in token of which the horse's footprints are shown to this day.'

Today the Cretan pigmy hippopotamus bears the Latin name *Phanourios minor* after the saint so commonly associated with its remains.

Fig. 3.3 *Imaginary dragon from the Island of Rhodes, from Kircher's* Mundus Subterraneus, *1678.*

Dragons

Traditions of dragons occur in many parts of the world, especially in Europe and China. In Europe, these animals were especially popular during the sixteenth and seventeenth centuries in Germany, Austria, Hungary and Switzerland. They were believed to be able to fly and to occupy the lonely heights and walls of the Alps, a concept illustrated in Kircher's *Mundus Subterraneus,* 1678. There was ample evidence for the supposed existence of such dragons (fig 3.3) in the form of fossil mammalian remains, especially those of cave bears, which occurred abundantly in caves throughout the area. A dragon skull from a cave in the Carpathians, illustrated by Johannes Hain in 1673, is clearly recognizable as that of a cave bear; and there are many hills and caves that have been named after dragons: for example the Drachenfels in the Siebengebirge, where Siegfried is supposed to have slain the dragon of the Niebelungen saga; and Drachenhöhle or Dragon's Cave near Mixtnitz in Austria, subsequently the finding place of remains of about 30 000 cave bears.

But cave bears are not the only mammals to have given rise to the dragon legend. In the main square at Klagenfurt in Austria is a fountain with a model of a magnificent winged dragon made by the sculptor Ulrich Vogelsang in 1590 (fig. 3.4). It seems fairly certain that the dragon is based on a skull of a woolly rhinoceros found in the neighbourhood about 1335 and which formerly hung on a chain in the Klagenfurt town hall. It is now in the local museum. The dragon fountain is believed to be the earliest attempt in the world of a palaeontological reconstruction.

Whereas the European dragon spread terror, the Chinese dragon was beneficial. It was one of four magic animals, the others being the unicorn, the phoenix and the tortoise. It signified wisdom; was a purveyor of rain at times of drought; and it guarded the Emperor. In consequence it has long been a favourite subject of Chinese sculptors, potters and artists, who usually show it to have a scaly body like a reptile, claws like a tiger and the antlers of a deer (fig. 3.5).

Vertebrate fossils (especially mammalian fossils) have long been regarded in China as bones and teeth of dragons, with great medicinal properties, a belief duplicated in a remarkable way by the later European beliefs about the magical properties of unicorn horn. Today, 'dragons' teeth' are still available, not only in China, but also in shops serving the Chinese communities in other parts of the world. In China, traditional Chinese medicine is practised alongside European medicine and drug stores are usually in two parts; one with European drugs, the other with Chinese drugs such as dried vipers, medicinal herbs, deer antlers and 'dragons' bones' and 'dragons' teeth' (fig. 3.6).

On examination, these 'dragons' bones and teeth prove not to be reptilian, as their supposed identity would suggest, but the remains of fossil mammals, of

Fig. 3.4 *The dragon fountain at Klagenfurt, Austria, 1590, believed to be the earliest attempt in the world of a palaeontological reconstruction.*

Fig. 3.5 *Chinese dragon, part of the famous 'Dragon Wall', Beijing (Peking). Note that it has the antlers of a deer, probably based on fossil material, these being common among fossil remains sold as 'dragons' bones' and 'dragons' teeth'. (Photo: H. D. Kahlke).*

Fig. 3.6 *Dragons' teeth street vendors in Kweilin, China. (Photo: H. D. Kahlke).*

various ages and in various states of preservation. Some of them fetch higher prices than others. Teeth are more costly than bones; and mineralized Tertiary specimens are more valued than those of Pleistocene age.

With ever-increasing demand for 'dragons' teeth', vast quantities of mammalian fossils have been ground up for medicine without ever being seen by palaeontologists and it is only during the last century that their scientific study has been undertaken. This massive destruction of fossiliferous sites, though lamentable, has nevertheless not been without advantages for it has helped lead palaeontologists to many localities that would otherwise never have been found.

There were many problems related to these 'dragons' teeth' that had to be answered. What was their true identity and where had they been found? What information did they provide about the past history of the fauna of China?

At first it was very difficult to obtain any background information about the fossils, as the localities were often kept secret by the local fossil hunters. In consequence, the first such remains to come to scientific notice, purchased from drug stores, were without details of provenance. Among those to make early studies of the drug-store specimens were Davidson (who, in 1853, recorded remains of rhinoceros, bear and the three-toed horse, *Hipparion*, from a drug store in Shanghai); and Owen, first Director of the British Museum (Natural History), who in 1870 based the first scientific description of such fossils on remains of *Stegodon*, hyaena, rhinoceros, tapir and chalicothere, purchased from a vendor of drugs in Shanghai by a Mr Swinhoe.

The first major account of the Chinese Pleistocene mammalian fauna appeared in 1903, when Dr Schlosser of Munich described sixty species of mammals from a large collection of teeth and jaw fragments that had been obtained from drug stores by the German scientist, Dr Haberer, about 1899–1902. He drew special attention to a human molar which (although subsequently shown to be only sub-fossil, probably of Mesolithic age) initiated the search for remains of fossil man in the vicinity of Peking, a search that later resulted in the discovery of the Choukoutien site where Peking man was found.

Nearly two decades were to pass before palaeontologists were able to start examining for themselves the localities where the dragons' teeth were being obtained. One of the first to do this was the Swedish palaeontologist Andersson, who, in 1918, started collecting fossil mammalian remains from crevices and caves at Choukoutien, near Peking. Shortly afterwards, in the autumn of 1921, Dr Walter Granger, palaeontologist of the Central Asiatic Expeditions of the American Museum of Natural History, made an expedition to the Szechwan province of China, where he had heard that fossil bones were being found in the neighbourhood of Wanhsien on the upper Yangtse. The fossils, he found, were being excavated from infillings in limestone caves by local farmers, who in the winter months had little else to do, and were stored by them until merchants from Wanhsien could come to collect the 'dragons' teeth'. By purchasing the best specimens from the farmers (though this made him very unpopular with the merchants) Granger was able to assemble for scientific study the largest collection of Chinese dragons' teeth of known provenance available up to that time. Species represented included *Stegodon*, rhinoceros, tapir, deer, hyaena, tiger, giant panda, monkeys, rodents and other mammals. Most complete of all the remains obtained by Granger was an almost perfect skeleton of a gaur (a relative of our domestic cattle) which, like many of the animals represented, had apparently fallen into a once open shaft where its bones had become buried among earth and stones.

But the most sensational discoveries were still to follow. Scientific excavations at Choukoutien led to the discovery of Peking man, then called *Sinanthropus pekinensis*. By the mid 1930s a dozen skulls had been found, together with other fragmentary remains representing about 45 individuals. The site proved to be a deep cleft (later excavated to a depth of over 50 metres) filled with stratified deposits, including ash layers, showing that Peking man had used the site for shelter and had known the use of fire. Other associated finds included primitive stone tools and abundant remains of mammals, including hyaenas. The skulls showed that Peking man was a primitive type with heavy brow ridges, closely resembling *Pithecanthropus* of Java. Both *Sinanthropus* and *Pithecanthropus* are now regarded by anthropologists as belonging to the same early species of man, *Homo erectus*.

Meanwhile a further field of research was being developed by the Dutch anthropologist, G.H.R. von Koenigswald. Arriving in Java in 1931, he started making a study of the mammalian teeth that were being sold in the drug stores outside China. At first he asked for 'animals' teeth', which led to little success, but when he asked for 'dragons' teeth' he was shown a great variety of imported Chinese remains. In 1935, while examining dragons' teeth in drug stores in Hong Kong, he made a sensational discovery – a molar of an immense previously unknown primate which he named *Gigantopithecus blacki*. The systematic position of *Gigantopithecus* was violently discussed. It was classified as a giant orang-utang; as a

Fig. 3.7 *Tower karst scenery near Kweilin, S. China. The opening of a fossiliferous cave can be seen in the limestone face of the central tower. (Photo: H. D. Kahlke).*

hominid – a missing link in the pedigree to mankind, coming immediately below the *Pithecanthropus*–*Sinanthropus* stage. Most striking were the dimensions of the teeth that were larger than those of any other primate, living or extinct. If the proportion of tooth size to height were the same in *Gigantopithecus* as in apes and man, this animal would have been about 4 metres tall.

By 1956, a further seven teeth of this creature had been found in Hong Kong drug stores, but still their origin remained unknown. In that year, however, a Chinese farmer, hunting for fossils in a cave in Kwangsi Province, found a mandible with teeth and took it to a drug store in Luichow. Fortunately the chemist recognized it as something special and sent it to the Institute of Vertebrate Palaeontology in Peking. The staff at the Institute at once began a systematic examination of the cave, which was found to be situated in very picturesque countryside with rivers winding between tall limestone towers of the sort so often illustrated by Chinese artists (fig. 3.7). The entrance of the cave, accessible only by bamboo ladders, opened 90 metres above ground level in the face of one of these towers.

Excavations produced more remains of *Gigantopithecus*, which is now represented in the scientific collections by three jaws and over 1000 teeth. There was also a rich associated mammalian fauna, which is interpreted as of lower Pleistocene age. No artefacts were found in the cave and there were no traces of fire. Subsequently *Gigantopithecus* has been found in two other caves in southern China; and an earlier species is now known from the Miocene of India.

Since no skulls or limb bones are yet known, *Gigantopithecus* still remains a subject for discussion among anthropologists, some of whom now suggest that the creature was an aberrant ape, adapted to living in open country, the disproportionately large molars of which reflect seed-eating habits. It may not have been as large as previously suggested, perhaps only the size of a large gorilla.

Today fossil mammal localities, which a century ago were being worked only for 'dragons' teeth', are being scientifically excavated and studied in many parts of China, especially under the auspices of the Institute of Vertebrate Palaeontology in Peking (fig. 3.8). There is now a clear picture of the types of deposit in which 'dragons' teeth' have been found. In addition to caves, they are also known from river and lake deposits and from loess. It is recognized that all the remains are not

Fig. 3.8 *Palaeontological studies in China today. Visitors viewing a recently excavated stegodont skeleton of lower Pleistocene age in the Museum of Natural History, Beijing (Peking). (Photo: Institute of Vertebrate Palaeontology and Paleoanthropology).*

of the same age. Two especially distinct assemblages are identified – the Pliocene *Hipparion* (three-toed horse) fauna; and the Pleistocene *Stegodon–Ailuropoda* (panda) fauna. Remains belonging to the *Hipparion* fauna (including rhinoceros, large giraffes, antelopes, deer, hyaenas, sabre-toothed cats, and mastodon) are mined from bone beds of northern central China. In the *Stegodon* fauna both mastodon and *Hipparion* are lacking, having been replaced by *Stegodon* and a horse. There are many pigs, deer, pandas, bears, tapirs, and rhinoceros; and primates, (including monkeys and a very large orang-utang) are abundant. Remains of this age, which often show gnawing by porcupines, come mainly from the rock fissures of southern China.

The geographical origin of the 'dragons' teeth' still being sold in drug stores in the various parts of China, is now fairly well understood. Fossils of the Pliocene *Hipparion* fauna are distributed mainly in the shops of northern China and do not generally appear in the south, where the market is supplied with fossils from the *Stegodon* fauna of the caves. Fossils from both regions continue to be distributed abroad, mainly via Hong Kong, though there is some control over the Pleistocene material, the most important part of which remains in China for scientific study.

Extensive geographical differences are now recognized between the Pleistocene mammalian faunas of south China and those of the north, which have Siberian affinities; and it is gradually becoming possible to subdivide the two faunal complexes described above into even shorter time units. At least three such subdivisions occur within the *Stegodon–Ailuropoda* complex. *Sinanthropus* and its associated fauna are believed to be of middle Pleistocene age, whereas *Gigantopithecus* is apparently earlier. The two animals do not, however, belong to the same lineage. The discovery in 1970 of three teeth of *Homo*, in association with remains of *Gigantopithecus* (i.e. ante-dating *Sinanthropus*) in Dragon Bone Cave, Jian Shi district, near Hopei on the Yangtse River, leaves no doubt that the unravelling of man's history in China has only just begun.

Although studies of the Choukoutien cave site suffered a great set-back when the early *Sinanthropus* collections were lost during the Second World War, new remains have since been excavated by Chinese anthropologists, who have declared the site a protected area and have set up a museum there.

Today scientists can attribute no special medicinal properties to these Chinese 'dragons' teeth' or to the 'unicorn horn' of Europe or to the supposed remains of St Phanourios on Crete; and there is no satisfactory explanation of how similar beliefs about the value of fossil bones and teeth could have arisen quite independently in so many different parts of the world.

Chapter 4 ***Looking for remains of Ice Age mammals***

We have seen how, as recently as three centuries ago, those who accidentally came across fossil mammalian bones and teeth were usually at a loss concerning the identity of the animals concerned and often attributed fanciful names to them such as dragons, unicorns and giants.

The palaeontologist of today, familiar with mammoths, cave bears, and many other Pleistocene mammalian species, is no longer confronted by problems of this nature; and his knowledge of the types of place where such remains can be expected to occur often enables him to find them in astonishing abundance. Most people, however, still pass their entire lives without ever encountering any mammalian fossils, which they see only in museums. Where does the Pleistocene palaeontologist look for such remains and how does he record and interpret them?

Nearly all fossils owe their preservation to having become buried in the ground, sometimes at great depth, where they are likely to remain obscured from human view unless exposed by erosion or by accidental or intended excavation. The banks of rivers and coastlines subject to erosion are sometimes especially productive of such finds. The late Mr A. C. Savin, who lived throughout his life on the Norfolk coast in eastern England, spent much of his spare time from the 1870s until 1940 looking for animal remains that had been washed by the sea, at times of storm, out of the soft Pleistocene Forest Bed deposits that form the cliffs there. The specimens were not numerous and were mostly collected individually, but during the long period of time that he was making observation of the cliffs he assembled, with associated geological information, over 4000 mammalian bones and teeth representing the Etruscan rhinoceros, elephants, bear, sabre-toothed cat, rodents and many other species. His finds, the greatest collection of Forest Bed mammals ever recovered, are now in the British Museum (Natural History). Similar remains are seldom found today since long stretches of the cliffs known to Savin have since been obscured by coastal sea defences.

Not all mammalian remains come to light as the result of natural processes of erosion. In populated areas the demand for clay for brick making, sand for cement, and gravel for concrete and road construction, combined with the practice of digging deep holes for basements beneath new buildings in central areas, provides a fruitful source of artificial exposures. We have already observed that, in cities such as London, where most of the buildings stand on Pleistocene deposits of the River Thames and where there are many outlying gravel pits, mammalian finds are frequent. Indeed, there can be few older workmen from building sites and gravel pits in the London area who have not come across teeth and bones at some time, although increased mechanization has led to a decrease in the number of remains recovered in this way. Excavator drivers may nevertheless still spot large remains, such as bison skulls and mammoth tusks, from their cab windows; and smaller remains commonly come to light at the washing plants of gravel pits.

Such methods of recovering fossil mammalian remains are unfortunately very crude. Large specimens are likely to get broken, small specimens are overlooked and the palaeontologist may have no means of determining whether all the remains are contemporaneous or if those from more than one level have been mixed together. Whether he is studying remains from a natural exposure, such as a sea-cliff or river bank, or from artificial excavations such as those mentioned above, his work only begins to become meaningful if he can make a detailed study of the deposits in which they have been found. For this he will first cut a vertical section at the side of the exposure, in order to show up any layering and, if circumstances allow, will then conduct more detailed excavation, sieving part of the deposits for remains of rodents, molluscs and other small animals and collecting sediment samples for mineralogical study and to be searched for pollen.

Often excellent co-operation has been established between the owners of building sites and quarries, on the one hand, and palaeontologists on the other, who would never have had the resources to open up such deep excavations on their own. Many important scientific deductions have followed from investigations connected with such commercial excavations (fig. 4.1).

So far we have considered only how mammalian fossils come to light but not how they become fossilized, a study of which can provide useful guidance about where to look for such remains, which do not occur everywhere.

For analogy, let us consider what happens to the carcass of a mammal when it dies at the present day. Most frequently, mammals die in the open, on the surface of the ground, where they are exposed to the attentions of carnivorous animals, insects and to atmospheric weathering. Their chances of fossilization are very remote. Even where bones escape the attentions of scavengers they are unlikely to survive the effects of sun, frost, rain and bacteria indefinitely, although the speed at which disintegration occurs varies greatly, depending on the climatic zone concerned. Dr Malcolm Coe, who studied the rate at which some elephant carcasses decomposed in Kenya found that, after six weeks, only the bones survived and that these too soon began to disintegrate. In the permafrost terrain of the Arctic, on the other hand (much of it snow covered for as long as 10 months during the year), the same processes may be greatly retarded. Remains of musk oxen, believed to have been killed by Eskimos at the time of the ransacking of McClure's abandoned ship, *The Investigator,* at Mercy Bay, Banks Island, Northwest Territories of Canada, more than a century ago, have been found with the horns and some soft parts still preserved. At Inuit archaeological sites in such regions (for example on Ellesmere and Bathurst Islands) bone artefacts are commonly found to have survived on the surface of the ground for more than a thousand years; whale bones (which are larger and of which there is more to decompose) on raised beaches, for three thousand years or more. Even such remains, if left long enough, would nevertheless ultimately decay.

Fig. 4.1 *Scientific excavation in a cave accidentally opened by quarrying at Yealmpton, Devon; examining the hippopotamus bearing deposits, 1961. (Photo: A. J. Sutcliffe).*

Fig. 4.2 *Fossil skeleton of a hippopotamus, about 1 800 000 years old, in old lake deposits (Bed I) at Olduvai Gorge, Tanzania. (Photo: A. J. Sutcliffe, 1960).*

Only under special conditions does fossilization take place. For example, a carcass of a hippopotamus on the shores of an East African lake will soon attract a group of carnivorous maribou storks, which will quickly strip the flesh from it, although they will probably cause little damage to the skeleton. If, at this stage, there should be even a small rise in the level of the lake, the bones would become covered by silt; and, protected from scavengers and from the air, the first step towards their fossilization would have been achieved. Subsequent sedimentation might lay down tens of metres of later deposits on top of the skeleton, which would remain obscured from sight until exposed once more by denudation or excavation.

The fossil hippopotamus skeleton shown in figure 4.2 was buried in this way about 1·8 million years ago, in lake deposits now re-exposed by erosion at Olduvai Gorge, Tanzania. The bones of this skeleton are heavily mineralized and of great weight. Such mineralization, by percolating ground water, is usually a slow process and can provide a rough indication of the age of the remains, but the speed at which it occurs varies from site to site and such evidence must not be regarded as precise. Sometimes bones of late Pleistocene age, which have lost part of their substance by leaching, but which have not yet become mineralized, are lighter than equivalent modern bones.

The Olduvai hippopotamus skeleton, illustrated above, is magnificent for its relative completeness and for the way in which the bones are still associated in their natural relationship. The earliest palaeontologists obtained special delight in studying such complete remains and in making reconstructions of the living animal. Moreover, when many such finds of different ages were compared together, evolutionary lines could be established. Such studies are still important, but frequently the only mammalian remains at a site are scattered and fragmentary and thus of limited value for anatomical study, though nevertheless associated with a wealth of information about the way in which they had become preserved and about the environment of the animals concerned. These fields of study are known as taphonomy (defined by Efremov, in 1940, as 'the study of the transition, in all its details, of animal remains from the biosphere to the lithosphere') and palaeoecology respectively; and today they take up as much of the time of the mammalian palaeontologist as do studies of the morphology and evolutionary relationships.

Among aspects which the taphonomist must consider are the relationship between the mammalian remains and the sediments in which they occur; and to any associated plant or other animal remains. For example, mammalian bones in a conical heap of angular rock fragments and earth beneath a shaft in a cave roof probably represent individuals which had fallen into the cave; whereas finely stratified, horizontally bedded clays, silts and sands in the same cave are likely to represent the deposits of an underground stream. The degree of disarticulation of bones and their orientation may provide further information. Thus a complete associated skeleton in a silt or peat deposit at an open site might be expected to have been buried on the surface of a marsh, or in a similar environment; whereas rolled and scattered bones in a coarse gravel are more likely to have accumulated in the fast flowing central channel of a river. The proportion of remains of aquatic and terrestrial mammals; of carnivores and herbivores; and of young and old animals provide further important evidence. Thus some caves occupied by bears or hyaenas have been found to contain almost exclusively remains of these animals which had died there, whereas pitfall deposits of the same age in other parts of the same caves contain more representative samples of the surface fauna, with a greater proportion of herbivores. Animal tooth marks on bones and cut marks made by man may indicate a lair or a habitation site. Much attention has been directed by archaeologists to distinguishing bone fragments broken by man from bones damaged by other processes. Associated plant and invertebrate remains, if preserved, may provide information about the climate prevailing at that time and whether there was woodland or open country in the neighbourhood of a site. Rarely preserved stomach contents sometimes show what a mammal had eaten for its last meal and give further information about climate (fig. 4.3).

Remains of flesh-eating insects, different species of which are characteristic of the various stages of decomposition of a carcass may show how long this had been lying on the surface before it was buried. Surface cracking of bones may also indicate the time of exposure before burial.

From diverse lines of study, such as these, the taphonomist and palaeoecologist are able to provide the palaeontologist with background information which, in turn, may help in the understanding of peculiar skeletal adaptations or changes of form in response to environment.

Now that we have considered the sort of evidence that the Pleistocene mammalogist must record with his finds, let us look in greater detail at how he knows where to search. Although, in some places, fossil mammalian remains occur in remarkable abundance, such sites are of restricted extent and would be unlikely to be discovered by casual exploration. In order to find the right place the palaeontologist must look for sites of Pleistocene sedimentation. Commonly these are situated close to places where equivalent processes are still taking place today, especially along the courses of rivers and near lakes and bogs; so that valleys and hollows are more likely to prove rewarding than the upper parts of hills (which are subject to erosion) unless the hill in question is itself made up of Pleistocene sediment or has caves in it.

There are few types of deposit in which fossil mammalian remains are likely to be found, as we shall now see.

Fig. 4.3 *Wood fragments from the stomach of a 13 000-year-old mastodon excavated from the site of a slightly saline spring at Taima-Taima, near Falcon, Venezuela by Drs J. M. Cruxent and A. Bryan in 1976. The crushed ends are believed to have resulted from chewing. Preservation was possible only because the ground had been continuously wet since the death of the animal. An associated projectile point, in the abdomen, suggests that it had been killed by man. (Photo: BM(NH)).*

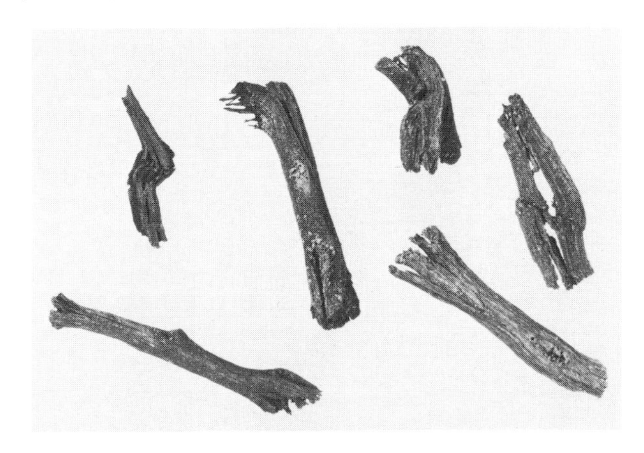

Fig. 4.4 *A spectacular marine fossil, now resting above sea-level a short distance behind the present-day coastline. Skeleton of the Pleistocene gray whale,* Eschrichtius *(skull in foreground, thorax beyond), being excavated in 1971 from old beach deposits of upper Pleistocene age at San Pedro, Los Angeles County, California. (Photo: Natural History Museum of Los Angeles County).*

Marine and coastal deposits

Since the Pleistocene is geologically relatively young, most marine deposits of this age are still below sea-level, where they are difficult to study. In consequence, although much information about sea bottom deposits has been obtained from bore hole studies, mammalian remains are unlikely to be found in any quantity except where such deposits are now situated above sea-level. Pleistocene marine deposits are thus a relatively unimportant source of mammalian fossils when compared with those that have accumulated on the land.

The Red Crag of Suffolk in eastern England (formerly worked commercially for phosphate nodules that were used for agricultural fertilizer) provides a fine example of a shallow water shelly marine deposit of Pleistocene age now partly above sea-level, though study of the occasional mammalian remains found in it is not without problems. Not unexpectedly, remains of whales and seals are more common than those of terrestrial mammals, which makes comparison with terrestrial faunas difficult; though such species as mastodon, elephant and deer, swept into the sea, are also represented. In addition, many of the fossils occur in a derived, sometimes rolled, condition, having been washed out of earlier terrestrial deposits by the Pleistocene sea and redeposited, already fossilized, in the same sands as contemporary remains. Distinction between these two groups of fossils has long presented a problem to palaeontologists. It is now accepted that, although the earliest mammalian remains from the Red Crag are of Eocene age (about fifty million years old) the Red Crag itself and the contemporary fossils in it are lower Pleistocene, less than two million years old.

Mammalian remains also sometimes occur in Pleistocene raised beach deposits, those of whales and seals again predominating (fig. 4.4).

River deposits

Probably most important of all the finding places of remains of Pleistocene mammals are the alluvial terrace deposits that occur along the flanks of rivers, sometimes at a considerable height above their present channels. Such terraces are of two main types – those of the upper part of a river, which slope gradually downstream at about the same gradient as the present river; and those of the estuary, which are controlled by the height of the sea. They have horizontal aggradation surfaces and terminate inland at the head of the tidal estuary of the time. Nearly the whole of London has been constructed on such sediments, and, as we have already seen, mammalian discoveries are frequent there.

Often these dry terrace deposits, which vary in constitution from coarse gravel laid down under torrent conditions (fig. 4.5) to silts and clays laid down in still water, contain rich accumulations not only of mammalian remains, but also of associated molluscs, plants and even insects, which give much valuable information about the climatic and other conditions under which the mammals were living.

Lake deposits

Some of the richest assemblages of Pleistocene mammalian remains have been found in old lake deposits. Lakes provide an especially favourable environment for fossilization, since not only do the carcasses of aquatic animals often sink to the bottom, where they soon become covered by sediment, but the remains of terrestrial mammals may be similarly buried upon the shore around the lake at the times of rising water.

Among the finest examples of such deposits are those of the East African Rift Valley (chapter 11), where Pleistocene faulting has elevated many of the old lakes to situations where they are now dry. These old land surfaces, with habitation débris of man, are commonly interbedded with lake deposits with remains of such aquatic animals as the hippopotamus (fig. 4.2), crocodile and catfish.

Marshes, peat and the effects of salt springs and bituminous substances

Both lakes and rivers are often bordered by marshes where, for example, mud accumulates at a lake head; or where a meander of a river, cut-off as an ox-bow lake, has become almost filled with sediment. Clays and silts are the characteristic deposits here. Often these have accumulated under oxygen-poor conditions so that they may contain well-preserved plant and insect remains. More rarely, there are the complete associated skeletons of animals that became bogged there. The Aveley elephants, previously mentioned, appear to be an instance of such entrapment.

In low-lying areas and in water-filled hollows on hillsides conditions may occur where concentrations of plant remains, prevented from decaying by lack of oxygen and by acid conditions, are preserved in bogs

Fig. 4.5 *A skull of a straight-tusked elephant preserved in Pleistocene gravels of the River Manzanares near Madrid, Spain, 1959. The skull has become separated from the rest of the skeleton and is lying with the upper molar teeth pointing skyward. (Photo: Keystone Press).*

Fig. 4.6 *The head of Tollund Man, (Iron Age, about 2000 years old) found in a Danish peat bog in 1950 and now preserved in Silkeborg Museum. (Photo: Silkeborg Museum).*

to become peat. Usually the greatest interest of the peat lies in its botanical content, but mammalian remains also occur from time to time. Sometimes the plant remains are mixed with other sediment in which case the deposit is known as 'detritus mud'. Some of the most astonishing peat bog mammalian remains, notably the Danish peat bog human burials, are of post-Pleistocene age but need to be included here for the sake of completeness. Since time immemorial, peat has been dug for fuel in many parts of the world. In Denmark, such excavations during the past two hundred years have led to the discovery of over 150 human bodies with various quantities of clothing; most in a remarkable state of preservation and many known to be of Iron Age date, about 2000 years old. In some instances, preservation was so perfect that the peat cutters, thinking that they had discovered a recent murder, called in the police. The best-known example is the Iron Age Tollund Man, found in 1950 (fig. 4.6). Not only was his hair preserved, but also his leather hat, a skin rope (apparently the cause of his death) fastened tightly round his neck; and the contents of his intestine, from which botanists were able to establish what he had eaten for his last meal.

Another well preserved body, also with a rope around its neck, was found by peat diggers in a bog in Cheshire, England, in August 1984. Although the soft parts of such a body may be most perfectly preserved, the acid conditions of a peat bog are less suitable for the preservation of bone. Of a similar body found at Damendorf, Germany, only the skin, leather belt and shoes survived; the bones having completely disappeared.

The peat bogs of Ireland have long been famous as the finding places of bones of the giant Irish Deer, *Megaceros giganteus.* In fact, the remains have mainly been found not in the peat itself, but in open-water mud deposits of late Pleistocene age, about 11–12 000 years old, underlying it. The peat, which is mostly post-Pleistocene, represents a later stage of sedimentation when the open water changed to bog. The largest of these giant deer remains are highly spectacular, some of the antler pairs having a span of as much as three metres. The Pleistocene mammalian fauna of Ireland will be further discussed in chapter 10.

Salt springs are a constant source of attraction to herbivores at the present day and the marshes around them sometimes contain accumulations of mammalian remains of Pleistocene age. Bones of hippopotamus and straight-tusked elephant, recovered during the course of motorway construction near Honiton in Devon, southwest England, in 1965, probably had some connection with a nearby salt spring, known as the Holy Shute Spring, long renowned for its supposed medicinal properties.

Better known is Big Bone Lick, Kentucky, USA, where immense quantities of remains of mastodon, bison and other animals have been found in a swampy area surrounding salt and sulphur springs. Big Bone Lick was one of the first fossil mammal localities to be investigated in North America, with finds dating back to the 1730s. The site is now preserved as a State Park.

At a number of localities in petroleum-producing areas, remarkable accumulations of remains of Pleistocene mammals, including mammoths, mastodons and ground sloths, have been found embedded in hardened asphalt. Best known of these is Rancho la Brea (literally 'tar ranch') in Los Angeles, California (see chapter 12). Other Californian examples include McKittrick and Carpinteria; and there is a further instance at Talara, Peru. At these sites fissures have allowed petroleum occurring at depth to reach the surface where, as a result of oxidation and the loss of volatile constituents, it has gradually consolidated to form asphalt. During Pleistocene times some substantial asphalt pools accumulated on the surface in this way, and the process still occurs on a reduced scale now. Whilst only partly hardened, such asphalt seepages present a serious hazard to animals (fig. 4.7).

Fig. 4.7 *Victim of a recent seepage of asphalt at Rancho La Brea, California; a ground squirrel (see chapter 12). (Photo: Natural History Museum of Los Angeles County).*

Asphalt is not the only bituminous substance in which remains of Pleistocene mammals have been found. At Starunia, eastern Europe, remains of a woolly mammoth and of a woolly rhinoceros of upper Pleistocene age with skin and flesh were found in 1907 during the course of excavation for a form of naturally occurring mineral wax known as ozocerite; and a relatively complete carcass of a woolly rhinoceros was found in 1929 as a result of a special excavation conducted to find more remains. The deposit is also highly saline and preservation appears to be the combined consequence of the oil and the salt. Remains of insects and plants were found with the rhinoceros carcass. The plant assemblage, which included dwarf birches and small-leafed willows, shows that tundra conditions, colder than at the present day, prevailed at the time of accumulation of the mammalian remains.

Glacial and periglacial deposits

So appealing to popular imagination are the frozen carcasses of woolly mammoths, bison (fig. 4.8) and other mammals that have been found in the arctic wastes of Siberia and Alaska (discussed at greater length in chapter 9) that mammoths are widely pictured as having lived on a terrain of snow and ice and even glaciers.

Let us consider more closely how such remains became buried and preserved. Although some present-day species of mammals, such as reindeer, collared lemming, polar bear and arctic fox can survive under conditions of seasonal snow and even ice, where there would appear to be little food available to them, all are dependent at the most northern periphery of their range on either the scanty vegetation of an arctic desert, or upon animals dependent on it, or on a carnivorous diet such as seals, fish or molluscs derived from the sea. The ice sheets themselves (and this applies also to the more extensive Pleistocene ice sheets) cannot support any mammalian life. Fossil remains may occur as erratics in Pleistocene glacial deposits (for example a glacially striated elephant tooth found on the Norfolk coast of eastern England) but they usually relate to a land surface pre-dating the glacial advance in question.

In the periglacial region, however, conditions for the preservation of mammalian remains have always been much more favourable. Such areas could support a limited fauna of cold-adapted species, the carcasses of which could sometimes be buried by gelifluction deposits before there was even time for them to decompose. Mudflows resulting from this process are potential traps for animals. It is recorded that, in 1947, about 25 of 150 reindeer that went to the beach at Nicholson Peninsula, Northwest Territories, Canada, became mired in a gelifluction flow. Herders managed to pull out 18, but the other seven were swallowed up in a short time. Probably they are still frozen in the permafrost today, potential frozen fossils of the future.

Although not all the carcasses of mammoths and other animals of Pleistocene age found in the frozen ground of Siberia and Alaska (see chapter 10) survived as the result of identical sequences of events, the process just described probably accounts for many of them. Further south, Pleistocene gelifluction deposits with remains of mammoth, woolly rhinoceros and other animals are common, though such deposits have thawed long ago so that only the skeletal remains are preserved.

Sand and gravel deposits laid down by water flowing from melting glaciers (glaci-fluvial deposits) provide a further environment where, more rarely, mammalian skeletal remains may become buried and preserved.

Volcanic and other wind-blown deposits

Since volcanoes sometimes erupt suddenly they have the ability to cause great destruction to plant and animal life, the remains of which may be quickly buried as potential fossils under thick layers of rock. This process can be of great value to palaeontologists since mammalian remains preserved in this way are likely to represent an entire cross-section of the fauna of

Fig. 4.8 *Frozen bison carcass from near Fairbanks, Alaska. A date of about 31 000 years was obtained on a piece of hide by the radiocarbon laboratory of the Geological Survey of Sweden. (Photo: T. L. Péwé).*

Fig. 4.9 In situ *plaster cast of the cavity left by one of the corpses at Pompeii. (Photo: Department of Antiquities, Naples).*

the time (i.e. a life assemblage) in contrast to, for example, caves and lake deposits where there may be a disproportionate over representation of remains of cave-dwelling carnivores and aquatic animals respectively.

Volcanoes produce two main types of deposit, fragmentary (or pyroclastic) material and lavas. Especially destructive are eruptions of the nuée ardente type, when clouds of incandescent dust roll down the slopes of volcanoes onto the plains below, burning up everything in their paths. During an eruption of this type from the volcano Mont Pelée in 1902, the 30 000 inhabitants of the town of St Pierre on Martinique and their domestic animals were annihilated by an asphyxiating blast which was so hot that it melted glass in the houses and set ships on fire in the harbour. In this instance, relatively little dust accumulated on the bodies.

The rapid mineralization that can be associated with volcanic eruptions sometimes leads to the preservation of the most unlikely soft parts of animals and plants. The Miocene ash deposits on Rusinga Island, Lake Victoria, Kenya, contain not only seeds and pieces of wood but also complete insects with even the moulds of the intestines preserved.

During the eruptions of Vesuvius which destroyed Pompeii in AD 79 most of the inhabitants apparently fled from the town, though some, who had remained, and a dog, were asphyxiated and buried beneath a deep layer of ash. This ash subsequently consolidated around the remains which decomposed, except for the bones, to leave cavities, some of which have been rediscovered during the course of excavations at the site. By pouring plaster into these natural moulds archaeologists have been able to reconstruct the forms of some of the people and animals that perished (fig. 4.9).

There is evidence that similar, potentially lethal, eruptions have occurred in many areas of the world during past eras and it is to be expected that some fossil mammalian remains in pyroclastic deposits result from similar processes. The finest known examples are of Tertiary rather than Pleistocene age. In 1978–9 excavations in Nebraska in a deposit of Miocene volcanic ash revealed the associated skeletons of over 200 rhinoceros, camels, horses and even birds, a tortoise and alligators buried ten million years ago by a volcanic eruption several hundred kilometres away from the site, which dwarfs by comparison the 1980 eruption of Mt St Helens in Washington State. Frequently, however, such deposits have been re-sorted by water or the bones represent a surface accumulation between ash layers; and the role of eruptions in the destruction of the animals concerned is difficult to assess.

It is unusual to find fossil mammalian remains in lava as the high temperature of eruption generally causes the destruction of organic matter by burning. For an example it is again necessary to refer to the Miocene. A natural hollow mould of a carcass of a rhinoceros, packed around by pillows of lava which had apparently flowed into water (where the dead and bloated animal was either floating or lying on the shore) was found near Blue Lake, Washington, USA in 1935.

Volcanoes are not the only source of wind-blown dust. Pleistocene loess (chapter 2) is another important source of mammalian remains. Since loess can accumulate very rapidly locally, as the result of drifting, bones lying on the surface of the ground have frequently been buried and preserved in it as fossils. In addition remains of marmots and other mammals are sometimes found in fossil burrows associated with old land surfaces between successive layers of loess. As we have already seen, some of the so called 'dragons' teeth', being sold in Chinese drug stores, have also come from loess deposits.

Cave deposits

Caves often contain extensive accumulations of fossil mammalian remains resulting from the animals concerned having fallen down shafts; having used caves as habitation places (for example bears and hyaenas); or having carried the remains of their prey underground. Once inside the cave, protection from outside climatic effects and, in limestone areas, alkaline conditions are highly favourable for the preservation of such remains. They will be discussed in greater detail in chapters 7, 8 and 12.

Chapter 5 **Dating Ice Age mammals**

We have considered the sorts of places where those concerned with Pleistocene mammals can find fossil remains, and how these originally became buried and preserved. After recording all the associated field evidence the palaeontologist must next try to interpret his finds. There are many lines of enquiry to follow. What did a mammal look like? What were its ancestors and was it ancestral to any living species? What other mammals coexisted with it and what were their ecological relationships? Under what climatic conditions did it live? What was its diet? Did it take part in any pattern of migration? If the mammal is now extinct, when did this happen and what caused it? Was it contemporary with human beings.

In such studies the need for accurate dating of the remains of the various mammals is especially important, since many of the lines of enquiry (for example studies of evolutionary changes, of migrations and extinctions) are dependent on a reliable time framework. Although the same general principles can be applied to the dating of deposits of all ages, the Quaternary had many peculiarities of its own and certain techniques can be used that are not applicable to earlier geological ages.

Three factors make the methods required for dating Pleistocene deposits rather different from those employed by geologists studying the rocks of the earlier eras. Often in earlier sediments the geologist is able to work on areas of tilted consolidated marine rocks that can be followed on land for hundreds of kilometres, allowing the examination of the relationships within long continuous sequences of strata. The Quaternary geologist, on the other hand, seldom can follow marine deposits for any distance as most are still below sea-level, where they can be studied only by means of bore holes. More often he has to rely on isolated terrestrial deposits – a peat deposit here; a cave deposit somewhere else – that cannot be correlated directly, because they are physically separated; and which may be difficult to relate on palaeontological grounds since chemical conditions might favour the preservation of, for example, only plant remains (but no bones) in the peat and of bones (but no plant remains) in the cave. A special difficulty of studying such fragmentary terrestrial deposits is that, even if all of them could be related in a correct time sequence they would still not represent the whole of Pleistocene time; indeed the formally recognized terrestrial time divisions may make up as little as 15–20 per cent of it. The palaeontologist must resist stretching his evidence to provide a continuous time scale for the Quaternary from terrestrial data that are obviously incomplete. Although sea bottom studies, previously mentioned, now provide a more nearly complete chronology for the Quaternary than can be obtained from the land, this method can only be indirectly applied to the dating of Quaternary mammals, remains of which are too rare in marine sediments to be encountered in bore holes.

Secondly, as we have already seen, the Pleistocene was unusual in being a time of great and repeated fluctuations of climate. These changes provide a valuable yardstick to which other Pleistocene events can be related; and those concerned with the dating of the Quaternary (unlike those working on most earlier eras) are heavily involved with matters of short term climatic change.

Thirdly, the Pleistocene is the most recent of the geological epochs, from which many of the plants and animals still survive, making possible study of their behaviour and ecological requirements as a basis for reconstructing the fossil evidence. In consequence, many of those concerned with Quaternary studies today are not primarily geologists but have entered the subject because of their knowledge of living plants or animals or of chemistry or climatology. Dating is a highly interdisciplinary operation and can only be achieved when all the specialists put their findings together.

In the remaining part of this chapter we will consider how the most important of the methods provided by these other disciplines assist in the establishment of a Quaternary time scale, to which we can then relate our mammalian studies.

Relative and absolute dating

Those concerned with dating geological events, including events during the Quaternary, make use of two different kinds of date: the so-called absolute date, which gives in years the time which has elapsed since a particular event; and the relative date, which tells only how one period stands in relation to another. Obviously, it is particularly interesting to obtain absolute dates where this can be done, but most methods for achieving this are developments of the last 40 years, so that the earlier workers had to make do entirely with relative dates. Absolute dating methods now make it possible, in favourable instances, to establish ages for particular horizons within such relative time sequences; but they are of restricted applicability and supplement rather than replace relative dating methods, which remain as important as ever in Pleistocene studies.

Relative dating

Almost any feature of the environment that changes with time and leaves some record in the ground can be used to set up a sequence of relative dates. The simplest situation is where a sequence of stratified deposits is superimposed at a single site. Figure 5.1 shows a section of an imaginary depression (a 'kettle hole') in the surface of a deposit of glacial clay, laid down by a former ice sheet, that has gradually filled with sediment, the uppermost layer of which contains remains of temperate plants.

We may infer that the glacial clay (layer 1), which is at the bottom of the sequence, is the oldest deposit and that layer 4 is the youngest. There may have been a break of sedimentation of unknown duration between beds 1 and 2 (an unconformity) and there was a change of climate from glacial to temperate between the time of deposition of beds 1 and 4. No absolute ages can be obtained from this part of the study but, with good fortune, it may be possible to correlate some of the beds with deposits of similar age elsewhere from other lines of evidence.

Unfortunately, Quaternary sedimentation is seldom so simple as this. Often the deposits show complex erosion or other disturbances that must be recognized if confusion is not to ensue.

In caves we are commonly confronted with the problem of partial washing out of Pleistocene infillings by stream action, leaving undermined consolidated deposits adhering to the walls and roofs and providing space for the accumulation of later deposits on the cave floor below. In this way the oldest deposits are higher than the later ones and the sequence appears to be inverted.

Figure 5.2 shows a section of an imaginary bone cave where such a situation has occurred. At one time the cave had been three-quarters filled with a bone breccia (layer 1) and this was sealed down by a flowstone (stalagmite) floor (2). The lower part of layer 1 was then washed out by stream action, leaving 2 forming an unsupported bridge across the cave with the consolidated upper part of 1 adhering to its lower side. Renewed sedimentation on the cave floor then led to the accumulation of layers 3–7, followed finally by a second flowstone floor, 8. Flowstone 8 is thus much younger than flowstone 2 which is situated above it;

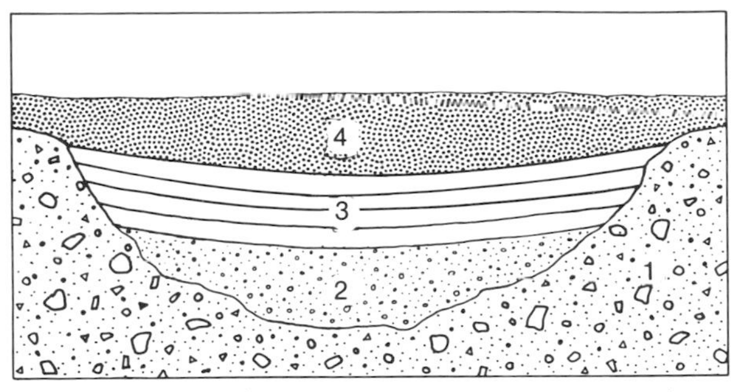

Fig. 5.1 *Relative dating. Section of an imaginary simple sequence of stratified Pleistocene deposits illustrating the age relationship of the various layers.*

Fig. 5.2 *Relative dating. Section of an imaginary cave passage showing the apparent inversion of strata as a result of the washing out of the lowest deposits.*

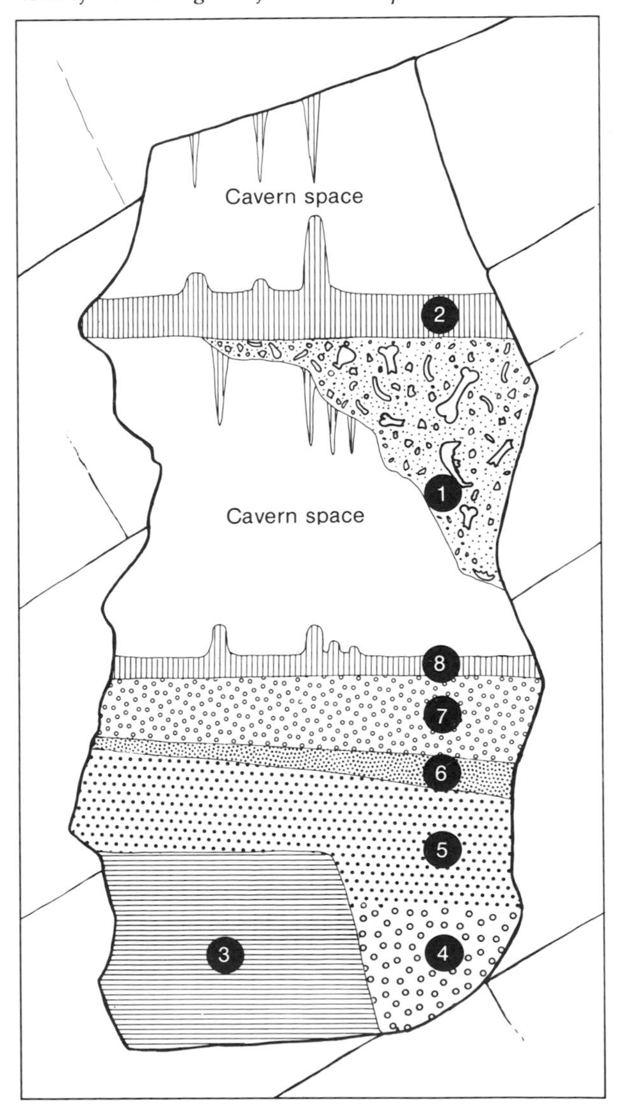

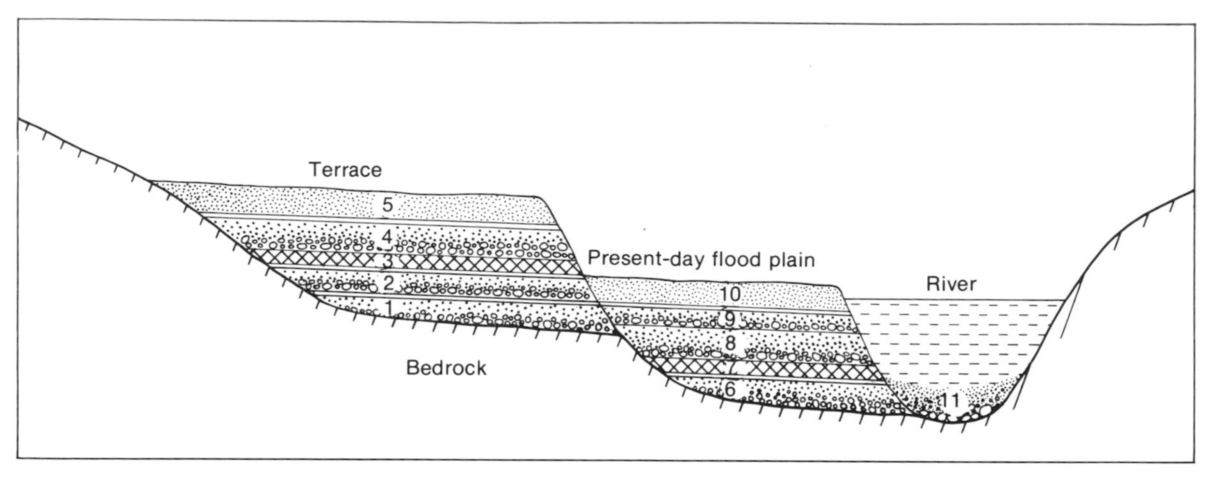

Fig. 5.3 *Relative dating. Transverse section across an imaginary series of river deposits flanking the course of a river.*

and the stalactites hanging from the lower surface of 1 and 2 are of similar late age. 'Inverted' sequences of this sort are very common in bone caves and special care must be taken to recognize their occurrence.

The structure of alluvial terrace sequences flanking the courses of rivers may be even more complex. Figure 5.3 shows a section across an imaginary series of river deposits in a situation where a river has experienced lateral displacement and deepening of its course. The oldest deposits (layers 1–5) now form a terrace above the general level of the river. Following their deposition there occurred a period of valley down-cutting and the cutting of a cliff against them by the river. The duration of this stage is uncertain but it could represent a period of many thousands of years while deposits, completely missing here, were laid down elsewhere. Then followed a further period of sedimentation and the accumulation of beds 6–10. Bed 10 is the top of the present-day flood plain and still receives sediment from the overflowing river at times of flood. In the bed of the river is deposit 11 which is swept along during every spate of the water and has not yet found a permanent resting place. It will be seen from the section that bed 2 (which is one of the lowest layers of the terrace) and bed 10 (which is the active flood plain of the river) occur at exactly the same altitude and could appear to be one continuous stratum, although a long period of time actually separates the two deposits.

Even more complicated than the study of caves and river terraces is the interpretation of the 'raised beach' and 'submerged forest' deposits, which provide evidence of past changes in the relative levels of the land and sea around our present-day coastlines. Such deposits, or those lying upon them, are an occasional source of fossil mammalian remains (fig. 4.4, p. 38), which may be locally very abundant in old sea caves associated with the raised beaches. If the age of the fossil shore line feature can be determined from sea-level studies, then the age of the mammalian remains can also be established.

Since the main oceans form one continuous sheet of water around the world, it should in theory be possible to correlate equivalent structures (such as old shore lines or coral reefs, even sea bottom deposits with evidence for a particular depth of water) at similar heights in all parts of the world, provided that altitude changes have resulted entirely from changes of sea-level and not from local elevation or subsidence of the land.

Sea-level specialists recognize three principal processes that can lead to relative changes in the level of the sea and the land, which must be clearly distinguished as a basis for all subsequent work.

Firstly, and most important, are the changes in the actual level of the sea, the principal causes of which, during the Quaternary, were the fluctuations in the size of ice caps. Low sea-levels correspond to glacial advances, high sea-levels to interglacials. This effect would have been compounded to a small extent, perhaps a metre or so, by contraction or expansion of the sea volume in response to temperature changes. Such changes are described as eustatic. Secondly there are local downwards and upwards movements of the land itself, in response to loading and unloading of superimposed weights. Such changes are described as isostatic. Their main cause during the Quaternary was the formation and melting of the great continental ice sheets. Thirdly there are instances of the relative changes between the level of the land and sea caused locally by earth movements (both mountain buildings and subsidence); and these are described as tectonic.

Another factor that must also be taken into consideration is the recent discovery from satellite investigations that the surfaces of the oceans do not, in fact, form a simple sphere but have quite extensive elevations and hollows. Any lateral movement of these structures will also cause an apparent change of sea-level.

It follows that considerable caution must be exercised when attempting to relate old shore line deposits in different areas on altitudinal grounds alone, especially where there has been tectonic displacement of the land. Early Pleistocene shore lines (which are more likely to have experienced tectonic displacement and which may subsequently have been re-submerged) are more difficult to interpret than those of the last 150 000 years or so which, being fresher, do offer reasonable scope for dating associated fossil remains. The possibility that sea-level stood at about the same height of successive occasions must also be taken into consideration.

Let us consider the combined result of eustatic and isostatic processes working together. Figure 5.4a shows a section across an imaginary land mass at a time of partial glaciation. The sea-level is low because much of the water is locked up on the land as ice. That part of the land which is covered by ice is isostatically depressed but the vertical extent of this decreases gradually towards the unglaciated part, where there has been no isostatic depression.

In figure 5.4b (which represents a situation such as we have at the present day) it is supposed that a warming of the climate has caused the ice sheet to melt, with consequent eustatic rise of sea-level, causing the submergence not only of the previous shore line but also of the most low-lying parts of the unglaciated area. In this way an old land surface, that may contain the fossil remains of mammals which lived on it, becomes the sea bottom.

In the glaciated area a different situation arises, since melting of the ice leads to an unloading of weight

Fig. 5.4 *Relative dating. Section of an imaginary area of land below, at time of partial glaciation and bottom, following the melting of the ice sheet, with consequent eustatic rise of sea-level and local isostatic re-evaluation of the land.*

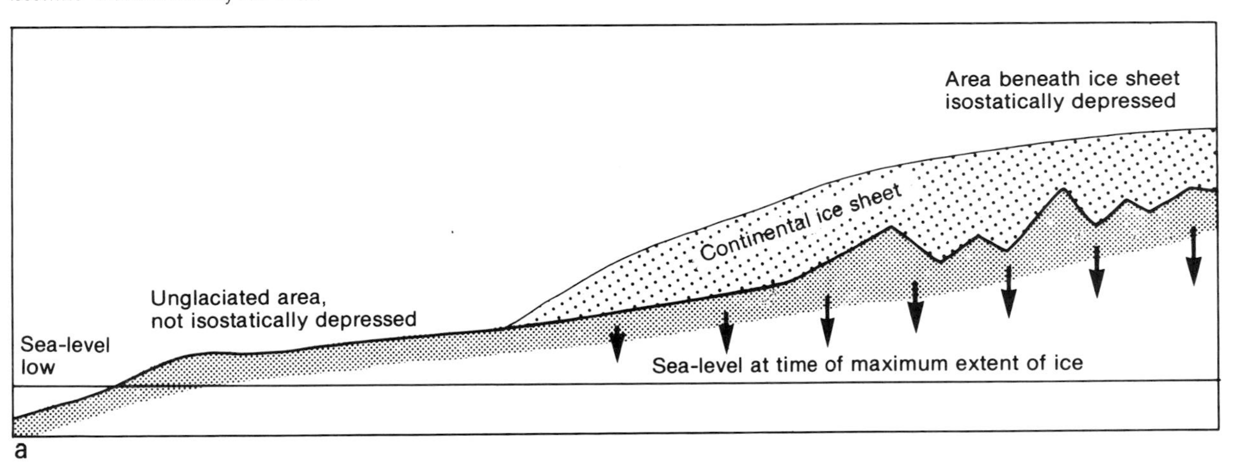

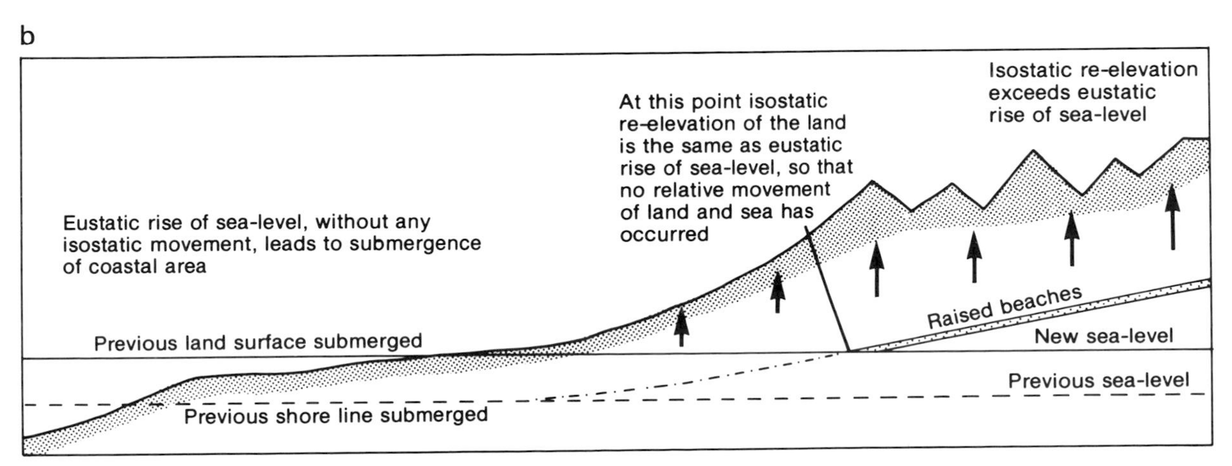

which is followed by isostatic re-elevation of the underlying land.

In regions where such isostatic re-elevation exceeds the eustatic rise of the sea the old shore line comes to lie above sea-level, where its cliffs, sea caves and beach deposits may be preserved with astonishing freshness. Such elevated shore lines slope downhill away from the areas of former ice cover towards the unglaciated areas, where they merge with submerged shore lines of the same age. At the junction of these two zones there is a short stretch of coastline where the eustatic rise of sea-level was the same as the isostatic re-elevation of the land. Here no change in the relative levels of the land and sea has occurred.

The British Isles (fig. 10.13, p. 143) provide a classic instance of a shore line structure of the type just described. Scotland is an area of recent isostatic uplift, with raised shore lines of very late date. The South of England and North Sea, on the other hand, are areas of eustatic submergence.

Marker horizons

So far we have considered only the spatial relationships of deposits at individual sites and regions. Let us now look at some of the criteria (other than absolute dating methods, which will be described later) whereby individual strata at different sites may hopefully be correlated with one another, allowing relative time sequences to be established over longer periods of time.

One of the simplest means of correlating two series of physically separated deposits that have a time overlap is to locate a marker horizon where the same readily identifiable event can be recognized at both localities. Layers of volcanic ash (the petrological composition of which varies from volcano to volcano and eruption to eruption; and which may settle over vast areas of land and in the sea, sometimes far from the original source) are of special value for this purpose. This method (tephrochronology) has been used with special success in East Africa where a recognizable ash bed, also shown by absolute dating methods to be 1·8 million years old, made it possible to relate important primate-bearing deposits at widely separated sites in Ethiopia and Kenya (see chapter 11). Unfortunately, however, volcanoes are restricted to limited areas of the world, and there are many regions (for example the British Isles) where this method cannot be applied.

Studies on palaeomagnetism can also be used to correlate strata from different sites. The earth's magnetic field has undergone a number of reversals in polarity in the past. A fossil record of these reversals is preserved in certain types of rocks, notably lavas and fine-grained sediment (including deep sea and lake

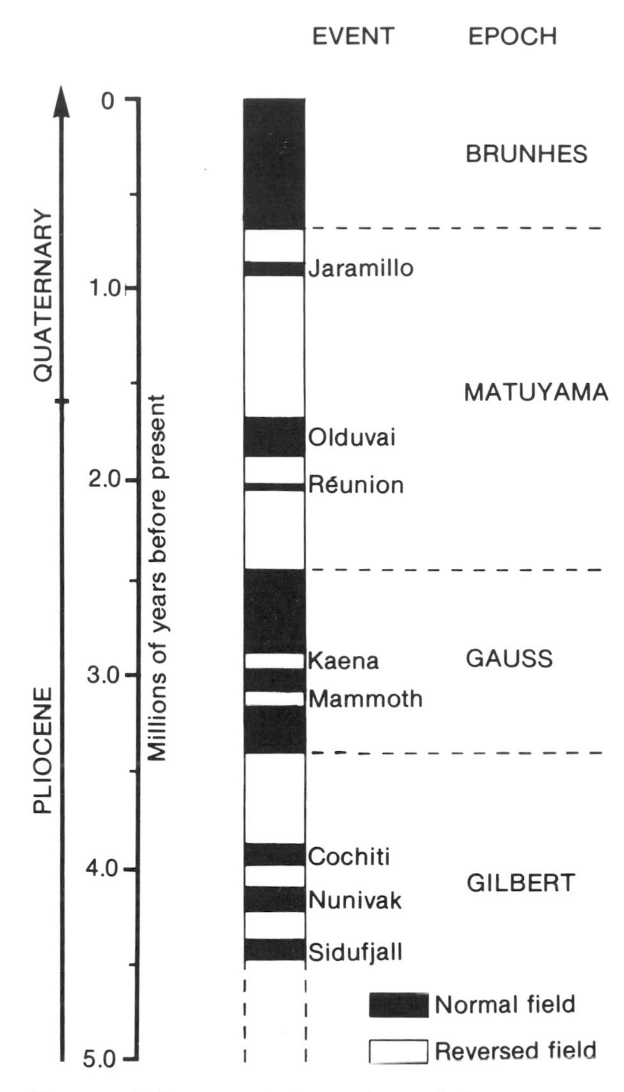

Fig. 5.5 *Palaeomagnetic changes during the last 4 500 000 years. Divisions are of two orders of magnitude: epochs, during which the polarity is predominantly normal or reversed over a relatively long period of time; and events, (mostly named after their place of discovery) which represent relatively short changes of polarity within the epochs. Absolute time scale established mainly by potassium argon dating, see p. 56–57. (After Berggren* et al.*, 1980).*

sediments) which take on the polarity of the earth's magnetic field at the time they are deposited. Using absolute dating methods, a time scale for these magnetic reversals has now been established that goes back far into geological time (fig. 5.5).

Since these reversals, which can be measured with a magnetometer, are a global phenomenon and the pattern of their history must everywhere be the same, this catalogue of magnetic events is useful in relating contemporaneous phenomena all over the earth. Evidence of a single event (i.e. a normal or a reversed polarity) is insufficient to allow recognition of the event concerned unless other lines of evidence are also available, but sequences of events are unique and can be related to the chronology shown in figure 5.5.

Sometimes deposits that are climatically diagnostic, for example glacial tills (which sometimes contain foreign rock fragments that may be characteristic of a particular glacial event), gelifluction deposits, loess and weathering horizons, make it possible to recognize the same climatic event at different, not too widely, separated localities. This method is nevertheless not without problems, because the number of climatic fluctuations that occurred during the Quaternary is so great.

Palaeontological evidence

Valuable though many unfossiliferous Pleistocene deposits may prove to be as climatic indicators, they usually do not have as great potential in chronological studies as those with fossil animal or plant remains, the individual species in which may be climatically diagnostic. In southern Britain, for example, remains of both hippopotamus and reindeer are found commonly, though never in the same deposit. The hippopotamus is usually interpreted as an indicator of warm conditions, reindeer of cold. But the mammalian evidence (including the evidence provided by man and his cultural remains – the province of the anthropologist) must not be considered in isolation. Remains of other animals and plants can be a source of similar information. Especially important among the remaining faunal groups are the other classes of vertebrates; marine, freshwater and terrestrial molluscs; insects; and from marine brackish water deposits only, foraminifera (protozoa). Although the possibility that the same 'hand of cards' may have been dealt twice must be constantly kept in mind, associated assemblages of animal and plant remains may sometimes be locally so distinctive that they can give an exact chronological position for the deposit in which they were found. Such assemblages can be considered as another type of marker horizon.

Those studying pre-Pleistocene fossil remains can often refer to continuous stratified sequences which accumulated over long periods of time, during which there were far-reaching evolutionary changes. The fossil species change progressively from layer to layer, with little likelihood of their changing back to those characteristic of earlier stages, and can conveniently be used for determining the relative ages of the various beds. Unfortunately this last mentioned principle can be applied to only a limited extent in Quaternary studies. This period of time was of relatively short duration, and provided little time for evolutionary change. Most of the molluscs, insects and foraminifera of the lower Pleistocene were species that are still living today (though not necessarily in the same region). Their value for dating purposes lies in their climatic significance or as constituents of diagnostic faunal or floral assemblages.

Some quite extensive evolutionary changes did occur among the mammals but, as we have already seen, the Quaternary was a time of massive geographical displacements of both fauna and flora as a result of changes of climate. In consequence it is seldom possible to follow mammalian evolutionary lines continuously at one place; and it is often difficult to distinguish evolutionary changes from changes resulting from immigration of closely related existing forms from some other region.

Among the mammals of greatest value for dating deposits of Pleistocene age are the rodents which, because of their short generation time, have greater potential for evolutionary change and for colonizing new environments than can be achieved by slower breeding mammals.

Such an evolutionary lineage, which has proved of special value in the interpretation of the Pleistocene sequence of Eurasia, including Britain, can be followed from the lower Pleistocene genus *Mimomys* to the present-day *Arvicola*, the water vole. *Mimomys* is characterized by rooted cheek teeth which wear down in old age. A gradual evolutionary transition can be followed to the condition in *Arvicola*, which has continuously erupting teeth which can be employed for chewing much coarser food without wearing out (fig. 5.6a). *Mimomys* is characteristic of the lower and lower middle Pleistocene; *Arvicola cantiana* of the

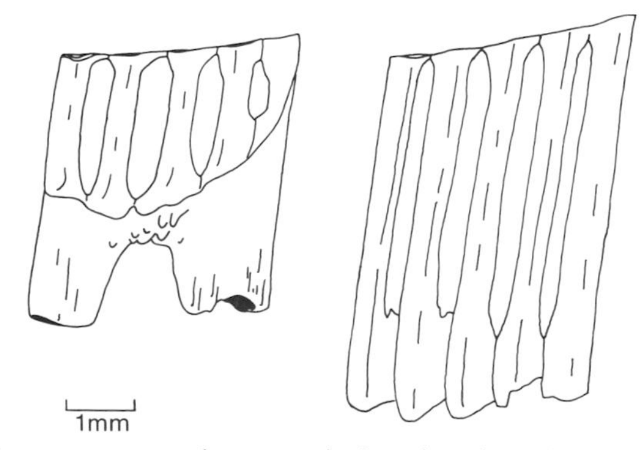

Fig. 5.6a) *Lateral views of cheek teeth of the rodent* Mimomys *(left) and of its descendant,* Arvicola *(right). Teeth of the former cease growth after eruption and develop roots; those of the latter continue to grow throughout the life of the animal and so lack roots. (Drawn by A. P. Currant).*

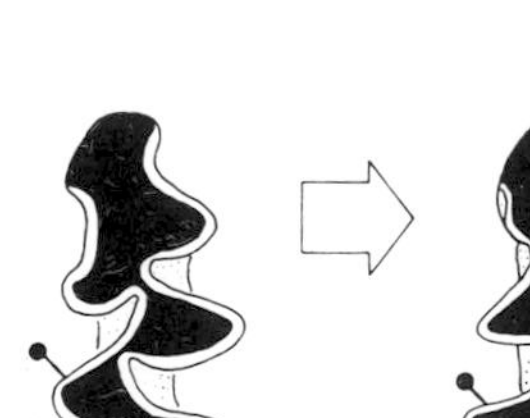

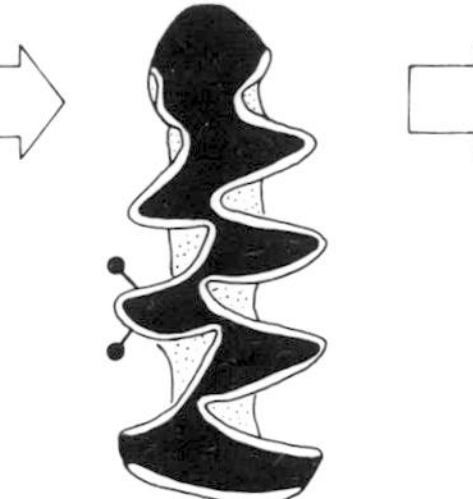

Fig. 5.6b) *Evolutionary changes observed by Von Koenigswald in the thickness of the enamel as seen on the biting surfaces of the cheek teeth of the rodent* Arvicola. *In the middle Pleistocene species,* A. cantiana *(left), the enamel is thicker on the posterior sides of the loops, as indicated by the sizes of the two circles. In the upper Pleistocene and present-day species,* A. terrestris *(right), the condition is reserved. There is a full range of intermediate forms.*

upper middle Pleistocene and *A. terrestris* of the upper Pleistocene.

The other vertebrate groups offer less scope than the mammals for establishing evolutionary series during the Pleistocene, though they do often provide valuable information about the prevailing climatic conditions, especially birds and reptiles. The European pond terrapin, *Emys orbicularis,* for example, is an indicator of warm summers. Although this animal appeared repeatedly during successive interglacials in the British Isles, it no longer occurs further north than southern and eastern Europe, owing to the summers normally being too cool and short to allow the eggs to hatch.

Similar evidence may be provided by the invertebrate groups. Climatically significant molluscs from the British Pleistocene include the freshwater bivalve *Corbicula fluminalis* (now extinct in Europe but occurring in the rivers of North Africa) from, for example, the interglacial mammal-bearing terrace deposits of Swanscombe, Kent; and the terrestrial gastropod *Columella columella* (which today has a high arctic and alpine distribution) from the cold climate of Ponders End – a well-known woolly mammoth locality – in the Lea Valley north of London.

Foraminifera can be useful in the study of associated mammalian remains for, not only are they sensitive to temperature changes and thus good climatic indicators in higher latitudes, but (their habitat ranging from brackish to fully marine) they can also provide us with valuable evidence about the height of the contemporary sea-level, which in turn may provide a guide to the age of the deposit concerned. They have also been used extensively in oxygen isotope dating of Pleistocene marine deposits. In a world-wide context, the planktonic foraminifer *Globorotalia truncatulinoides,* which evolved from *G. tosaensis* at approximately the Plio–Pleistocene boundary, and

Fig. 5.7 *Foraminifera indicative of Pleistocene or later age, from the earliest Pleistocene of Italy. Left,* Globorotalia truncatulinoides; *right,* Hyalinea balthica. *Highly magnified stereoscan photographs: greatest diameter 0·45 and 0·36 mm, respectively.*

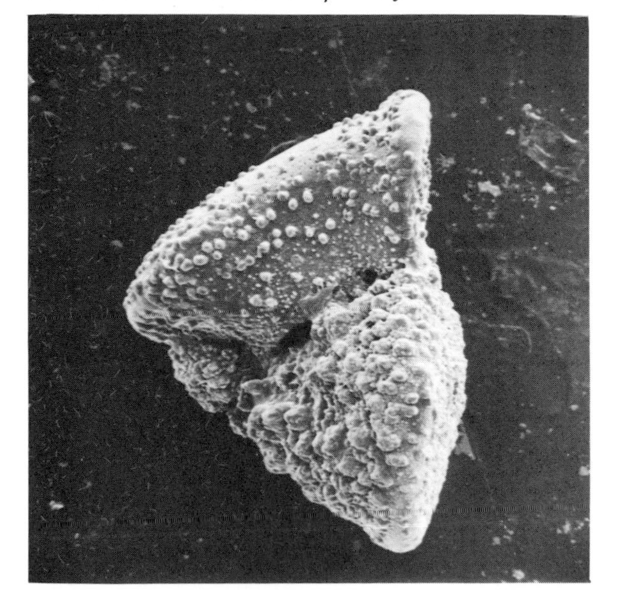

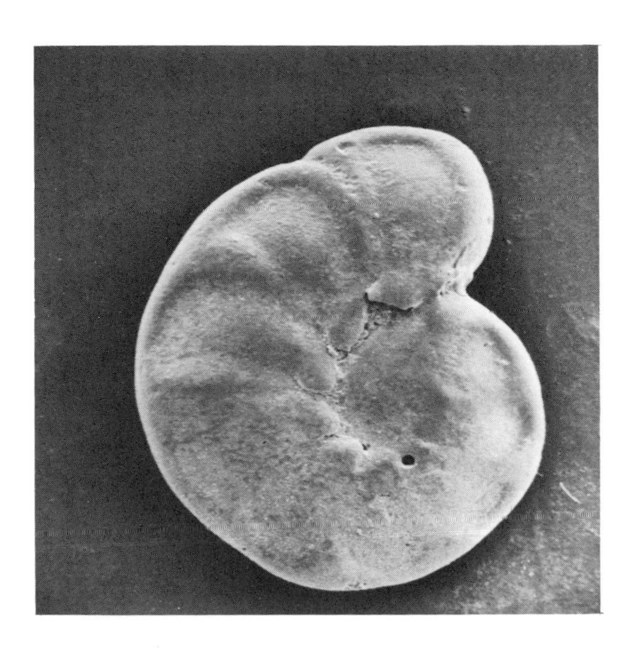

the benthic species *Hyalinea balthica*, which made its first appearance at about the same time, are useful indicators of Pleistocene or younger age (fig. 5.7). The sudden global extinction of *Stylatractus universus* (one of another group of single-celled marine organisms – the radiolaria), about 410 000 years ago provides a further useful marker horizon in marine Pleistocene sediments.

Of special value in the dating of mammalian finds from Pleistocene sites is the information provided by the study of associated insect remains, especially those of beetles which are particularly robust and make up the greatest part of any fossil insect assemblage. Many of the insects represented in Pleistocene deposits are species that still survive at the present day (although often not in the same region) and their present-day distribution and ecological requirements indicate with great accuracy the likely climatic conditions formerly prevailing. Not only do such studies provide information about temperature but also whether the climate was continental or oceanic and even about the most likely vegetation of the neighbourhood.

Among the insects represented in the 120 000-year-old hippopotamus-bearing interglacial river terrace deposits of Trafalgar Square, London, is the marshland beetle *Oodes gracilis*, no longer found in Britain and which today has a more southerly distribution, ranging across central and southern Europe (fig. 5.8). In contrast, a beetle fauna from a river terrace deposit with remains of woolly mammoth, woolly rhinoceros, horse and bison at Fladbury, Worcestershire, England, dated by carbon[14] (see absolute dating, below) as about 38 000 years old, was found to include many species no longer present in the south of Britain, but having a much more northern and even arctic distribution at the present day – evidence of the greater severity of the British climate 38 000 years ago.

Plant remains of Pleistocene age, like those of insects, also provide important climatic information which in turn may provide evidence about any associated mammals.

The study of Pleistocene plants is a much older discipline than that of insects. Under favourable conditions entire tree trunks and seeds and leaves may be preserved and the occasional survival of plant remains in stomach contents (fig. 4.3, p. 37) and fossil droppings is of even greater interest. Such remains accurately reflect the diet of the plant-eating mammals concerned and many indicate whether they were living in a warm or cold, dry or wet environment.

Today the study of Pleistocene plants is a sophisticated discipline. Samples of plant-containing sediment are usually studied in two parts. One part, which is of the greatest volume, is sieved for leaves, twigs and seeds (the macroscopic remains) (fig. 5.9); whilst a

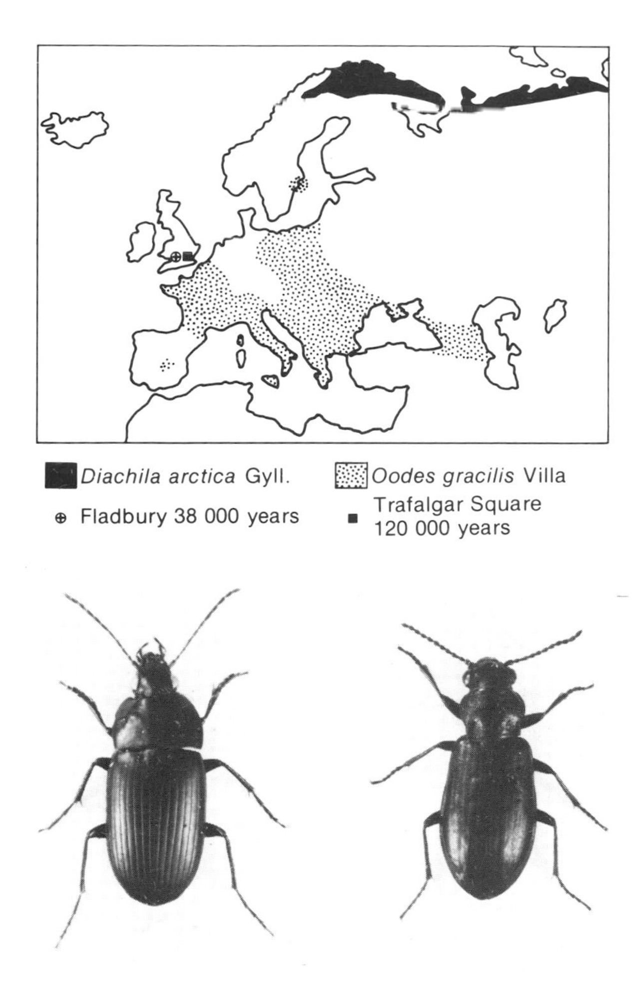

Fig. 5.8 *Top, present-day distribution of two species of beetle represented in British Pleistocene deposits.*
Above, left, Oodes gracilis, *remains of which were found in the hippopotamus bearing deposits of Trafalgar Square, London (c. 120 000 years old). The southern distribution of this and other beetles from the same deposit supports the supposition that the British hippopotamus was an interglacial species.*
Above, right, Diachila arctica, *one of the beetles from the mammoth bearing deposits of Fladbury (c. 38 000 years old). Its northerly distribution, like that of many of the associated species, indicates climatic conditions more severe than at present. Data supplied by R. Coope. (Photos: British Museum (Natural History)).*

Fig. 5.9 *Some macroscopic plant remains found associated with bones and teeth of hippopotamus and other warm climate mammals in terrace deposits of the River Thames at Trafalgar Square, London. Top left, fruit of the cocklebur,* Xanthium *sp; top right, fruit of water chestnut,* Trapa natans; *centre, three hazel nuts,* Corylus *sp; bottom left, half of the winged fruit of a southern European maple,* Acer monspessulanum; *bottom right, a hawthorn twig still bearing its thorns. The temperate nature of the flora (*T. natans *and* A. monspessulanum *are southern European species no longer occurring in England) confirms the interglacial data supposed from the mammalian evidence. Associated insect remains (Fig. 5.8) are also of warm climate species.*

much smaller sample is processed for pollen and spores (which are highly resistant to decay) which are then examined under the microscope (fig. 5.10) This last mentioned study is known as palynology.

When all the plant species from a particular locality have been identified consideration can then be given to their climatic significance. In Britain, plants of the temperate forest stages include the hemlock spruce, *Tsuga* (now extinct in Europe); oak, *Quercus*; hornbeam, *Carpinus betulus*; elm, *Ulmus*; lime, *Tilia*; wingnut, *Pterocarya faxinifolia*; hazel, *Corylus avellana*; water chestnut, *Trapa natans*; and the grapevine *Vitis*. Plants of the open country cold stages include mountain avens, *Dryas octopetela*; purple saxifrage, *Saxifraga oppositifolia*; dwarf birch, *Betula nana*; polar willow, *Salix polaris*; and dwarf willow, *Salix herbacea*.

Whereas such contrasting floras as those described above make the distinction between warm and cold stages relatively simple, it is far less easy to distinguish successive interglacials or successive cold stages from one another on the evidence of plant remains, as many of the same species reappear repeatedly. Oak and elm, for example, are characteristic of the warmest part of all the British interglacial stages and they provide no means of determining which one is represented. Other plants, such as the hemlock spruce, *Tsuga*, of the lower Pleistocene, and the fir, *Abies*, of

Fig. 5.10 *Greatly enlarged stereoscan photographs of Recent pollen grains showing the sort of differences that permit specific identification. Right, oak,* Quercus *sp. a temperate species; left, dwarf willow,* Salix herbacea, *today restricted to montane and arctic habitats and interpreted from Pleistocene localities as evidence of severe climatic conditions. Size of grains 0·033 and 0·025 mm respectively.*

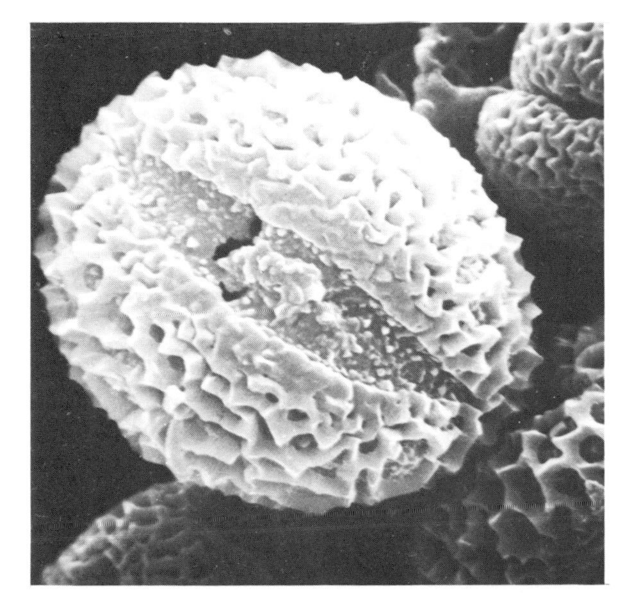

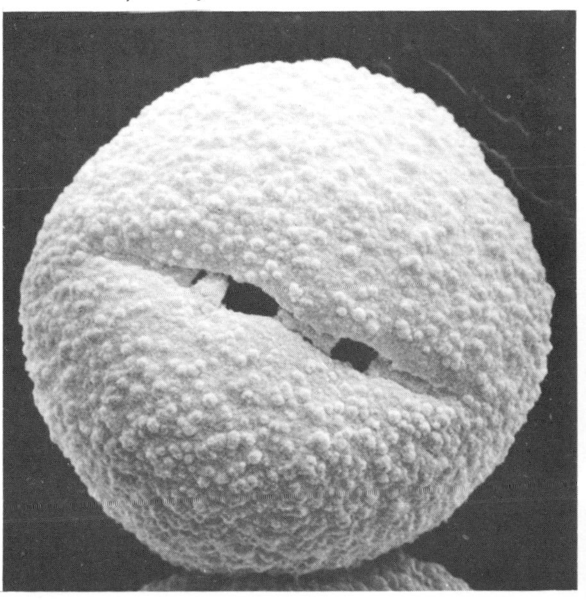

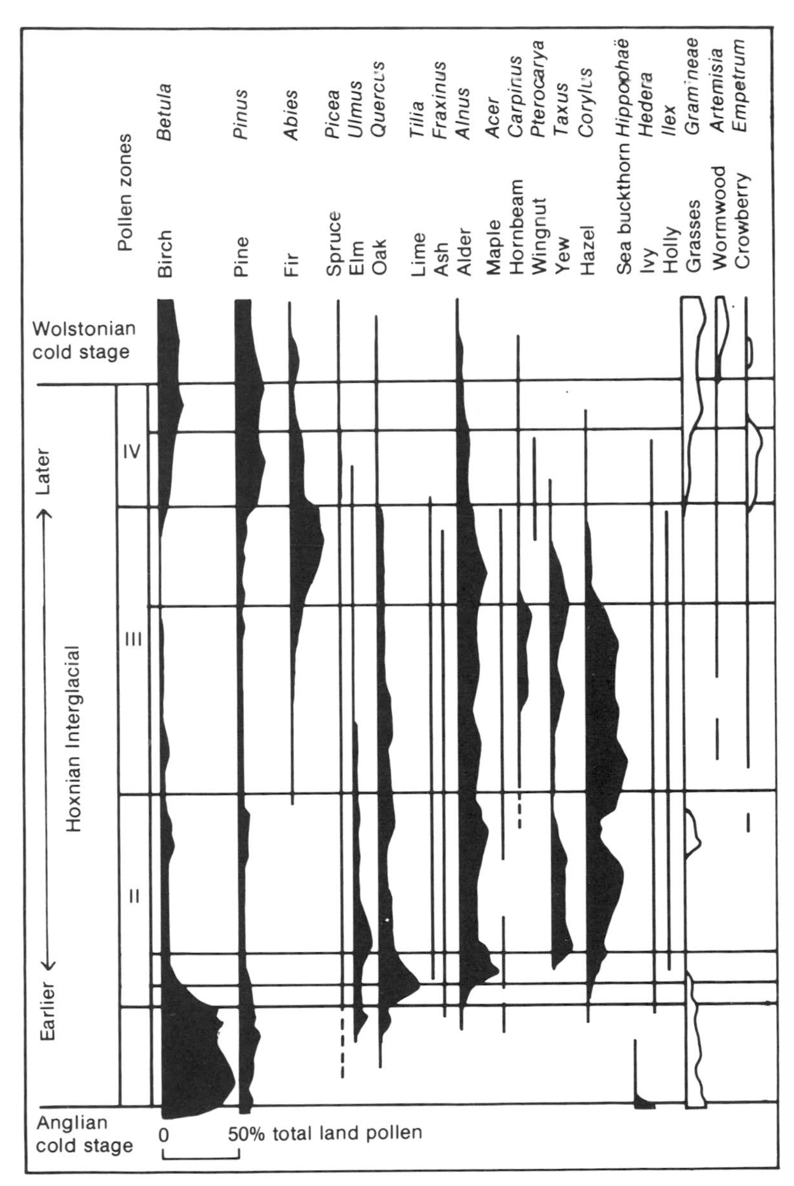

Fig. 5.11 *Pollen diagram from an English interglacial site. This example from Marks Tey, Essex, is of Hoxnian Interglacial age (see chapters 6 and 10). The vertical divisions, numbered pollen zones I–IV, represent stages of time (older at the bottom); the irregular black areas represent the relative abundance of pollen of the various plant species during these stages. A complete interglacial cycle is represented, with birch and pine abundant during the early and late zones I and IV. During the warm zones II and III there was an abundance of elm, oak, alder, yew and hazel (after Turner, 1970). Pollen diagrams can be constructed to illustrate different data. This example (a composite summary diagram from several closely situated auger holes) shows land pollen percentages—trees black, dwarf shrubs and herbs white, aquatic plants omitted. Other schemes may show, for example, percentages of arboreal (tree) pollen or pollen concentration within a sedimentary unit.*

the middle Pleistocene, have restricted distribution and are important aids to distinguishing earlier and later interglacials.

In areas beyond the limits of the Pleistocene ice sheets changes of rainfall can similarly be reconstructed from the fossil plant evidence.

In the early days of development of the studies of Pleistocene plant remains it was customary to produce a list of species found at a particular site; and this was used to establish the environment at the time the deposit was laid down. Such a method does not make the best use of the evidence, however, and today it is more usual to calculate the abundance as a percentage of the total number of pollen grains from samples collected at 10-cm intervals throughout the deposit, and to plot these on a diagram such as that shown in figure 5.11. When interpreting such pollen spectra, however, it cannot be rigidly assumed that the relative abundance of pollen grains of particular species of plants necessarily accurately reflects the abundance of the plants in question. Some plants produce more pollen than others; some produce their pollen at different times of year from others; some pollen can be carried by the wind further than that of other plants; that of some plants is less prone to decay and more likely to become fossilized than that of others; and trampling by large mammals along river banks can cause the destruction of forest and its replacement by herbaceous plants. The interpretation of pollen diagrams is a highly specialized discipline.

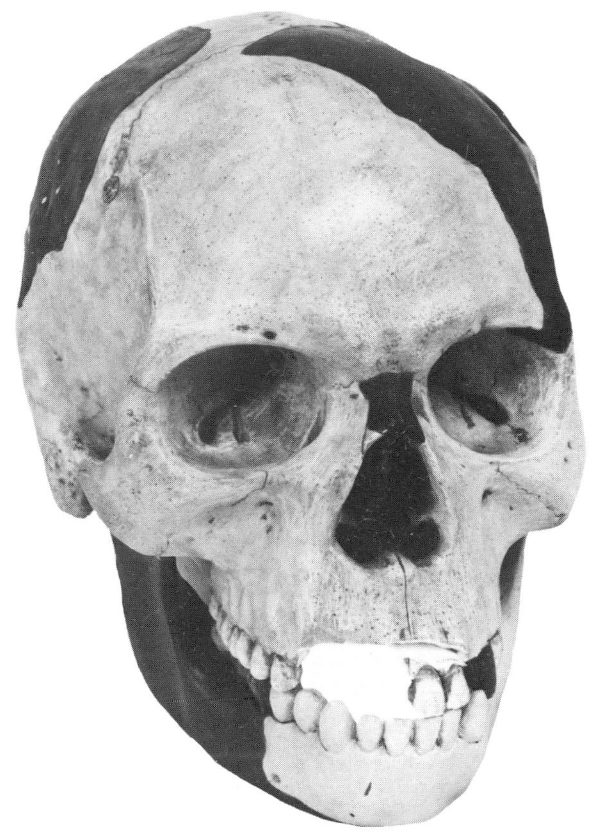

Fig. 5.12 *The Piltdown skull, as reconstructed by Dr A. Smith Woodward, about 1920; original specimens dark, reconstructed parts white. Although at first acclaimed as a lower Pleistocene 'missing link' it was subsequently shown by chemical and other tests to be based on fraudulently planted human and ape remains of no great antiquity.*

Chemical methods of establishing relative dates

The relative ages of mammalian remains at individual sites can sometimes be established by quite simple chemical methods. Bones and teeth are composed of two main parts – the protein collagen and a mineral called hydroxy-apatite, which is a calcium phosphate. On burial, two processes of chemical change begin. The collagen breaks down and the soluble products are washed away, with the result that the amount of nitrogen decreases with time. The mineral part of the specimen picks up small traces of certain elements from the ground water, in particular fluorine and uranium. Fossil bones and teeth of early date are likely to have a lower nitrogen content and a higher fluorine and uranium content (all of which can be determined by analysis) than those of more recent age. The rates at which these changes occur are dependent on conditions of burial, so that such methods of dating cannot be used for correlating remains found at different sites, but when applied to individual sites they can nevertheless be of the greatest value.

One of the most spectacular applications of the methods just described was in the exposure of the Piltdown skull hoax in 1953. Between about 1911 and 1915 a series of fragments of a very thick primate skull, a canine tooth and part of a lower jaw were found in association with parts of teeth of mastodon, stegodon, hippopotamus and beaver and artefacts in gravel deposits at Piltdown, Sussex, England. The finds were interpreted as being of lower Pleistocene age and the skull was thought to be a missing link in the ancestry of man, for which a new genus, *Eoanthropus* or 'dawn man' was established (fig. 5.12).

As the years passed, and other remains of early man were found in China and Africa, it became increasingly difficult to accommodate Piltdown man in any rational scheme of human evolution. In 1949, Dr K. P. Oakley of the British Museum (Natural History) carried out fluorine tests on a variety of finds from Piltdown, to find whether the skull, jaw bone and associated mammalian remains were all of the same age. The results were astonishing for, whereas the other mammalian specimens showed an appreciable fluorine content (some of them as much as 3·1 per cent) the skull, jaw and canine tooth had less than 0·4 per cent. The *Eoanthropus* remains could no longer be accepted as of the great antiquity previously claimed.

Soon it was realized that the assemblage of fossils was not a natural one but had been fraudulently brought together from diverse localities of different ages. The stegodon and mastodon teeth were probably lower Pleistocene; the hippopotamus and beaver upper Pleistocene; and the *'Eoanthropus'* remains were not even fossil. It was announced that Piltdown man, which had been accepted as genuine by anthropologists for over 30 years, was a carefully prepared hoax.

The Piltdown hoax had been exposed by relative dating methods at a time when methods of absolute dating (which will be our next topic for discussion) were still in their infancy and could not yet be applied, particularly because of the large amount of material required for such tests. For a carbon^{14} date it would have been necessary to destroy the Piltdown specimens completely. By 1959, it had become possible to establish dates from much smaller samples, and parts of the skull and jaw were sent to the radiocarbon laboratory at Groningen in Holland. The results were: 500 ± 100 years old for the jaw; 620 ± 100 years for the skull. The startling conclusions that had been based on the earlier relative dating methods were fully vindicated. The specimens chosen for the hoax must nevertheless have been acquired as antiquities, for they were not entirely modern.

Absolute dating

So far we have considered mainly methods of obtaining relative dates, which already provided a basis for working out the chronology of the Pleistocene long before the advent of absolute dating methods, most of which have become available to us only since the 1940s. We have also encountered some of the problems of having to work with a relative time scale, the most notable being that our climatic framework for dating events can only come into its own when we are quite sure how many climatic fluctuations actually occurred; and that stage in Pleistocene studies is only just beginning to be reached. We do not have a continuous record of Pleistocene sedimentation on land and, even if all the known terrestrial sequences of deposits could be placed in relative order, there would still be periods of time of unknown duration for which there is no sedimentary record. How can we recognize such gaps in the fossil mammalian evidence? Clearly, the availability of only a few absolute dates within such relative sequences would be of the utmost value.

Today there are several methods of absolute dating available to us. They can be divided into two categories. Firstly there are those which derive a time scale by counting sequences which result from annual events such as the deposition of layers of silt in certain types of lakes or the formation of rings in trees. The thicknesses of both such silt layers and of tree rings vary slightly from year to year, in response to external factors, permitting overlapping parts of sequences to be correlated locally from site to site and from tree to tree. Sophisticated computer programmes have been introduced to assist with such studies. In this way records of events over long periods of time can be built up which, if they extend up to the present day, can be given an exact age in years. In Scandinavia, layered or 'varved' sediments in glacial lakes have been used as a basis for dating the successive margins of the Pleistocene ice during the last 15 000 years.

In the same way, changing thicknesses in tree ring series can be used for correlating from one tree to another. In the southwest North America a tree ring sequence for the Bristlecone Pine has been built up which goes back 7000 years.

Varves and tree rings have proved valuable locally as a means of dating Pleistocene and later events, but few mammalian remains have been found at sites where these methods can be applied. In consequence, they have so far been unimportant in mammalian studies.

Secondly, and more important, there are the absolute dating methods that involve the measurement of decay of radioactive substances, which takes place with the passage of time; and these have been of world-wide importance in establishing the ages of mammalian remains. Many elements have different isotopes, some of which are unstable and will break down at predictable rates. These are the radioactive isotopes. Different radioactive substances have different rates of decay, which are unaffected by outside factors such as temperture and pressure. If one knows the rate of decay of the parent isotope and can measure the proportion of it which remains, it is possible to work out how much time has elapsed since the decay process started. It follows that radioactive isotopes that break down slowly are most suitable for determining the age of the earliest rocks in the geological time scale, whereas more quickly decaying elements are employed for dating events during the Quaternary. The earliest presently known rocks on earth (of Pre-Cambrian age), from near Godthaab in Greenland, have been dated as 3860 million years old on the basis of decay of rubidium to strontium. In Pleistocene studies, three isotopic breakdown sequences are of special importance: carbon→nitrogen; potassium→argon; and uranium→thorium (a short-lived isotope in breakdown series uranium–lead).

The rate at which these various radioactive isotopes decay is measured in terms of what is known as their 'half lifes' – that is the length of time required for half of the 'parent' isotope to decay to its daughter form. The end of one half life interval is the beginning of a new one so that after the second half life only a quarter of the parent isotope remains; after a third half life only an eighth; and so on (fig. 5.13).

The half life of potassium^{40} (which decays to argon^{40}) is estimated as 1 310 000 000 years; whereas that of carbon^{14} (which decays to stable nigrogen^{14}) is only 5730 years. The former method cannot be

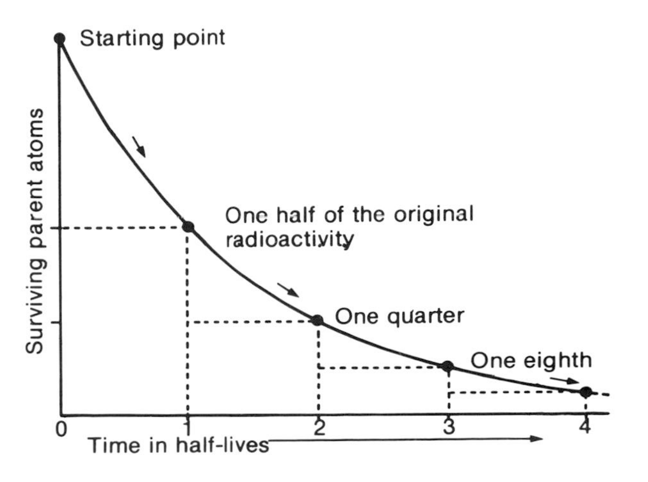

Fig. 5.13 *Half life curve for a radioactive element. Note that there is a gradual flattening of this curve towards infinity. Decay does not proceed at a uniform rate (like the burning down of a candle) until nothing is left.*

applied to rocks younger than about 100 000 years old, whereas the $carbon^{14}$ method has a maximum age limit of about 50 000 years, though this may soon be extendable with new equipment and great care, to 70 000 years. The time gap (in the upper Pleistocene), between these two methods is conveniently filled by the $uranium^{234}/thorium^{230}$ method, which potentially covers the period from 10 000 to 350 000 years ago.

Let us now consider the applicability of these three important methods of radioactive dating in slightly more detail, especially where they can be applied to the study of mammalian remains.

The radiocarbon method, devised by W. F. Libby in the USA in 1946, depends on the uptake by living organisms of $carbon^{14}$, which is being continuously produced in relatively small quantities in the earth's upper atmosphere, as the result of bombardment by cosmic rays.

Atmospheric nitrogen and free neutron→ $carbon^{14}$ + proton

The newly produced carbon atoms are rapidly oxidized to carbon dioxide. About 0·03 per cent of the earth's atmosphere is carbon dioxide, in which the carbon is mostly composed of the stable $isotope^{12}$, with smaller amounts of the stable $isotope^{13}$. Only about one atom in every million carbon atoms in the atmosphere is $carbon^{14}$, so that atmospheric carbon dioxide has very low radioactivity. When green plants take carbon dioxide into their systems during the course of photosynthesis a small amount of $carbon^{14}$ is incorporated. Green plants form the base of the terrestrial food chain, so that all higher living organisms (plants and animals), whose tissues are ultimately built up from plant material, are also in equilibrium with the atmospheric $carbon^{14}$. When death occurs no more $carbon^{14}$ enters the tissues and the only way in which the amount present can change is by radioactive decay to nitrogen.

$Carbon^{14}$ →nitrogen + beta particle

If the amount of $carbon^{14}$ remaining in a sample of once living material is measured, then the time since the death of the organism can be calculated. Since suitable material for radiocarbon dating (which includes animal bones as well as plant remains) is widely available at many localities, this method is of immense value in mammalian studies during the latest part of the Quaternary. Two components of bone have been utilized in such studies – collagen and bone apatite. The collagen fraction is the most dependable for dating purposes, though weathering may lead to its removal from otherwise potentially datable bones. Only if no collagen survives is bone apatite utilized, but this method is considered less satisfactory because organic carbon in ground water may replace the original carbon in the apatite, causing dating error.

Radiocarbon dating is not without other problems, most notably the large quantity of each fossil which, using conventional methods, must be destroyed during analysis. In the case of bones ideally 100 g is needed, which is more than can usually be spared from specimens of any importance. Fortunately it is now possible to date bones not amenable to conventional techniques by means of new accelerator methods, which utilize amino acids specific to bone. Not only are specimens of as little weight as 1 g or less sufficient but it should be possible in the future to extend the method beyond its present limit of about 50 000 years, optimistically to around 70 000 years.

It has been found that various corrections must be made during the calculations if discrepancies, which may amount to 500 years or more, are not to appear in the results. Several causes of such discrepancies are now recognized. Firstly the basic method described above assumes that the rate of formation of $carbon^{14}$ in the upper atmosphere has always been the same. There is some evidence that this is not true and that 'radiocarbon' years do not correspond to absolute years through the whole of their range. Secondly marine food chains are not in equilibrium with atmospheric carbon dioxide, so that remains of whales and other marine animals sometimes have apparent $carbon^{14}$ ages which are too great. Lastly the amount of $carbon^{14}$ now present in the atmosphere has been changed by the burning of fossil fuels since the Industrial Revolution and more recently by the detonation of thermonuclear devices. Thus living material is no longer suitable for use as a standard for comparison when calculating the age of fossil material. Only wood dating from before the Industrial Revolution can be used in the provision of such standards. These problems are now well understood and, with caution, the $carbon^{14}$ method provides an accurate means of dating mammalian remains from the latest part of the Quaternary.

The potassium–argon method, developed during the late 1950s has now provided a great number of dates for deposits with Pleistocene mammals, though unfortunately it cannot be universally applied, as it is dependent on the availability of potassium-rich volcanic lavas and ashes and, less reliably, glauconitic sediments. This method has been especially successful in the East African Rift Valley, where it has provided an absolute time scale for richly mammaliferous australopithecine sites back to more than four million years ago (chapter 11).

The uranium–thorium method is also of great

importance since it can be applied in non-volcanic areas, where potassium–argon dating cannot be attempted; and, with a potential age range going back to about 350 000 years, it can be used for determining the age of mammalian remains too old to be dated by carbon^{14}. There is the additional advantage that suitable material for dating is widely available in naturally occurring forms of calcium carbonate such as stalagmite formations in caves, travertine deposits around springs, molluscs, corals in Pleistocene coral reefs and in foraminifera buried in the sea floor. These types of deposit are often found to have trapped, at the time of their formation, traces of uranium; and the quantity of daughter isotopes present can be used to determine the time since their deposition. This method has been especially successful in the dating of bone deposits in caves, which often occur interbedded between datable flowstone floors.

The same method has also been applied to the dating of actual bone specimens. Whilst some of the results obtained by this method correspond well with those obtained independently from other lines of evidence, great discrepancies occur in other instances so that such dates must be treated with caution. This difficulty appears to stem from the assumption, necessary for the method, that once uranium has been taken up by a bone on initial fossilization no further addition has occurred (a 'closed system'), whereas such is the attraction of the bone for uranium that often this has not been the case. In theory it is possible to determine whether a system has remained closed by measuring proportions of not only the thorium but also of another daughter product of the decay of uranium, protoactinium. Research into this problem is still continuing.

Chemical methods of obtaining palaeotemperatures

As previously observed, one of the greatest problems in the way of establishing a continuous chronology for the Quaternary lies in the fragmentary nature of the terrestrial deposits and the difficulty in recognizing intervals of time unrepresented by any sediment. The identification of deposits that accumulated under known climatic conditions (which might be employed as marker horizons) is a valuable aid to chronological studies, but there is the constant risk of confusing climatic cycles that were close in time. Those studying the Quaternary have long looked for a method that would allow them to obtain a record of the entire sequence of climatic fluctuations in a less complex manner than by trying to tie together the fragmentary and discontinuous continental evidence.

It was not until after the Second World War, at the time that sea bottom investigations were beginning to make possible the recovery of long cores of undisturbed sediment from the ocean floor and when absolute dating methods were also first being applied, that a method was found of resolving this problem. Pioneer of such studies was Cesare Emiliani who, in 1955, following the theoretical work of Urey, published a generalized climatic curve based on measurements of the relative proportions of the various isotopes of oxygen that occur in the calcium carbonate shells of foraminifera in the successive layers of cores obtained from the Caribbean, Pacific and Atlantic ocean floors. This method depends for its application on the slightly different behaviour of the two main isotopes of this element during equilibrium exchange reactions. Proportions change as evaporation takes place, just as distillation separates substances of different volatility. There are three isotopes of oxygen – 16 (normal oxygen), 17 (the proportion of which is so small it can be ignored in the present discussion) and 18. In water about one oxygen atom in 500 is oxygen^{18}, the remainder normal oxygen. Both isotopes are stable so that, once incorporated into calcium carbonate or ice, the proportions do not change with time, and they provide a permanent measureable record of the relative amounts locally present at the time of formation of these substances. Since the proportions of the two isotopes of oxygen in calcium carbonate are related to the temperature prevailing at the time of precipitation, this method was initially developed as a potential means of reconstructing past sea temperatures. It was soon realized that variations in the proportion of the two isotopes of oxygen in the sea water itself also played a major role in these studies. When water evaporates, oxygen^{16} goes off at a proportionately greater rate than oxygen^{18}. As we have already noted, during the glacial episodes much water was transferred from the sea (the level of which fell on occasions at least 100 metres) onto the land, where it contributed to the build up of the great ice sheets. This lowering of the sea-level had the effect of changing the isotopic content of the remaining reservoir of sea water, which became slightly enriched with oxygen^{18}. It is said to have been isotopically 'heavy'. At the same time the water vapour carried from the sea, subsequently to be precipitated elsewhere as rain or snow, was isotopically 'light'. The accuracy of this method has subsequently been independently confirmed by studies of the present-day temperature tolerance of the species of foraminifera found in the deep sea cores, which provide a parallel means of reconstructing past temperatures (see pp. 50–51 and chapter 6). The oxygen isotope method can also be applied over shorter ranges of time to the calcium carbonate of fossil corals, to stalagmite formations in caves and to fossil wood.

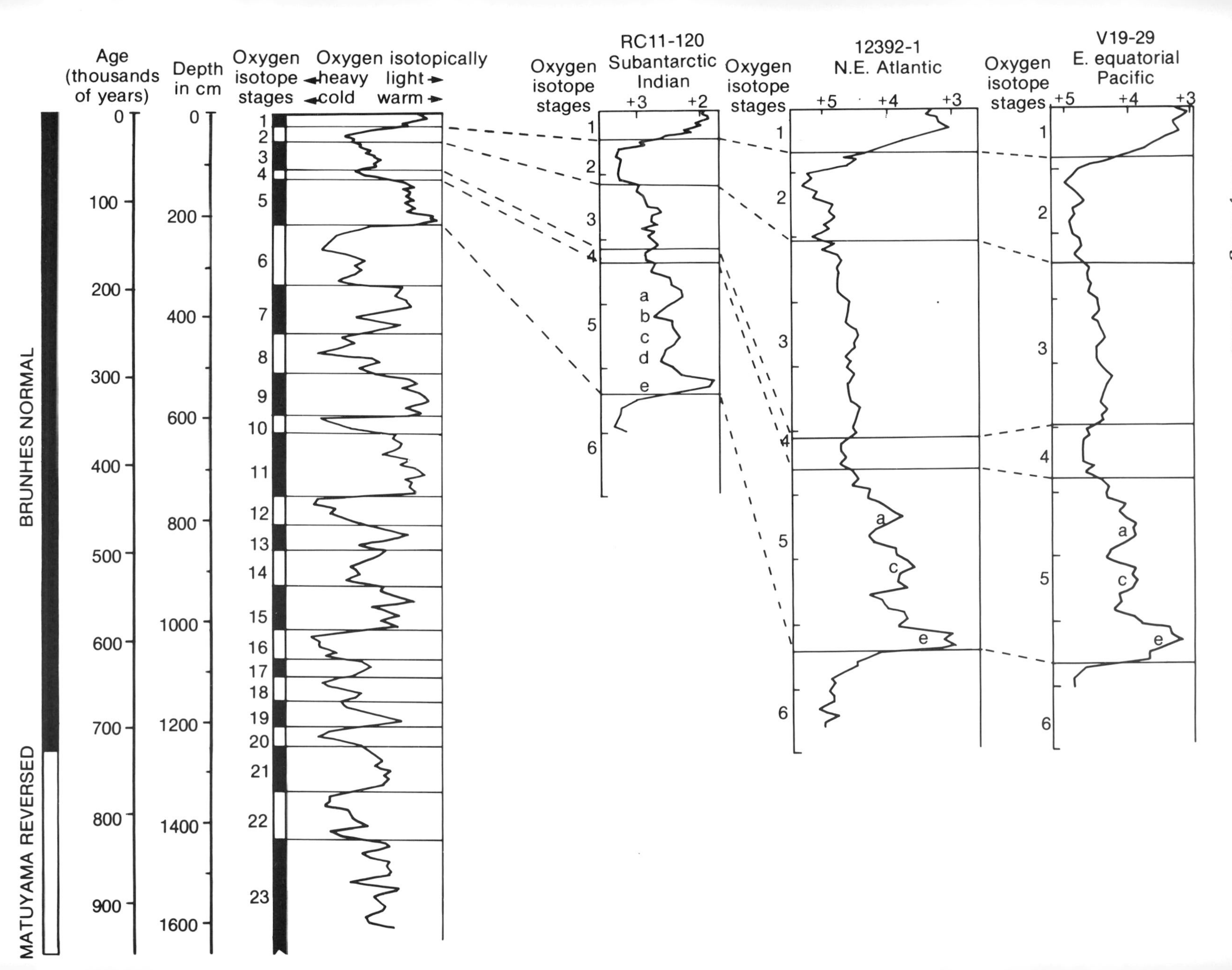
Age
(thousands
of years)
Depth
in cm
Oxygen
isotope
stages
Oxygen isotopically
heavy light
cold warm
BRUNHES NORMAL
MATUYAMA REVERSED
RC11-120
Subantarctic
Indian
12392-1
N.E. Atlantic
V19-29
E. equatorial
Pacific
Oxygen
isotope
stages

It is good fortune for those concerned with the reconstruction of past climates that oxygen isotope ratios vary in response to physical conditions and can be interpreted as a fossil record of past fluctuations of climate. Whenever there have been such fluctuations, so they were recorded in the continental ice, with isotopically light ice accumulating on land during the glacial stages; whilst, at the same time, foraminifera and other calcareous organisms were incorporating proportionately more isotopically heavy oxygen into their skeletons from the sea. Basically the changes recorded are a measure of the changing volume of the earth's continental ice. They also provide an indication of the amount of water in the ocean and thus of times of high and low sea-level. Organisms from the tropics would have had different isotopic composition from those living at the same time in more temperate latitudes, but, at any individual locality, a record of changing ratios in the different layers of deposits can be interpreted as a record of climatic change.

From the sequence of climatic fluctuations so established, Emiliani devised a chronology of oxygen isotope stages (odd numbers being given to interglacials even numbers to the glacials (fig. 5.14) as far back as stage 14. This scheme was further elaborated by Skackleton 1969 who subdivided interglacial stage 5 into five isotope sub-stages a–e with three warm peaks – a, c and e – separated by cold sub-stages b and d (fig. 6.3 a, p. 64).

Since these cores also provide a record of palaeomagnetic events (the date of which is known) it is also possible to establish an absolute time scale for the climatic events. By 1973, on the basis of core V28–238 from the Pacific, the story had been taken back 870 000 years (stages 1–22, with the Brunhes–Matuyama boundary of 700 000 years ago shown to be in stage 19); by 1976, (on Pacific core V28–239, which measured 21 metres in length and which recorded also the Jaramillo and Olduvai magnetic events) back to the top of the Pliocene, more than two million years ago. The sequence of events so established – at least seventeen major cold–warm cycles averaging only about 100 000 years each during the Quaternary – surprised Quaternary workers, who had previously been accustomed to thinking in terms of four or five. The deep sea evidence now provides the most complete record of Quaternary climatic change yet available by any method. This scheme subsequently received great acclaim from those trying to establish a chronology for the Quaternary from continental evidence and, used in conjunction with absolute dates from continental sites, is now widely accepted as being the most satisfactory yardstick to which studies should be related. Writing in 1979, John and Katherine Imbrie appropriately described its upper part as the 'Rosetta Stone' of late Pleistocene climate.

New dating methods

Study of the rates of racemization of amino acids in Pleistocene mollusc shells provides a relatively new dating method, although this depends critically on the history of temperature at the sites being investigated. The method has already proved valuable in permitting (sometimes very surprising) chronological separation of deposit at different sites, including mammalian sites, and it has good prospects, after calibration, of providing a means of absolute dating in the future.

Other new methods of dating Quaternary deposits are constantly being devised. Fission track and electron spin resonance dating are other potentially useful new methods, likely to be of greater importance in the future.

Tying the dating evidence together

We have considered the great variety of methods now available for dating Quaternary deposits. All are not applicable, however, at any single locality, and often it is necessary to follow a very indirect course of reasoning. A fossil mammal bone, by itself, may provide no information about its age; but if it is associated with remains of molluscs, insects and plants, all of which indicate interglacial conditions, then there is good reason for believing that the mammal lived during an interglacial period. One of the many other lines of study may then perhaps make it possible to identify the interglacial event in question; sometimes even to determine an absolute date for it. It is from such fragments of evidence that the chronology of the Quaternary is being reconstructed.

Fig. 5.14a) *Oxygen isotope record from foraminifera in deep sea core V28–238 from the west equatorial Pacific: with depth in centimetres; palaeomagnetic stratigraphy; and boundary ages calculated from assumed uniform rate of sedimentation. The troughs (even stage numbers) are periods of enrichment of oxygen isotope 18 and are interpreted as times of increase in the volume of the continental ice mass; the peaks (bars on left side, odd stage numbers) as periods of ice reduction. (After Shackleton & Opdyke, 1976).*

b–d) *Climatic fluctuations of the last 160 000 years as shown from the oxygen isotope records from different bore holes in the Pacific, Atlantic and Indian Oceans. Although the various cores are of different lengths, resulting from different rates of ocean sedimentation; and there are absolute differences in the range of oxygen isotope ratios (figures across tops of columns, measured as ‰, represent differences from a mean ocean standard) resulting from temperature differences at different latitudes, the general picture of climatic change is almost identical in each instance. The division of stage 5 into three warm parts (5e, c and a) and two cold parts (5d and b) is clearly shown. (After Shackleton, 1977).*

Chapter 6 ***Drawing the threads together***

By employing dating methods such as those described in the previous chapter, those working on the Quaternary have accumulated a vast amount of evidence that is gradually being pieced together to provide an overall chronology for this period of time. We have considered especially, by way of an example in chapter 2, the conditions prevailing at the time of the last main advance of the great ice sheets which, together with the hot deserts of the low latitudes, spread to their greatest extent about 18 000 years ago (fig. 2.6, p. 19). We cannot, however, recreate any single comprehensive picture of conditions prevailing throughout the Quaternary (itself so often spoken of in broader terms as 'the Ice Age') as the climate was changing all the time. A series of reconstructions, representing different stages of time, would be necessary for this purpose. Let us now look at the rest of the Quaternary time scale, which provides the framework for our mammalian studies, in greater detail.

Classical Quaternary chronologies

Before the advent of deep sea studies, all attempts to construct a chronology for the Quaternary were based on piecing together the fragmentary continental evidence previously described, with its associated hazards of overlooking time intervals unrepresented by any sediment.

As the result of such work during the past 90 years, climatically based Quaternary chronologies were established for most continents of the world and, until recently, these provided the principal basis for associated mammalian and other studies. In Europe, the earliest widely accepted of these chronologies was Penck & Bruckner's four-fold glacial subdivision of the Alps (in ascending order Günz, Mindel, Riss, Würm) in 1909; and other local chronologies were established for other continents. Some of the more important of these are shown in Table 2, since further discussion of the Quaternary would not be possible without reference to them.

All these sequences became well established in the literature. The four-fold Alpine subdivision seemed for a while to be matched by four-fold divisions in other areas too, but, although such equivalence was suggested by some writers, it could not be firmly demonstrated except in the case of the last glacial episode, and it became increasingly uncertain for the earlier events.

As time passed and further studies were carried out, it became apparent that the continental chronologies were greatly over-simplified and did not give a full record of all the climatic fluctuations that had actually occurred. Firstly there was the deep sea evidence, suggesting many more than four or five glacial – interglacial cycles. In 1977, Kukla, who had been studying the palaeomagnetically dated loess sequences of Europe, concluded that the classical Alpine sequence of four glacial–interglacial cycles in fact covered eight such cycles, going back to marine stage 22 (800 000 years ago); the supposed interglacials, originally based on studies of river terraces, actually representing periods of accelerated crustal movements and not climatic events. He appealed for the classical terminology to be abandoned in all inter-regional correlations, which, he argued, should instead in future be based on the oxygen isotope sequence of the deep sea sediments.

Meanwhile, in the British Isles, palaeobotanists and others working on early Quaternary deposits were finding evidence for additional cycles before the Cromerian; and mammalian palaeontologists, studying the later deposits, were finding it increasingly difficult to accommodate all their mammalian faunas into such a simple chronology.

By this time the establishment of absolute dates was also adding fuel to the argument for a climatically more complex Quaternary. A radiometric date of 0·6 million years for a marker horizon of volcanic ash (Pearlette Ash, type '0') found in deposits with a very late Kansan or early Illinoian mammalian fauna in the Great Plains area of the USA demonstrated a considerable discrepancy between the continental climatic chronology (from which an antepenultimate glaciation or penultimate interglacial age would be inferred) and the deep sea chronology, with evidence for about six major climatic cycles after this date. A date of 1·2 million years for Pearlette Ash, type 'S', in deposits regarded as of earlier Kansan age, suggests that the Kansan itself, as it had come to be applied, covered a time-span of several such cycles.

Meanwhile, in East Africa, potassium–argon dating of volcanic ash beds and lavas alternating with the mammal-bearing sediments of the Rift Valley had provided a continuous absolute time scale to which the mammalian faunas could be related, not only for the whole of the Pleistocene but for part of the Pliocene and Miocene too, a period of more than 20 million years (chapter 11). As a result of this work the interpluvial–pluvial chronology of the Quaternary has long been superceded, as a framework to which other events could be related, by this absolute time scale. This, in turn, now provides a much better key to which these fluctuations of precipitation (which did indeed occur) can themselves be related.

Today there is an increasing tendency among Quaternary geologists to attempt to tie their terrestrial evidence to the continuous palaeomagnetically dated chronology provided by the deep sea cores, rather than to continue using the continental chronologies tabulated above (which nevertheless continue to occupy considerable prominence in their studies) as their yardstick.

On first consideration these two methods might be seen to be providing results that are so conflicting that it would be difficult to reconcile the differences, but this need not be the case if consideration is given to some of the problems of making comparison. In most cases these are problems of terminology. Specialists in different disciplines must be able to name the same thing in the same way.

Firstly, if we are to continue using the term 'interglacial', its meaning must be clearly defined. In geological terminology it has become widely employed to describe the units of time between the glacial advances. Glacials and interglacials have tended to become rigid 'boxes' into which all climatic events have to be fitted. It would nevertheless be more accurate to see a gradual geographical gradation from the glacial state at high latitudes to the tropical state at low latitudes, the relative areas of which are being constantly adjusted in response to climatic change, with associated shifting of the various vegetational zones of the world. The most striking evidence of glacial–interglacial fluctuations is found in mid latitudes, most notably in Europe. In high latitudes, for example in parts of Siberia, are areas that have been continuously tundra throughout the Quaternary, with no significant ameliorations of climate; at the equator areas where there has been no glaciation, except at very high altitudes, and the predominant manifestation of climatic change has been in the amount of rainfall.

Further problems arise in connection with defining the magnitude of particular glacial events. By convention the world is said to be in an interglacial state at times of minimum ice cover and maximum temperatures, and in a glacial state at times of maximum ice cover and minimum temperatures. The term 'interstadial' has been widely employed to describe minor warm episodes within the glacials, not warm

	EUROPEAN QUATERNARY STAGE NAMES			NORTH AMERICA	SIBERIA	CHINA	EAST AFRICA
	ALPS	N. EUROPE	BRITISH ISLES				
HOLOCENE	Postglacial	Holocene	Flandrian			Holocene	Nakuran and Makalian pluvials; un-named interpluvials
PLEISTOCENE	Würm glacial	Weichselián glacial	Devensian glacial	Wisconsinan glacial	Sartanian Karginian Zarianian	Tali glacial	Gamblian pluvial
	Riss-Würm interglacial	*Eemian interglacial*	*Ipswichian interglacial*	*Sangamonian interglacial*	*Kasantsevian interglacial*	*Lushan-Tali interglacial*	*Interpluvial*
	Riss glacial	Saalian glacial	Wolstonian glacial	Illinoian glacial	Tasovian	Lushan glacial	Kamasian pluvial
	Mindel-Riss interglacial	*Holsteinian interglacial*	*Hoxnian interglacial*	*Yarmouthian interglacial*	*Schirtinskian*	*Taku-Lushan interglacial*	*Interpluvial*
	Mindel glacial	Elsterian glacial	Anglian glacial	Kansan glacial	Samarovian glacial	Taku glacial	
	Günz-Mindel interglacial	*Cromerian complex*		*Aftonian interglacial*	*Tobolian*	*Poyang-Taku interglacial*	
	Günz glacial			Nebraskan glacial	Oruchanian	Poyang glacial	
	Penck and Bruckner, 1909	Zagwijn, 1975	Geological Society, 1973		USSR Ministry of Geology 1983	Duan Wanti *et al.*, 1980	Leakey, 1931

Table 2 *Quaternary chronologies, based on continental evidence for Europe, North America, Siberia, China and Africa*.*

*The European, Asian and North American sequences are based on glacial–interglacial cycles, that for Africa on pluvial–interpluvial cycles. No lateral time equivalent is implied, except in the case of the last glacial event (Würmian, Devensian, Wisconsinan, Sartanian) which, being the most recent of the glacial advances with the best preserved evidence, can probably be reliably correlated on a world-wide basis.

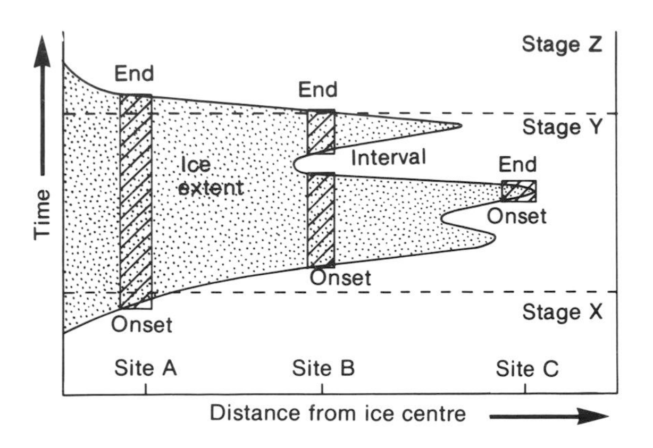

Fig. 6.1 *Although the terms 'glacial' and 'interglacial' have commonly been employed, with world-wide implication, to distinguish major climatic fluctuations during the Pleistocene, such a concept is greatly over-simplified. In this diagram at site A (near the centre of glaciation) there is one long glacial episode during the period of time shown. At site B, further from the centre, there were two glacial stadia divided by a warmer phase of interstadial status. At C there was only one brief glacial event and beyond that conditions were for most of the time of interglacial status. (After Andrews, 1979).*

enough to be called interglacials; stadials for minor periods of cold within the interglacials, not cold enough to be ranked as glacial episodes. But an interglacial in France might be of only interstadial status in Britain, and the same climatic event could be defined as being of different intensity and duration at different places (fig. 6.1). Our problem is partly one of trying to apply the terms 'interglacial' and 'interstadial' too rigidly. In response to this some workers favour defining these climatic events on their mid points, rather than assigning to them a beginning and an end.

Most serious of all the problems is that of specialists in different disciplines, working at different terrestrial sites, having insufficient overlap of data to enable them to relate their findings. Each of the named climatic stages in the various countries is based on evidence from one particular field locality. This is known as the type locality (stratotype). Not all of the evidence from the type localities is from the same range of disciplines. In the British Isles, for example, all the interglacials are based on pollen sequences; most of the glacials on unfossiliferous glacial deposits. Usually the type localities are many kilometres apart, so that only sometimes can direct chronological relationship can be established; and intervals of time, unrepresented by any deposits, are difficult to identify. Long distance correlations can be claimed where, for example, a glacial deposit is of great lateral extent or where apparently identical sequences of fossil plant assemblages occur at more than one place; but the last mentioned scheme requires the assumption that the same 'hand of cards' could not have been repeated more than once and, in light of the deep sea evidence, it is no longer feasible to assume this. The same uncertainty would be associated with any attempt to identify a particular climatic event on, for example, its mammalian or insect faunas alone.

Gradually, a more interdisciplinary approach to this problem gains momentum. For any specific climatic episode, many different disciplines may be applied to its study – plants, molluscs, insects, mammals, sea-level and other geomorphological studies, uranium series dating of travertine, racemization studies of amino acids, as well as other methods. If we apply too few of these disciplines to the study of a particular site, we run the risk of confusing two climatic episodes that were close in time, but, if a variety can be taken together, then the signature of that episode is likely to be unique and no confusion can occur. Of course, there are no sites that provide evidence for all these disciplines, but when all the evidence from different sites of the same age is taken in combination it becomes increasingly feasible to recognize particular climatic episodes and to spot the 'odd man out', which has been misplaced.

As new evidence comes to hand, so the gap between the terrestrial and deep sea evidence is gradually diminishing and it is seen that both are parts of the same climatic story. No longer, however, can it be inferred that a series of five or six climatic cycles in one region corresponds to five or six such cycles in another region, if it is not known which five or six of the many cycles, that actually occurred, are involved. Climatic episodes apparent from terrestrial evidence can be reliably compared and hopefully correlated over very long distances only where absolute dates are available. It can be expected that new glacial and interglacial stages will be formally defined within the terrestrial chronologies in the future.

Tying the marine and terrestrial records together; the last 160 000 years

With the concept of a more complex glacial–interglacial chronology now firmly established, the need to try to correlate deep sea and terrestrial evidence has intensified.

Such correlation is relatively straightforward for the latest part of the Quaternary, which has the best preserved evidence, but becomes increasingly difficult as we progress back into time; although radiometric dates already being obtained (especially the superb series from the African Rift Valley – see chapter 11) provide a preview of the more complete world wide chronology to come.

The 1970s were a period of exceptional progress in studies of the period back to about 160 000 years ago, with the deep sea and continental evidence now showing close agreement in a whole series of disciplines. That the results should be so consistent can be no coincidence; and there is good reason to believe that, even though much scope for elaboration still remains, a reliable and detailed record of events during this period is already beginning to be obtained.

The deep sea oxygen isotope record

We have already examined the deep sea oxygen isotope record at the end of chapter 5 and clearly recognizable in its upper part (fig. 5.14, p. 58) are the Holocene Interglacial (oxygen isotope stage 1); the Last Glaciation (stages 2–4) with the glacial maximum of 20–15 000 years ago (2), already described in chapter 2, and a minor warm stage of interstadial status (3); three warm peaks of interglacial or interstadial magnitude (stage 5); and an earlier very cold stage (6). The details of stage 5 are of special interest as there is clear evidence of three warm sub-stages (a, c and e) where only one (the Last Interglacial) had previously been recognized from terrestrial evidence. The earliest of these, stage 5e, is of greater magnitude than c and a; and there are two cold stages (d and b) of glacial magnitude.

Palaeontological evidence of past fluctuations of the temperature of the sea

From a study of assemblages of foraminifera (using especially the polar species *Globigerina pachyderma*) and coccoliths (the calcareous sheaths of marine algae) of known climatic sensitivity from a series of North Atlantic deep sea cores, McIntyre & Ruddiman demonstrated in 1972, on palaeontological grounds, a series of fluctuations of ocean temperature during Quaternary times that closely matches the oxygen isotope evidence. Not only were they able to demonstrate changing temperatures at the sites of the individual bore holes but, from this evidence, also to reconstruct past migrations of polar and tropical waters, which provide a direct oceanographic analogue of the equivalent ice sheet movements on land (fig. 6.2).

Fig. 6.2 *Maps of the North Atlantic Ocean, showing distribution of polar and tropical water at three stages during the late Quaternary. Left, present day; centre, c. 18 000 years ago (Last Glaciation, oxygen isotope stage 2, note polar water along the west coast of Portugal); right, c. 120 000 years ago (Last Interglacial, oxygen isotope stage 5e; sub-tropical water extending almost to the British Isles). The various temperature areas, being based on palaeontological evidence, are described as 'ecologic water masses'. The significance of the currents shown is discussed on page 69. (After Ruddiman & McIntyre, 1976).*

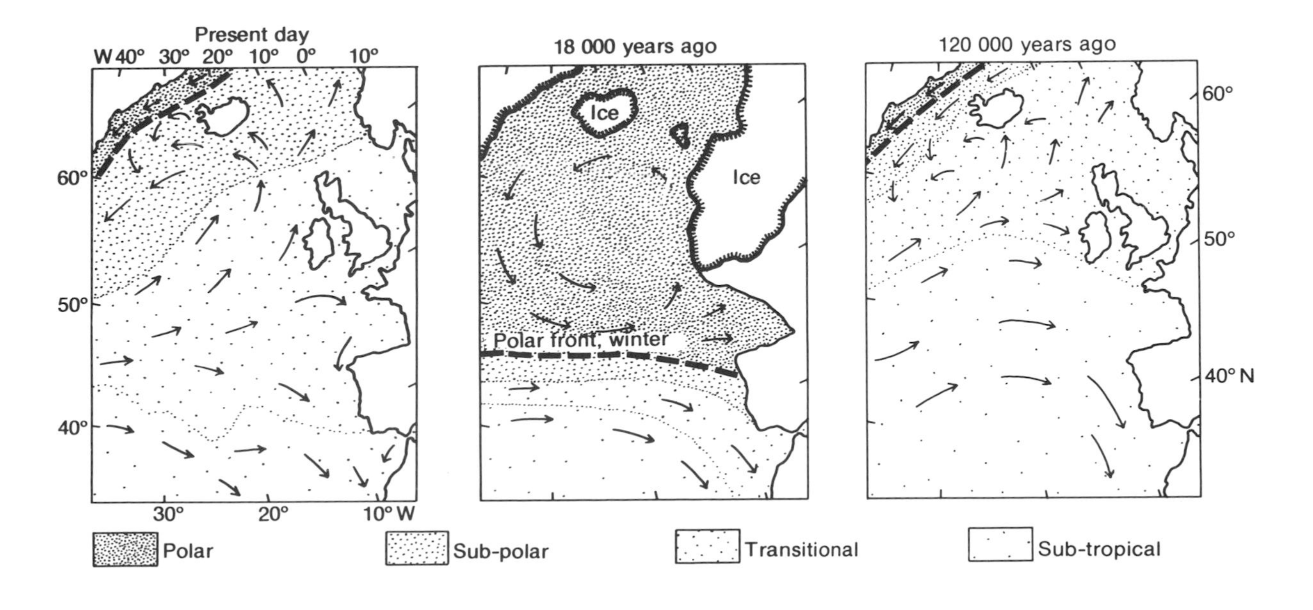

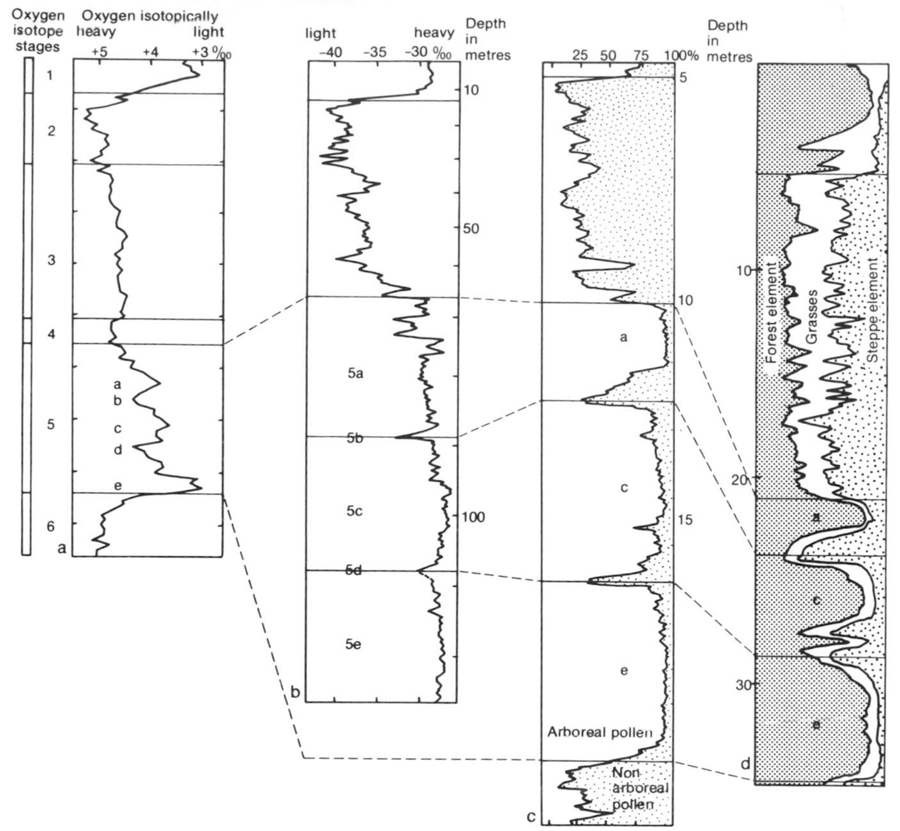

Fig. 6.3 *Climatic change during the upper Quaternary, based on the evidence of several disciplines at widespread localities. In each instance the general picture is very similar.*

a) *Oxygen isotope ratios in equatorial Pacific core 1239 2-1. (part of fig. 5.14 repeated).*

b) *Oxygen isotope ratios in a 1390-metre-deep stratified ice core from Camp Century, Greenland. The Holocene (stage 1), the Last Glaciation (2–4) and three peaks of interglacial status divided by cold stages (stage 5), are clearly apparent. Note that whereas, in the deep sea record isotopically light oxygen is interpreted as evidence of warmer conditions, isotopically heavy oxygen as evidence of colder conditions, here the situation is reversed since isotopically light oxygen is preferentially transferred in water from the sea to the land. (From Dansgaard* et al., *1971).*

c) *Vegetational changes during the last 140 000 years from an analysis of plant remains from the deep peat bog at Grande Pile, Vosges Mountains, France. The simplified graph shows the relative proportions of tree and other plant species throughout this period. The flora varies from forest (the peaks) to tundra (valleys). Although based on evidence totally different from that forming the basis of figure b, the similarity of the two diagrams is astonishing. (After Woillard, 1978).*

d) *Vegetational changes during the last 120 000 years, from an analysis of plant remains from Tenagi Phillipon, Greece. The diagram shows changes between forest and steppe, basically indicating changes of precipitation at this Mediterranean locality. (After Wijmstra & van der Hammen, 1974).*

Recent Studies of a long core of sediment, its earliest part going back to 3·5 million years ago, from Lake Fuguene, Columbia, South America, have demonstrated fluctuations of forest and mountain turrock grass during the last 110 000 years at this high altitude locality, with a remarkable similarity to the vegetational sequence for the same period of time at Tenagi Phillipon.

The record provided by past fluctuations of sea-level

As we have already seen, times of glaciation are characterized by low sea-levels, whereas, during the interglacials, the sea rose once more as the result of melting of the ice sheets. At the height of the last glacial advance (oxygen isotope stage 2) sea-level was at least 100 metres lower than at the present day and extensive areas of present day sea bed were dry land with associated terrestrial flora and fauna. The high sea-level events are represented by fossil beaches, coral reefs and other shore line structures at various altitudes above present sea-level in many parts of the world. In Barbados, three former high sea-levels, dated by the uranium–thorium method on coral, have been shown to correspond to the three warm stages of oxygen isotope stage 5 (Barbados III, the highest, average age 122 000 years = 5e; Barbados II, average age 103 000 years = 5c; and Barbados I, average age 82 000 years, the lowest = 5a). It is known, however, that Barbados has also been affected by uplift of the land so that all these shore lines occur at a higher altitude than can be explained simply by glacial–interglacial exchanges of water. Three high sea-levels there were indeed, though probably only 5e (the warmest of the three) stood much above that of the present day. There are many other well-dated instances of this last mentioned high sea-level from diverse parts of the world.

The record provided by the oxygen isotope composition of the continental ice sheets

Like the sediments of the ocean basins, the ice of the world's ice sheets (having accumulated as a series of annual snowfalls) is also commonly finely stratified; and a study of the oxygen isotope ratios from the various layers provides a further important method of reconstructing past climatic conditions, although it cannot be extended earlier than about 130 000 years ago. A core obtained by drilling through the ice at Camp Century in Greenland (fig. 6.3b) has demonstrated a sequence of events closely agreeing with the later part of the deep sea evidence. Comparable results have been obtained from an ice core drilled from the Antarctic ice sheet.

Palaeobotanical evidence of past fluctuations of climate on the land

Plant remains from deep stratified peat bogs, lakes and similar deposits have provided additional corroboration of the climatic sequence which is now becoming so familiar. Plants respond rapidly to climatic change and a change from, for example, forest to tundra vegetation, can be interpreted as evidence of climatic cooling. Changes of continentality can be similarly detected. During the earliest stages of such studies only the most recent part of geological time could be investigated in this way because of the difficulty in finding sufficiently deep peat deposits, but recently the record has been taken back over 120 000 years at a number of sites, notably at Grande Pile in the Vosges Mountains (fig. 6.3c) and at Echets Bog near Lyon, France and at Tenagi Phillipon in Macedonia (fig. 6.3d).

The basic principle of this method is to study the proportions of pollen of the various plant types from successive layers and to reconstruct climatic conditions for each from the floral assemblages represented. All these sites show a sequence of events closely resembling that from the deep sea and ice sheet oxygen isotope records. Similar palaeobotanical studies that have recently been made in the southern hemisphere, notably in Colombia and Australia, have also produced evidence about climatic events in these regions during late Quaternary times.

The record of past climatic events provided by dated glacial deposits

In instances where absolute dates can be established for glacial advances, so these events can also be placed within the overall chronology. One of the most important studies of this nature is that of Richmond who, writing in 1976, reported the relationship between glacial and volcanic deposits, accurately dated by the potassium–argon method, around the area of the Yellowstone National Park, Idaho and Wyoming. He was able to demonstrate extensive glacial advances 140 000, 114 500 and 88 000 years ago which he correlated with the marine oxygen isotope stages 6, 5d and 5b. During the non-glacial interval between the last two events, a large lake occupied the basin of the present Yellowstone Lake.

Direct land–sea correlations

Although the deep sea oxygen isotope record, because of its completeness, is now widely accepted as representing probably the best framework to which the fragmentary terrestrial data can be related, such correlations are seldom straightforward. As we have observed previously those working on marine terrestrial deposits are usually working in different disciplines, with no overlap of data. Mammalian remains are unlikely to be found in cores of deep sea sediment. Direct correlations are commonly impossible and time relationships of sequences of deposits under comparison can then only be established in instances where absolute dating methods can be applied.

An important exception to this generalization is to be found in situations where terrestrially derived dust and pollen have been blown out to sea and mixed with marine sediments. Greatly increased inputs of dust derived from the Sahara into the sea off West Africa, from the Arabian and Indian deserts into the Indian Ocean and from Australia into the Pacific, about 18 000 years ago, confirm the hypothesis previously outlined that this glacial episode was also a time of expansion of many of the world's desert areas. From studies of similarly derived pollen the regional vegetation of land areas can sometimes be directly related to the deep sea oxygen isotope and foraminiferal sequences; although such evidence must be interpreted with caution, since the pollen of some plants travels further than that of others. Of special interest are the findings from core SU8132, recently obtained from the Bay of Biscay, 100 kilometres west of the Spanish–Portugese coast (fig. 6.4). Studies of pollen from the part of this core spanning oxygen isotope stage 5 permit direct correlation with the equivalent part of the sequence at Grande Pile, in France, and gives precision to a correlation that was previously hypothetical.

A further important observation emerges from the study of this core. Comparison of the isotopic

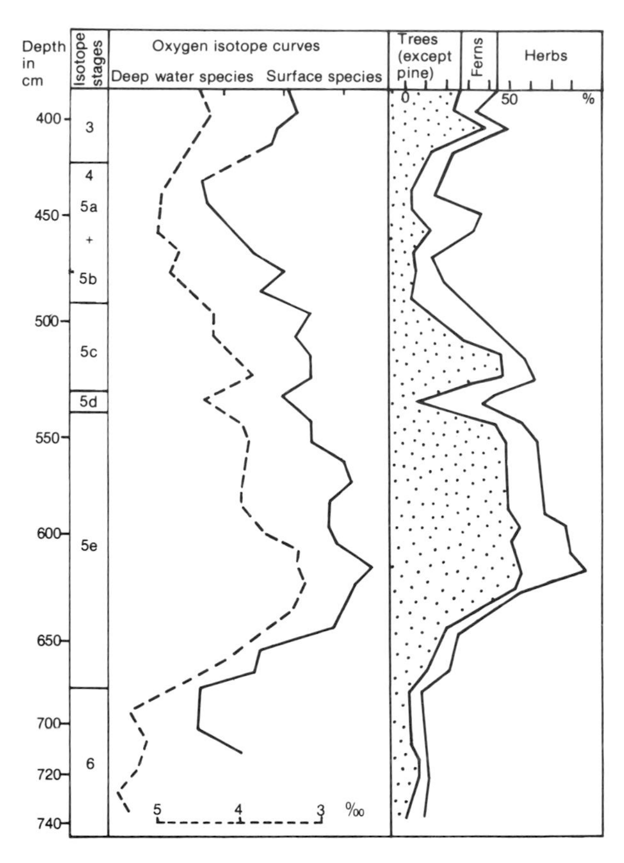

Fig. 6.4 *Direct correlation between the deep sea oxygen isotope chronology and a terrestrial pollen sequence (from the Iberian Peninsula), in core SU 8132 from the Bay of Biscay. Two oxygen isotope curves are given, for both deep water and surface foraminifera. Although the ratios differ, since the two groups have different oxygen isotope uptake, the resulting curves correspond closely to one another. Compare with the pollen sequence from Grande Pile, figure 6.3c. (After Turon, 1984).*

evidence (which reflects the state of world ice cover) and of the palynological evidence (which reflects the vegetation of the Iberian Peninsula) demonstrates that northern hemisphere cooling at the end of the Last Interglacial did not occur simultaneously everywhere; temperate conditions continuing on the mid-European continent for some time after the initiation of renewed glaciation on the high latitude land areas.

Diachronism

Prior to our discussion of core SU8132 we have broadly assumed that the climatic fluctuations which we have been describing, though manifesting themselves differently at different latitudes, were events of world-wide synchroneity. Certainly all the climatic zones of the world must, at any one time, be interrelated to one another; and the last major glacial episode, which reached its climax about 18 000 years ago, appears to have been accompanied by approximately synchronous glacial expansion in Scandinavia, North America, Antarctica, the Andes and on the mountain tops of many other parts of the world. Only a few decades ago the belief of many workers was that this expansion reached its furthest limits everywhere at about the same time. With improved dating methods, however, and with new field studies, it has become increasingly apparent that this was not so and that there was much local variation in what actually happened. The geological term for this is diachronism. A sedimentary formation is said to be diachronous when it becomes laterally younger in the direction in which deposition was being displaced, for example at a moving ice front or shore line. Even today there is much local variation in the behaviour of glaciers; for instance on the islands of the Canadian high arctic, where individual examples of both advance and recession can be observed in the same region. As we have already seen, the Quaternary ice sheets began to build up in Antarctica earlier than in the northern hemisphere. Recent work by Boulton has demonstrated that 18 000 years ago, when the North American and Fennoscandian ice sheets were at their most extensive (fig. 2.6, p. 19), large parts of the Arctic were unglaciated. There was, however, extensive expansion of glaciers within the Arctic about 11–8000 years ago, by which time the last mentioned ice sheets had almost disappeared. He attributed this to starvation of the Arctic of precipitation at the time of build up of the great ice sheets, followed by penetration of moisture into the Arctic as these retreated and the northern part of the North Atlantic became unfrozen once more. However cold the climate, ice sheets cannot form where there is no moisture. The surface of a warm interglacial sea is potentially an excellent source of water vapour for the formation of ice at high latitudes (fig. 6.5).

Recent studies by the working group known as CLIMAP have shown from deep sea evidence that, although the ocean temperature at the time of minimum ice volume of the Last Interglacial (oxygen isotope stage 5e) was in general not significantly different from that of the present-day ocean, the mid-latitude part of the North Atlantic was slightly warmer (which is of special interest to us as it includes Europe), the Gulf of Mexico cooler. Surprisingly they also found that different parts of the world oceans, and presumably of adjacent maritime portions of continents, registered the full warmth of the Last Interglacial at times differing by as much as several thousand years.

As in the glacial areas, so in the hot desert areas of the world, there is evidence that the times of maximum and minimum precipitation were not synchronous everywhere. Although, as we have already seen, there was an almost world-wide expansion of hot deserts centred on the time of the glacial maximum of 18 000 years ago, the currently desert areas of western North America experienced a high rainfall episode. Of shorter duration, in 1982–83, a change in the Pacific weather pattern, known as 'El Nino', resulting from temporarily changed sea current patterns, brought equatorial sea temperatures above normal, heavy flooding in western North America, the Gulf Coast and Cuba; while severe droughts affected Central America, South Africa, Indonesia and Australia.

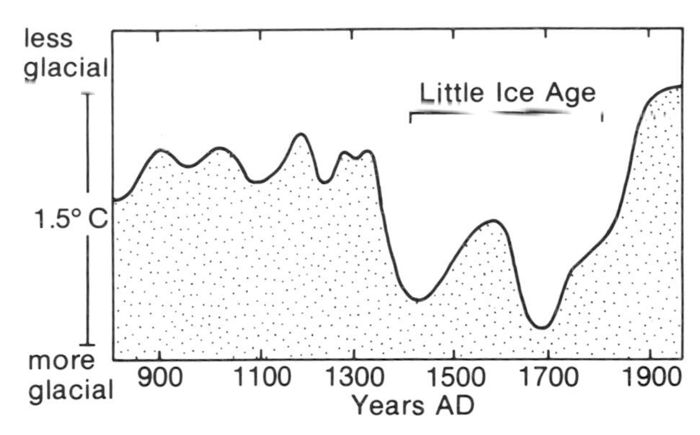

Fig. 6.6 *Climate of the past 1000 years, including the 'Little Ice Age'* c. *AD 1450–1850 from manuscript records. (After Imbrie & Imbrie, 1979).*

The 'Little Ice Age'

Best documented of Quaternary climatic events is one of the most recent – the cold period which continued from about AD 1450 to 1850 which became known as the 'Little Ice Age' (fig. 6.6).

During the thirteenth century it had still been warm enough for Vikings to colonize southern Greenland, on one occasion apparently reaching the east coast of Ellesmere Island, only 1200 kilometres from the North Pole. With decreasing temperatures and increase of sea ice they were forced to abandon their settlements, while many northern hemisphere glaciers advanced beyond their present limits. Crop failures in Scandinavia led to extensive emigration and depopulation. In England the cultivation of vines, which had been an important industry in the twelfth and thirteenth centuries was largely abandoned. The frozen lakes and rivers of continental Europe were immortalized by paintings of winter scenes by artists such as Avercamp (fig. 6.7) and Brueghel. By the end of the nineteenth century the climate had ameliorated once more. A slight cooling has taken place since 1940.

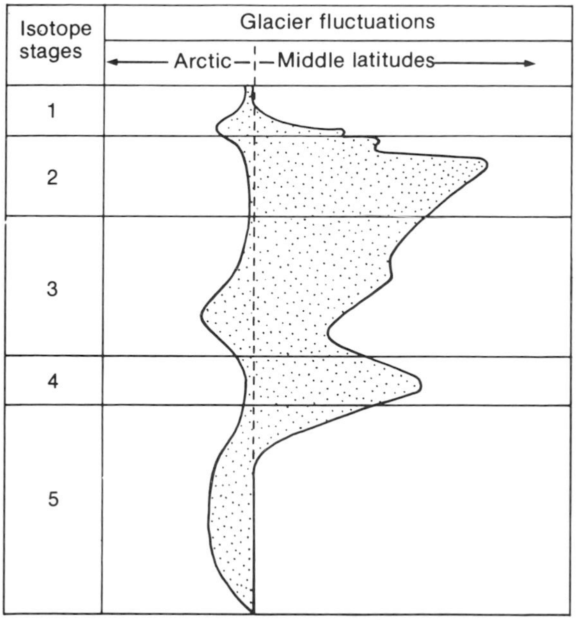

Not least of the consequences of man's study of the 'Little Ice Age' has been his increased awareness of the spectacular changes of agricultural productivity that can result from quite small fluctuations of climate and his increasing concern about what the future may hold for him.

Fig. 6.5 *Suggested time relationship of glacier fluctuations around the Atlantic sector of the Arctic and at the southern margins of the Fennoscandian and North American ice sheets during the last 130 000 years. Note ice advance in Arctic at times of retreat in mid latitudes. For explanation of numbering see fig. 5.14. (After Boulton, 1979).*

Isotope stages
Glacier fluctuations
← Arctic – – Middle latitudes →
1
2
3
4
5

Causes of climatic change

Although the causes of climatic change lie outside the scope of this book we cannot wholly overlook this topic since our studies of the Quaternary are centred almost wholly around such changes which have occurred in the past, the understanding of which in turn provides data for the climatologist concerned with the future.

Basically, all that has to happen to bring on an ice age, is for more snow to fall in winter than is lost by ablation the following summer. There is still no final agreement between scientists about the detailed combination of processes that actually causes this to happen, although sufficient is now known about the

Fig. 6.7 Winter landscape with ice skaters. *Painted by Avercamp during the 'Little Ice Age' when such scenes were commonplace in the long, cold winters. (With acknowledgement to the Rijkmuseum, Amsterdam).*

various factors involved to provide a basis for reasoned thought.

Firstly there is the possible effect on the earth's climate of changes in solar activity. A further suggestion is that ice ages may have been caused by the interaction of the sun with dense interstellar clouds. According to this theory, solar luminosity is temporarily increased when the sun passes through such a cloud, which leads to increased precipitation and ice accumulation on earth. The known periodicity of glacial events on earth, it has been suggested, may be explained by the passage of the solar system through the spiral arms of such clouds.

In addition, there are the processes, within the earth itself, that could lead to climatic change. The presence of large land masses at sufficiently high latitudes to catch and hold masses of snow, seen in the context of continental drift, has been suggested as a contributing factor towards the oncoming of ice ages, including the pre-Quaternary ice ages, throughout geological time.

Then there is the possible effect of variations of the earth's orbital characters. Most important of these are:

a) the angle of the ecliptic (the tilt of the earth's axis of rotation from the plane on which the earth travels around the sun). At present it is 23·5° and is known to have varied between 22·1° and 24·5° approximately every 4000 years. The obliquity produces the seasons and is one of the factors modifying climatic zones.

b) the eccentricity of the orbit. Since the earth travels around the sun in an ellipse there is a time of year when it is nearer to the sun (the point is known as the *perihelion*) than during the remainder of the year. At present this occurs during the winter of the northern hemisphere. The eccentricity fluctuates with periods of 93 000 years. A decrease in eccentricity reduces the differences between the lengths of the seasons on earth.

c) the precession of the equinoxes. This is a slight conical movement of the earth's axis. It results in a slow shifting of the spring and autumn equinoxes, and summer and winter solstices, which delimit the seasons; and it has a periodicity of 21 000 years.

There have been various attempts to calculate the combined effects of these three variable factors on world climate, most notable being that of Milankovitch who, in the 1920s, prepared graphs of the overall variation of radiation that would be expected to have reached various latitudes of the earth's surface during the last 600 000 years. The graph showed many peaks and valleys of various amplitudes; more than could be accounted for by the Alpine sequence of four glaciations that formed the most popular basis for Quaternary chronology at that time. Although claims were made that observed fluctuations of Quaternary climate corresponded with those postulated by Milankovitch, this hypothesis was subsequently largely abandoned, mainly because dating methods had not yet progressed sufficiently far to allow meaningful comparison. With the availability of the more detailed deep sea record it was possible to reassess once more the Milankovitch theory. The similarity of the climatic fluctuations postulated by Milankovitch and that demonstrated by the deep sea evidence is very close; and once more it is widely accepted that changes in the earth's orbital geometry are a fundamental cause of the succession of Quaternary ice ages. (fig. 6.8).

Finally, there is the possible effect of terrestrial processes on world climate. Increased dust content of the atmosphere following volcanic eruptions, for example, can interrupt the amount of insolation received from the sun and have a cooling effect. The potential cooling effect of dust thrown up by large meteorites has also been proposed as a triggering mechanism for the oncoming of glacial conditions.

It is also possible that, once commenced, a glacial phase can become partially self-perpetuating. More insolation is reflected back into space from the surface of the ice sheets (increased albedo); and the fall of sea-level which accompanies ice sheet expansion has the effect of increasing the altitude of the land relative to the sea, and so further compounds the effects of cooling.

So far we have considered mainly primary causes to explain the dramatic fluctuations of climate known to have occurred during the Quaternary. But the world's weather pattern is a complex manifestation of the interactions between continents, the oceans and the atmosphere; and it is to local conditions that we must look to explain the smaller details. Secondarily caused changes in the behaviour of an ocean current, for example, can cause extreme and sudden changes of climate along the margins of a neighbouring land mass, whereas the same event might hardly be noticeable further inland, in the continental interior.

Nowhere is the effect of marine currents on the climate of an adjacent land area better demonstrated than in the instance of the North Atlantic and Europe. Although the British Isles lie at about the same latitude north of the equator that South Georgia (an island which today carries permanent glaciers) lies south of it, waters of Gulf Stream origin provide Europe with anomalous warmth as far north as Iceland and the Norwegian sea (fig. 6.2). At the time of the glacial advance centred upon 18 000 years ago this pattern was greatly altered by an anticlockwise current of polar water, originating in East Greenland and extending as far south as northern Portugal and New York. Along the east–west line where this current met the warmer waters from the south at about 42°N, was a zone of abrupt temperature change known as the 'polar front'. The position of this front has moved repeatedly during the Quaternary, with warm currents almost reaching Greenland at the height of the Last Interglacial, at about 120 000 years ago. North of 42° sea temperatures have varied by as much as 18C° between full glacial and interglacial extremes, whereas south of this line they differed as little as 3C° in some areas. In Europe the passing of the polar front appears to have been the dominant thermal event within each climatic cooling or warming, causing temperature changes which were not only extreme but also very rapid. In western Scotland an ice cap over 100 km long formed and melted again in only 800 years between 11 000 and 10 000 years ago. The pollen evidence from Grande Pile suggests that during the period of cooling at the end of oxygen isotope stage 5e, about 115 000 years ago, a change from temperate forest to pine–spruce–birch taiga may have occurred in only about 150 years. Fossil insects in Europe suggest that climatic change may sometimes have occurred even more suddenly. We will examine the effects of these changes on the mammalian faunas of the British Isles in greater detail in chapter 10.

During the last century another potential cause of climatic change has gained importance – the accidental effects of man himself, who is burning up more and more fossil fuel and pouring ever greater quantities of effluvia into the atmosphere. Although smoke has a cooling effect on the climate, the increased carbon dioxide which man is creating may have a greenhouse effect on the world and make it warmer. It has been calculated that between the middle of the nineteenth century and today the concentration of atmospheric carbon dioxide has risen from about 270 to 340 parts

Fig. 6.8 *(With acknowledgement to the* New Scientist*).*

per million and that it could double by the middle of the twenty-first century, causing a possible rise of temperature of 2–3°C, shifting of the world's agricultural belts by many degrees of latitude, reducing continental ice and raising sea-level. So great could be the effect of such changes on the future quality of life for man that the balance between free carbon dioxide and the earth's carbon reservoirs, such as forests (currently being rapidly cut down by man or killed by acid rain) and the oceans (which can absorb carbon dioxide) has become a major topic for concern and scientific study.

Also important in such studies is work currently being conducted to establish atmospheric carbon dioxide concentrations in the geological past – information now available from analyses of deep sea sediments and of gas bubbles in the ice sheets of Antarctica and Greenland. Early results indicate that, during the last glaciation, there were only about 200 parts per million. The cause and effect relationship between climate and carbon dioxide concentration is nevertheless little understood and remains a subject for continuing research.

Man's poor agricultural practices present a further potential cause of climatic change, especially in marginal regions such as the edges of hot deserts, where the removal of vegetation can cause loss of soil, increased albedo and decrease of rainfall. The droughts currently occurring in the Sahelian regions of Africa (fig. 1.4, p. 13) may in part have been caused by man's activities, although it is difficult to distinguish the relative importance of natural processes of climatic change and human interference in such instances.

The possibility that material thrown up by an exchange of nuclear weapons could set off an ice age now also receives serious consideration. Recent studies suggest that, beyond a certain threshold in the magnitude of the exchange, the world could be so enveloped in smoke that radiation from the sun would be shut out, freezing would ensue and any survivors would die from cold and resulting crop failures. Conditions might be expected to return to normal after perhaps three months or a year, although the effects of other processes, such as restructuring of the earth's atmosphere (which might prolong the cooling) are quite uncertain and, as pointed out by Turco and others, writing in 1984 'this is not a subject amenable to experimental verification – at least not more than once!'

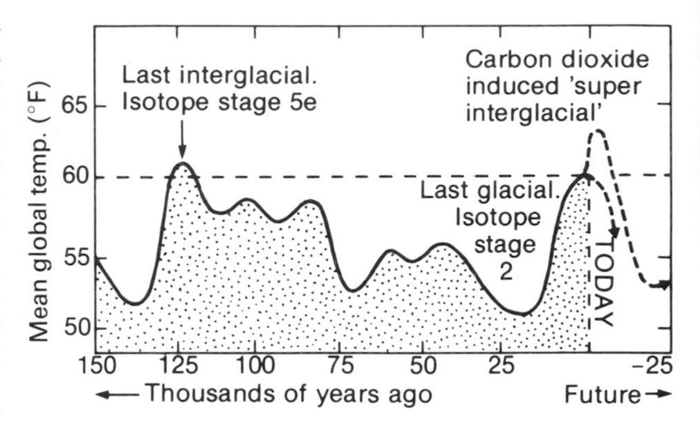

Fig. 6.9 *Climatic forecast for the next 25 000 years. The lower dashed line predicts an expected cooling if man-created effects are not taken into consideration; the upper line a possible temporary warming resulting from the 'greenhouse effect' caused by man's burning of fossil fuels. (After Imbrie & Imbrie, 1979, 80).*

The future: greenhouse or refrigerator?

What does the world's climate hold for man and the other mammalian species in the future? There are two contrasting approaches to this problem. On the one hand there is the evidence of palaeoclimatologists who point out that the present interglacial has already continued for 10 000 years and that, if the frequency of previous glacial–interglacial cycles is any guide, we should go into the next glacial episode very soon, possibly quite quickly. Indeed, if the calculated effects of variations of the earth's orbit are projected into the future, then an increase in the size of the world's ice sheets is to be expected. On the other hand, if carbon dioxide production continues at the present rate, the current interglacial might become even warmer. What will actually happen? An interesting compromise has been suggested by John and Katherine Imbrie, who predict that the cooling trend expected to lead to the next ice age may be delayed until the present period of carbon dioxide induced warming has run its course, perhaps 2000 years from now (fig. 6.9).

There are others who point out that, even though the atmospheric concentration of carbon dioxide is known to be increasing, it does not seem to have been accompanied by the expected increase of temperature. Another possible effect of carbon dioxide increase might be hotter summers and colder winters throughout Europe. But there are other aspects about which we can be more certain. For perhaps 90 per cent of the last million years the climate has apparently been colder than today, with more extensive ice development. The present interglacial is not typical and the existing locations of many of the world's most prosperous industrial and farming areas, especially on the continents of the northern hemisphere, can be but temporary. In Europe, although a return to glacial conditions might occur quite suddenly, monitoring of the movements of the polar front should nevertheless enable man to anticipate the change.

The climatic background to Quaternary mammalian studies

From findings, such as those outlined above, a clear picture emerges of a Quaternary composed not only of many glacial–interglacial cycles but, within each of these, many minor subdivisions which were not exactly synchronous everywhere. Local changes in the amount of rainfall and degree of continentality of the climate were all part of the same general process of climatic change. It is against such a background that our fossil mammal studies are set (fig. 6.10).

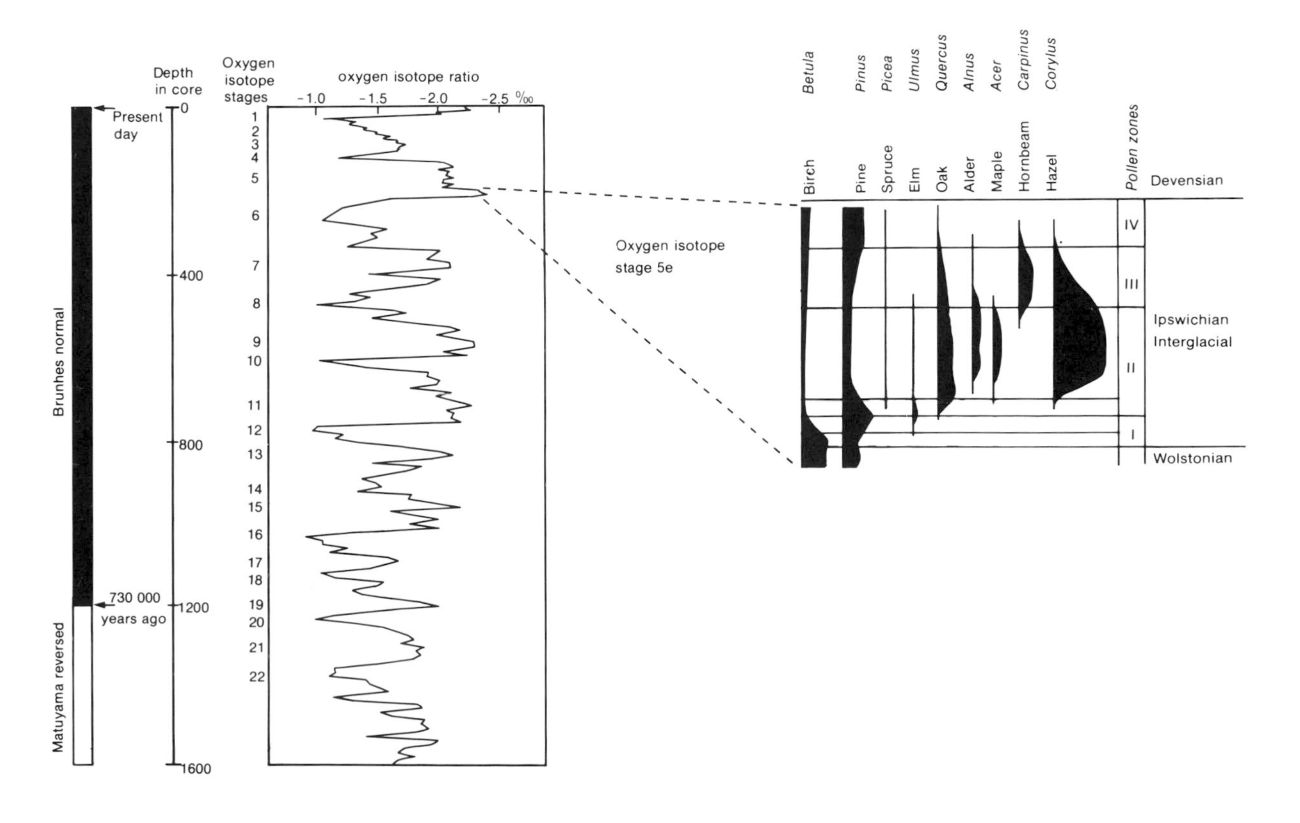

Fig. 6.10 *The climatic background against which mammalian faunal changes need to be considered. Left, the deep sea evidence; oxygen isotope record of equatorial Pacific core V28-238. (After Shackleton & Opdyke, 1976). There could have been many climatically induced changes in the mammalian fauna, which were quite sudden and of relatively brief duration. Right, the pollen zones of a single interglacial (in this instance the Ipswichian of the British Isles: a composite diagram; isotope stage 5e). There could have been significant changes of mammalian fauna from zone to zone within the interglacial. (After Sparks & West, 1977).*

Chapter 7 **Fossils in caves**

Fig. 7.1 *Underground stream in Ogof Ffynnon Ddu (Cave of the Black Spring), South Wales. The stratification of the Carboniferous limestone in which the cave is formed can be seen dipping gently to the left. A bedding plane forms the ceiling. (Photo: T. D. Ford).*

For the palaeontologist looking for fossil remains of animals and for the archaeologist seeking evidence of early man, caves provide one of the richest fields of research. Much of our existing knowledge of extinct fauna and of man has come from a study of remains, sometimes preserved in amazing abundance, found in caves. How are caves formed and why should there be such great concentrations of bones and teeth and artefacts in some of them? How do the palaeontologist and archaeologist set about investigating caves and what sort of information do they hope to obtain?

Firstly, let us consider the way in which caves may be formed, which varies with the nature of the rock in which they occur.

Cave formation

Limestone caves

The commonest type of cave, which includes most of the world's largest (for example the Flint Ridge–Mammoth Cave system, Kentucky, USA, with over 200 km of interconnecting passages) is that formed by solution of limestone by slightly acid ground water. Such caves often begin as small cavities along planes of weakness in the bed rock below the water table, which gradually increase in size over tens of thousands of years to form chambers and entire cave systems. If there is a lowering of the water table (for example as a result of valley down-cutting) they are drained of their ground water, although enlargement may continue as the result of invasion by surface streams and falls of rock from the roof. Finally parts of the cave become accessible through openings from the surface, which allow entry by animals and man (fig. 7.1).

Lava caves

A second type of cave is that formed in lava flows on the flanks of volcanoes. When a low-viscosity lava stream cools its surface may sometimes solidify first, forming a roof and preventing the further cooling of the lava underneath. This liquid lava, continually replenished while the volcano is erupting, moves in a channel further and further downstream with the 'roofing over' process following after it. If, after the eruption ceases, the lava drains away, a cave will remain (fig. 7.2).

Fissure caves

Although relatively unimportant, this third cave type is mentioned here as it has occasionally become the receiving place for remains of Pleistocene mammals. Tension in hard rocks will sometimes lead to the formation of vertical fissures which, if not open to the surface, will develop into true caves. Such fissures are common in the rocks of volcanic areas (for example in Iceland, East Africa and America) and in dissected areas of hard sedimentary rocks overlying clay. At Ightham in Kent, in southeast England, many mammalian remains have been found in infilled fissures of this last mentioned type.

Sea caves

Sea caves, formed by wave action and the battering of pebbles and boulders thrown against cliffs by the sea, may be found in any type of rock sufficiently robust to resist collapse. They most frequently develop along planes of weakness such as faults or volcanic intrusions and may have a large entrance chamber, although, unlike terrestrial limestone and lava caves, they usually terminate after a relatively short distance. Sometimes the force of waves entering a cave breaks through the roof to form a 'blow hole' from which water is thrown

Fig. 7.2 *Lava cave on Mount Suswa, Kenya. The pendant structures on the ceiling and walls are festoons of cooled lava and are not to be confused with the stalactites of limestone caves (fig. 7.6) which are deposited from solution in water. A rockfall has brought down the nearest part of the ceiling together with the festoons on it; and has obscured the floor. (Photo: A. J. Sutcliffe).*

onto the cliff top above at times of storm. Caves into which the sea still enters are unsuitable for habitation other than by seals and bats. Any mammalian remains in them are likely to be swept away. Caves left high and dry by a fall of sea-level may contain bone deposits similar to those in terrestrial caves (see fig. 10.12, p. 142).

Minor cave types

Nearly all the bone caves of the world belong to one of the types described above. For completeness two other rare types must also be mentioned.

On Mount Elgon (an extinct Miocene volcano, on the border of Kenya and Uganda) are many large caves in consolidated volcanic ash. Their origin is unproved, although some of them may have formed, like limestone caves, by solution by water of the relatively soluble salts holding the ash together. Some of the caves have been secondarily enlarged, possibly entirely formed, by sodium-deficient elephants going into them to mine these salts (see pp. 75–76 and fig. 7.3).

Around the lower slopes of the Pleistocene volcano Fantale in Ethiopia are several hundred beehive-like rock mounds with blind circular caverns inside them. These unique structures are the product of a nuée ardente eruption when a cloud of incandescent ash was erupted from the volcano at a very high temperature. This settled on the ground around the volcano, locally to depths exceeding ten metres. There the particles became fused together and there was a massive escape of gases from, and associated blister formation in, the surface of the cooling 'welded ash-flow tuff' so formed. These blisters (of various sizes, mostly about 15–20 metres across, though sometimes approaching 100 metres) were a product of this process. Those that are still intact cannot be entered from outside, though many others have holes in their sides or summits that permit access for man and animals.

No mammalian remains more than a few hundred years old have yet been found in the caves of either Mt Elgon or Mt Fantale though the rapid accumulation of bones that is taking place in some of them at the present time, as the result of diverse processes, is of special interest to palaeontologists trying to interpret fossil deposits in caves.

Cave fossils

True cave fossils are found in secondary infillings within caves and are consequently younger than the caves in which they occur. They are not to be confused with fossils already in sedimentary rocks forming cave walls.

The quantity of mammalian remains found in caves and their fine state of preservation is sometimes astonishing. Two factors contribute to the accumulation of such deposits. Firstly, caves are places where remains tend to become concentrated by natural processes. Some caves are dangerous, with shafts in their roofs down which mammals may fall. In addition, some animals use caves as breeding places or eating or sleeping shelters and they may die there or leave the bones of their prey there. Secondly, though not invariably, caves are places where remains are likely to survive as fossils, once they have been deposited there. They are protected from weathering by the cave roof and the alkaline conditions prevailing in limestone caves favour the preservation of bone. Bones found in lava caves may be less well preserved.

Although throughout the world, for many thousands of years, man has been attracted to caves as places where he could shelter or dig minerals, the study of fossil remains from caves is a relatively young science. Already, however, by the early nineteenth century, it was becoming apparent to those who were beginning to excavate in caves that the processes that had led to the accumulation there of sediment and bones in the past could, in favourable situations, still be observed taking place at the present day. Studies of this fossilization in action, together with the development of excavation techniques that separate fossils from different layers, have since developed into the sophisticated methods being applied by palaeontologists today.

Let us now examine these present-day processes in greater detail as a basis for comparing some fossil examples.

The accumulation of sediment and bones in caves

Since bones are part of the sediment of a cave the processes of accumulation of both can be considered together.

Cave sediment may be either of two sorts – that originating inside a cave and that which has come from outside. The former (the insoluble residue from the rock of limestone caves and rock falls from cave roofs) are unimportant as a source of mammalian remains although crushed skeletons of animals and man have sometimes been found beneath rock falls; for example the mummified body of a Pre-Columbian gypsum miner from Mammoth Cave, Kentucky, USA.

More important are the sediments that have found their way into caves from outside: talus cones beneath shafts; mudflows; stream deposits; and organic matter taken in by animals, including man himself.

Natural open shafts are common wherever there are caves. Although most of these are quite shallow, in limestone areas it is not unusual to find shafts 100 metres deep leading down to vast cave systems below (for example Gaping Gill, Yorkshire) and sometimes they exceed 300 metres. Such open shafts act as collecting places for débris falling in from above. If there is no cave stream below to wash it away, this builds up to form a conical talus which may finally block the shaft to ground level. Such shafts are of great danger to animals walking nearby, which may fall into them, their bones then becoming incorporated in the talus deposit below. A skeleton of a rhinoceros recently found beneath a 'sky light' in the roof of a lava cave on Mount Suswa in Kenya, was of an animal that had apparently fallen in from above.

One of the finest examples in the world of a talus cone with remains of Pleistocene mammals beneath a once open shaft was discovered in 1939 in Joint Mitnor Cave, Buckfastleigh, Devon in southwest England. Among the animals that had fallen into this were the hippopotamus, narrow-nosed rhinoceros, straight-tusked elephant, bison, giant deer, red deer, fallow deer, wild boar, cave lion, spotted hyaena, wolf, fox, wild cat, badger, brown bear, hare and rodents. The fauna is characteristically warm and is believed to be of Last Interglacial age, about 120 000 years old. It is of special significance that, although there is a fair representation of carnivorous species, most of the bones are of herbivores, especially bison, which are the animals which make up most of any mammalian population on the surface and which are statistically most likely to become entombed in this manner. There is no evidence of the cave ever having been a den (though hyaenas may have been attracted there by decaying carcasses); and the relatively few carnivore remains are probably of animals that fell or climbed in and were unable to get out.

Occasionally such deposits may be so saturated with water that they turn into mudflows and progress along cave passages carrying incorporated bones with them. In this way, bones can be transported for considerable distances underground without the aid of running water.

Although water-laid sediments, both those laid down along stream beds (fig. 7.1) and in still water lakes, are very common in caves, the frequency with which bone deposits have been explained away by 'washing underground' greatly exceeds the number of instances where this has really happened. Without doubt, some isolated bones and even associated skeletons which have been found in cave stream deposits were carried underground by water; but it would require an unusual disaster on the surface for most streams to collect more than a few bones. Considerable concentrations of mammalian bones do nevertheless sometimes occur in Pleistocene stream deposits in caves; and here it is necessary to consider whether any other process has also taken part in their assembly.

The hippopotamus-bearing deposits found in the cave near Yealmpton, Devon, (fig. 4.1, p. 36) were composed of horizontally bedded layers of gravel, sand, silt and clay that had clearly been laid down by a former cave stream; though how the bones had found their way into these deposits was not at all clear when the cave, which was filled to the roof with sediment, was first accidentally broken into by quarrying operations. Further quarrying resulted in the opening of a large cave chamber with a series of small talus cones beneath shafts. The animals had apparently fallen into the cave, in the same manner that occurred in Joint Mitnor, and their remains were then secondarily picked up and carried away by the cave stream flowing through the chamber, for redeposition elsewhere along its course. Groups of hyaena coprolites (droppings) spread along the streamway suggest that live hyaenas also had access to the cave.

Most important of all the processes of bone accumulation in caves is the activity of cave dwelling animals, which may die underground or leave the remains of their food there. Man may also leave his implements and the ashes of his fires and bury his dead in caves.

The variety of animals that go into caves is very great. Herbivores, though not usually considered to be cave dwellers, sometimes go into caves for shelter and may perish there if there are unsuspected holes in the cave floor.

The caves of Mount Elgon, previously mentioned, are a great attraction to the local antelopes, buffalo and even elephants – not for shelter but for the salts (notably mirabilite or glauber salt – sodium sulphate) which these sodium-deficient animals can obtain there. Especially astonishing is the behaviour, previously mentioned, of the elephants, animals which, in the open, do not like climbing over rocks; yet they are prepared to scramble over rough ground and to go deep inside the caves, beyond the limit of daylight, in search of salt which they break from the cave walls with their tusks (fig. 7.3). Thick deposits of elephant dung occur locally on the floors of these caves.

Sometimes accidents occur to these animals while underground. The mummified remains of antelopes which have fallen into spaces between boulders are not uncommon; and two baby elephants are known to have perished in one cave, during the last decade, in a similar manner. Collapse of the roof of this cave in 1981, as the result of undermining by elephants, occurred on such a massive scale that it was not

Fig. 7.3 *Elephants inside Kitum Cave, Mt Elgon, Kenya, which they have entered in order to obtain salt. (Photo: I. Redmond).*

possible to determine whether any of these animals had been buried under the rock fall.

Most universal of all the groups of mammals which inhabit caves are the bats, of which there are about 900 living species (though not all of them go underground). With highly developed echo-location those which are cave-dwellers can fly deep into the dark zones of caves, where they shelter, have their young and, at medium latitudes, hibernate during the winter. Often they die underground and their skeletons are incorporated as potential fossils into the floors of the cave; and there are few caves, where conditions are suitable for the survival of fossil remains of small mammals, which do not contain some bones and jaws of these animals. In fact, very few fossil bat remains have been found outside caves, which provide nearly all the fossil evidence about this group of mammals.

The animals mentioned so far often leave their own remains in caves, but they are not responsible for carrying those of other animals there. Sometimes, large accumulations of mammalian and other bones may be taken underground by animals living there, especially birds of prey, some carnivores, porcupines and man.

Birds of prey commonly use caves as roosting or nesting places, penetrating underground to various distances. Two sorts of food débris are left there by such birds of prey – discarded parts of carcasses fed to the young and pellets regurgitated by both adults and young in order to get rid of fur, feathers, bones and anything else indigestible that they have swallowed. Vast piles of bones of small mammals, birds and even frogs and toads may accumulate in this manner and are common in Pleistocene cave deposits throughout the world.

Various carnivores inhabit caves at the present day as others have done in the past. Current examples include the African leopard, which commonly carries underground the remains of baboons it has killed; and the bobcat (lynx) and jaguar of North America. In Europe, during upper Pleistocene times, two carnivorous/omnivorous cave dwellers were of special importance – the spotted hyaena and the bear (great cave bear of the continent and brown bear of the British Isles, fig. 7.4). We will return to these animals on p. 80. In North Africa, Asia and America other species of bear were cave dwellers. The lion, lynx, wolf, red fox and wolverine (fig. 7.5), also represented among the cave fossils, are other northern hemisphere mammals that apparently went underground of their own free will. In Australian caves (chapter 13) the carnivorous marsupials, native cat, Tasmanian devil and Tasmanian wolf (the last two now extinct on the Australian mainland, the last (fig. 13.8, p. 196)

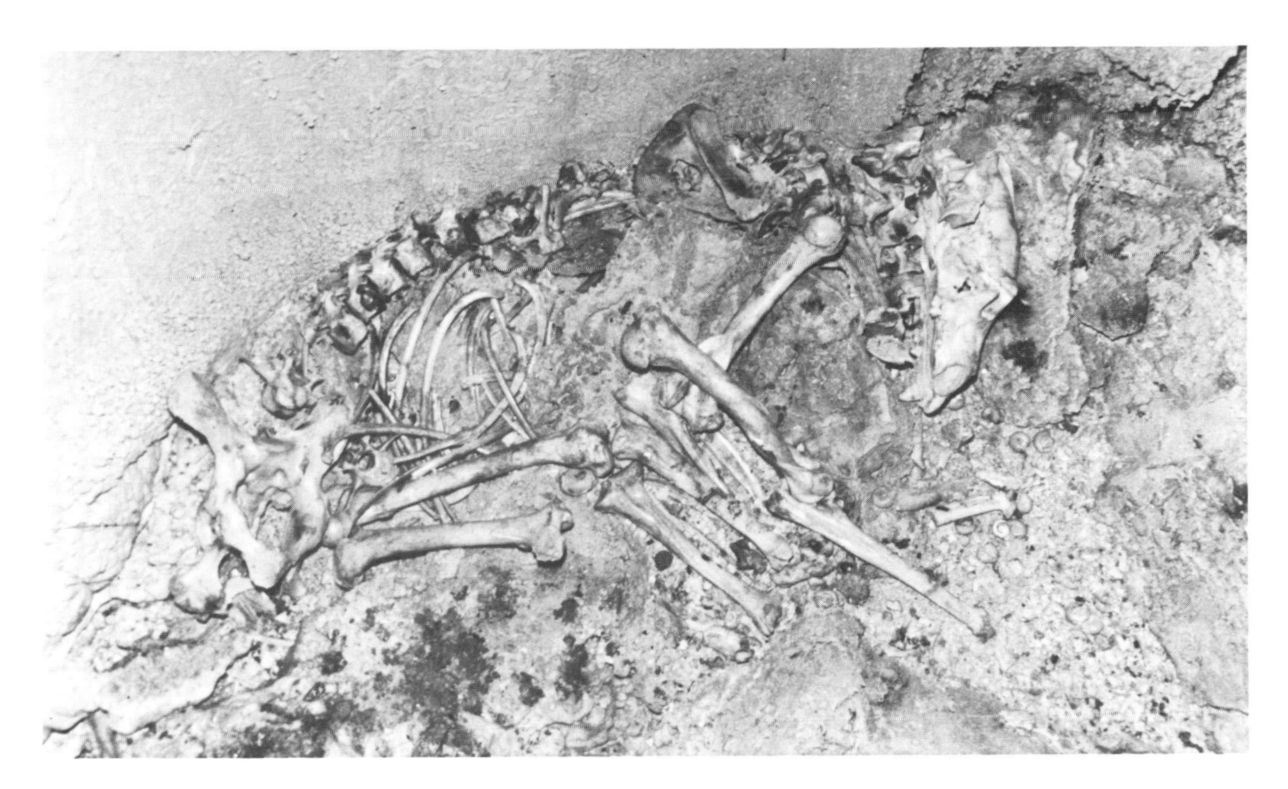

Fig. 7.4 *Skeleton of a brown bear (pelvis on the left, skull to the right, limb bones below, which apparently died during hibernation, in the cave Urkizetako Leiza III, Pais Vasco, N. Spain. (late Pleistocene or Holocene). (Photo: J. Telleria).*

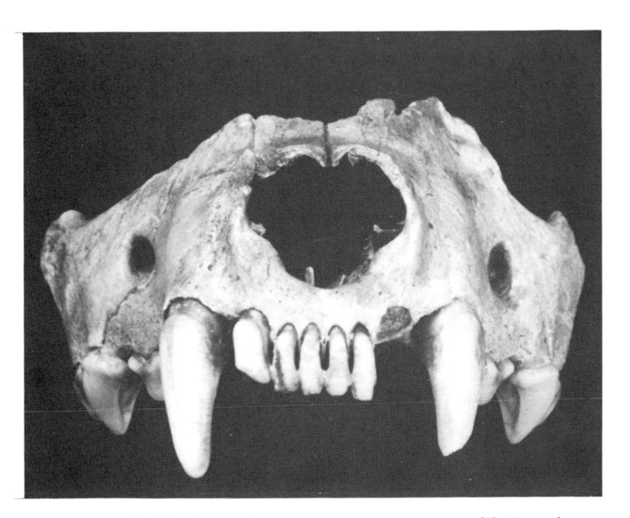

Fig. 7.5 *Skull of a wolverine, c. 85 000 years old, found in Stump Cross Cave, Yorkshire. The animal apparently entered the cave intentionally and died there by mischance. (Photo: BM(NH)).*

probably extinct everywhere) may also have been responsible for the accumulation of bone fragments, partly in their droppings, in caves.

Most complex of all the types of occupation débris associated with cave-dwelling animals is that resulting from human habitation of caves. Even today, in Africa and other parts of the world, caves are sometimes used for shelter by man and his domestic animals. Commonly they are simply adapted for habitation by the construction of woven brushwood barricades across their mouths and there may be similar screens inside for keeping domestic animals separate from the living areas. In Europe, there are many examples of quite recent stonework extensions to the entrances of natural caves to create houses or religious buildings or even pigeon roosting places. Some of these are still intact today.

Human occupation of caves can be a major cause of sedimentation; with ashes from fires, food débris, the waste material from tool making, earth carried into the cave from outside and even human excrement jointly building up to form great thicknesses of stratified deposits. Such accumulations are especially common in the entrance regions of caves though they are sometimes also found further underground; and man is not always averse to lighting fires in parts of caves with poor or little ventilation.

Important examples of accumulations of remains of Pleistocene mammals carried into caves by Palaeolithic man include a pile of bones and skulls of woolly mammoth and woolly rhinoceros found during the Cambridge University excavations in the Cotte de St Brelade, Jersey, in the 1960s and 70s; and great quantities of bones of horse and reindeer excavated over a century ago in the Cave of Bruniquel, Tarn et Garonne, France. Among the Bruniquel finds were numerous fragments of reindeer antlers which had been cut by Palaeolithic man, using sharp flint knives, for the removal of slivers of antlers that were to be fashioned into other tools. The waste fragments of antler had then been abandoned on the cave floor.

There are many examples of caves being used as human burial places. The upper Palaeolithic skeleton with its associated rods and rings of mammoth ivory found in Paviland Cave, South Wales, in 1822 was apparently an interment; and other burials of diverse ages have been found in caves throughout the world.

So far we have considered mainly bone accumulations which have become buried under cave floors as further sedimentation has taken place. But sedimentation is not always a continuous process and occasionally caves have been discovered with floor deposits unburied and unchanged after being sealed away from the outside world for thousands of years. Under such circumstances not only may any skeletal remains stay unburied (fig. 7.4), but even the footprints and scratch marks of such cave dwelling animals as bears, man and (in America) the jaguar, have sometimes been preserved.

Fig. 7.6 *Shatter Cave, Mendip, Somerset. This cavern chamber illustrates some of the forms of speleotravertine (stalactites, straws, columns, organ pipes, curtains and floor deposits) commonly found in limestone caves. The joined stalactite-stalagmite column is about 2·5 metres high. (Photo: A. E. Mc R. Pearce).*

Speleotravertine

There is one further common type of deposit that does not fit into either of the categories described above which must be discussed separately here – the hard often crystalline material that is precipitated in caves from drops and films of water containing dissolved matter which have entered cavern chambers from above.

This varies in composition according to the nature of the cave, although in limestone caves it is almost always calcite or aragonite (two forms of calcium carbonate, the substance of which limestone is mainly composed). It is then known as travertine and the term speleothem has been devised as a general term for the diverse shapes. There are three main forms – flowstone, which is deposited from thin films of water on floor and walls, dripstone (stalactites and stalagmites) and crystals in cave pools (fig. 7.6). If crystalline and pure, these formations can sometimes be dated using the uranium series method (see chapter 5); and where mammalian remains occur between dated flowstone horizons, their date can also be established.

By its mere presence, travertine provides an indication of the climatic conditions prevailing at the time of its deposition, since its formation (fig. 7.7) is typically a phenomenon of wet interglacial episodes, growth ceasing under arid or glacial conditions. New methods now also make it possible to determine the prevailing temperature. Usually growth is very slow, some layers having taken many thousands of years to form. In instances where more rapid deposition can take place, for example in pools, plant leaves and even insect wings (so important in the interpretation of surface sites, though unusual in cave deposits) have

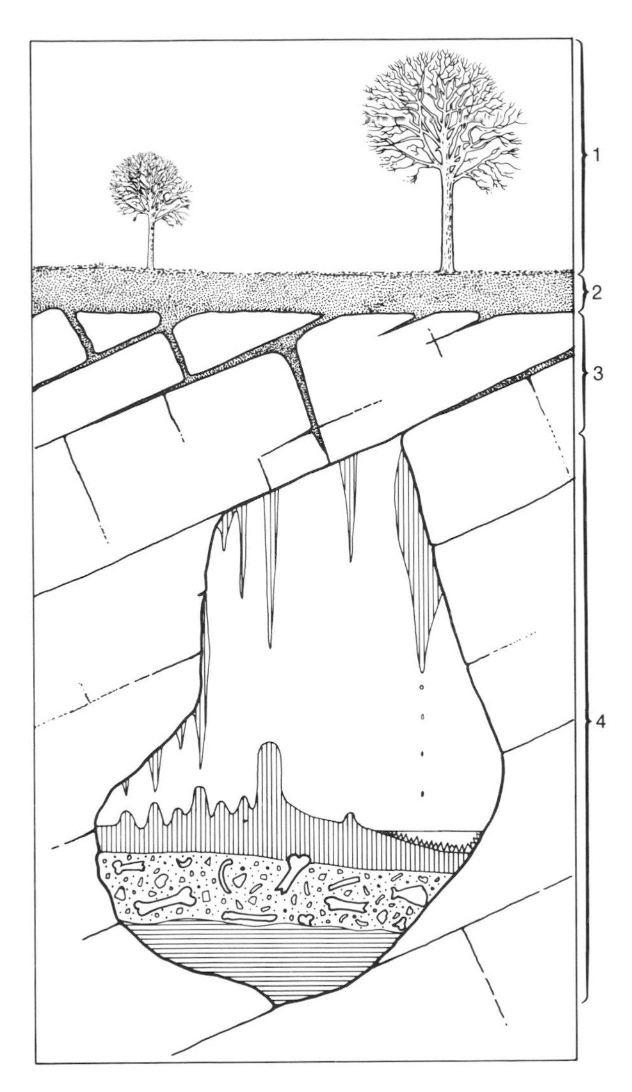

Fig. 7.7 *The chemistry of flowstone and dripstone formation.*

1 **The atmosphere**

Carbon dioxide from the air dissolves in rain to form carbonic acid:

$$\underset{\text{water}}{H_2O} + \underset{\text{carbon dioxide}}{CO_2} \rightarrow \underset{\text{carbonic acid}}{H_2CO_3}$$

There is generally very little carbon dioxide in the atmosphere, so that rain-water falling on the ground contains only a small amount of carbonic acid.

2 **The soil**

For any appreciable amount of limestone to be dissolved, a more concentrated solution of carbonic acid is needed. Biological processes in the soil maintain a higher concentration of carbon dioxide there than is present in the atmosphere. The concentration of carbonic acid in rain-water passing through the soil is therefore increased.

3 **The limestone above the cave**

The carbonic acid solution, coming into contact with the calcium carbonate of which limestone is formed, reacts with it to form soluble calcium bicarbonate:

$$\underset{\text{carbonic acid}}{H_2CO_3} + \underset{\text{calcium carbonate}}{CaCO_3} \rightarrow \underset{\text{calcium bicarbonate}}{Ca\,(HCO_3)_2}$$

The solution is carried through cracks in the limestone until it reaches the cavern space. As the carbonic acid becomes used up by reaction with the limestone its erosive force weakens and thus many vertical cavities become rapidly narrower with depth. So long as the cavities are completely filled with solution the carbon dioxide cannot escape.

4 **The cavern space**

The concentration of carbon dioxide in the cavern space is usually approximately that of the free air outside. The solution of calcium bicarbonate dissociates on reaching the cavern space; the carbon dioxide acquired in the soil is gradually lost, and the insoluble calcium carbonate is precipitated in various crystalline forms:

$$\underset{\text{calcium bicarbonate}}{Ca(HCO_3)_2} \rightarrow \underset{\text{calcium carbonate (calcite and aragonite–flowstone and dripstone)}}{CaCO_3} + \underset{\text{water}}{H_2O} + \underset{\text{carbon dioxide}}{CO_2}$$

The loss of carbon dioxide takes place slowly so that precipitation continues as the water drains through the cave. When all excess carbon dioxide has been given off no more calcite is precipitated.

sometimes been preserved as moulds, providing valuable background data for associated mammalian studies.

Disturbance of cave sediments and their contained mammalian remains

In most caves, stratified sequences of deposits with mammalian remains can be excavated and studied in the same manner as stratified fossiliferous deposits elsewhere. The lowest stratum can be interpreted as the oldest; the uppermost (which forms the present-day floor of the cave) as the youngest. Such deposits are nevertheless very susceptible to disturbance which, if unrecognized, could make detailed stratigraphic work valueless. We have already seen how washing out of the lower deposits in a cave, followed by further sedimentation, can lead to an apparent inversion of strata (fig. 5.2, p. 45). Other common disturbances are collapse pits and burrows, where a burrowing animal may dig up fossil bones from an earlier stratum and throw them out onto the cave floor; or it may die in a burrow where its skeleton can be confused with earlier fossil remains in the deposit forming the burrow wall. More rarely, disturbance may be caused by the interment of human remains. Earthquakes and frost disturbance sometimes cause the shattering of travertine in caves.

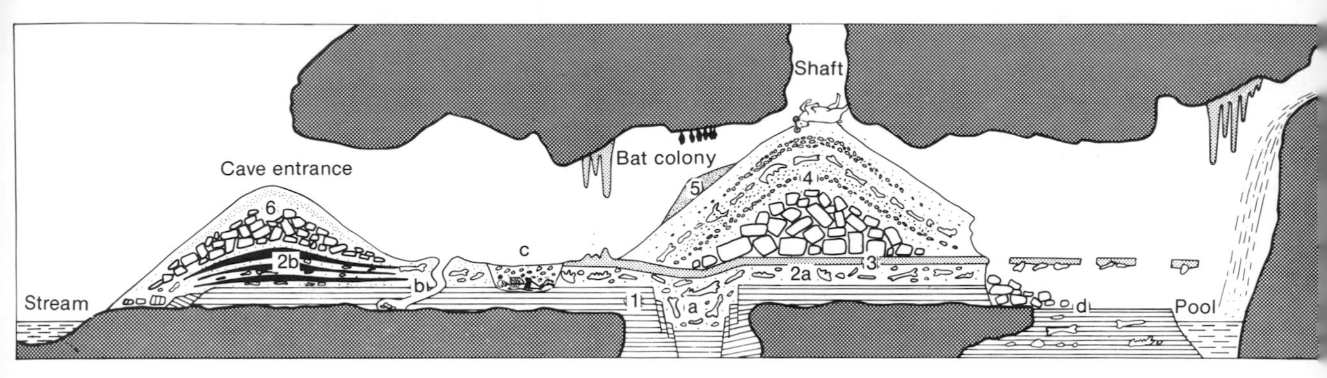

Fig. 7.8 *Vertical section through an imaginary bone cave, illustrating some important types of cave deposit. 1, water-laid sands and clays; 2a, deposits of an animal lair; 2b, hearths of fires made by man; 3, stalagmite floor; 4, talus cone with bones of animals which fell down a shaft; 5 bones and dung of bats; 6, second talus cone at cave mouth. The deposits show disturbance at several places: a, collapse pit; b, burrow; c, human burial; d, washing out and redeposition by a stream.*

A section of an imaginary bone cave

Figure 7.8 shows a vertical section of an imaginary bone cave in which the main types of bone deposit described above and the disturbances which may affect them are shown together. It is unlikely that, in reality, any one cave would show so many different features. Consideration of deposits such as those shown in the section emphasizes some of the problems associated with reconstructing the sequence of events leading to their accumulation. Only careful examination of the burial (c) can show that it is later than deposit (2), which is at the same level. Similarly, the bones of the burrowing animal, which has died in the burrow (b) are surrounded by deposits of layer 1 and (unless the burrow is carefully cleaned out before 1–2 are excavated) could be thought to be older than is really the case. The collapse pit (a) and the undermined flowstone floor (d) create similar problems. A further hazard lies in the excavation of layer 4 since the divisions are sloping and finds from them will become mixed if this deposit is excavated in horizontal layers.

Bone caves of the world

Although the same processes of bone accumulation take place in caves in all parts of the world, the mammalian species whose remains can become fossilized in them vary both regionally and in time. The diversity of assemblages is very great. Let us look briefly at this diversity.

Western Europe

Nowhere in the world is there greater variety of mammalian faunal assemblages in caves than in Western Europe, where the repeated effects of changes from glacial to interglacial conditions are so strongly marked. Many of these sites are also of great archaeological importance. There will be some elaboration of these topics in chapters 10 and 8 respectively.

Of the animals known from fossil remains in European caves, probably the great cave bear of Western Europe has attracted the greatest attention. The Drachenhöhle, a cave near Mixnitz in Austria, is estimated to have contained the remains of over 30 000 cave bears, which accumulated as a result of a small number of animals continuing to occupy the cave over a long period of time.

The Pleistocene spotted hyaena of Europe, remains of which have so often been found in caves, has also interested palaeontologists and archaeologists because of a controversy about its behaviour which arose in the 1950s. Was it a cave dwelling animal, which also left the bones of its prey in caves, as was suggested by William Buckland in 1823, following his excavations in Kirkdale Cave, Yorkshire in northern England; or, as there were also signs of human habitation in some of the caves concerned (for example, in Kents' Cavern, Devon in southwest England), could man have been the agent of bone accumulation?

In order to resolve this problem it is necessary to consider the behaviour of present-day African spotted hyaenas. Until recently it was thought that because of their greater size, the spotted hyaena remains found in the Pleistocene deposits of Europe represented a distinct subspecies. A study of the size of both present-day animals and of the fossil material, however, has shown a continuous geographical size gradient (smallest at the equator, larger at high latitudes, reaching its peak in Europe – an example of

Bergmann's Rule). No such sub-specific difference can be accepted any longer. Direct palaeoecological comparison between Pleistocene European and living East African spotted hyaenas, if undertaken with caution, is therefore feasible.

The living spotted hyaena is both a hunter and a scavenger that can break quite large bones, leaving its characteristic tooth marks on them. Some British caves in which such remains have been found, for example Kirkdale Cave, are much too low to have been places of human habitation and today it is generally accepted again that the European spotted hyaena was indeed a cave dwelling animal that carried the remains of its prey underground, as claimed by Buckland over a century and a half ago.

Island faunas

Islands often have unusual faunas that have evolved in isolation from those on the nearest mainland, sometimes in absence of competition from other species, sometimes under adverse conditions of food supply. In cave deposits of Pleistocene age on some of the Mediterranean islands have been found remains of dwarf and giant and other peculiar mammals which evolved in this way. Mammals represented include pigmy hippos and pigmy elephants on Malta, Sicily and Cyprus; a pigmy deer on Crete; a giant dormouse on Malta; and the peculiar goat-like *Myotragus balearicus* on Majorca. Madagascar, long separated from the mainland of Africa, has its own distinct fauna, reflected in fossil remains found in its caves. Of special interest are the fossil lemuroids, first described from the Cave of Androhomana, near Fort Dauphin in the southern part of the island. These remains are probably of relatively recent date.

Africa

Among the most important African cave sites are the famous lower Pleistocene Australopithecine localities of South Africa; Makapansgat and Sterkfontein. The Sterkfontein fauna consists almost exclusively of extinct animals, including a giant pig, *Notochoerus*, almost twice the size of a warthog; *Libytherium*, a giraffid with antler-like horns; *Hipparion*, the three-toed horse; carnivores; *Dinopithecus*, a giant baboon; and many other genera.

Asia

There are many fossiliferous caves in China, most notable being the middle Pleistocene site of Choukoutien, the finding place of Peking man. The discovery of 'Dragons' teeth' in other Chinese caves has already been described in chapter 3.

In the SE Asian region, one of the most important cave excavations has been that in Niah Cave, Borneo. A rich fauna of late Pleistocene and post-Pleistocene age was found associated with important archaeological remains. Mammalian species include various bats and monkeys, orang-utang, pangolin, rodents, bear, tapir, rhinoceros and deer. Most of these animals still survive in the region at the present day.

Australia

Fossil remains have been found in many Australian caves, some of the most important sites being situated in the Nullarbor region of Western Australia. The mammalian fauna of Australia (which will be discussed in more detail in chapter 13) is peculiar in that this continent became separated from the rest of the world in Cretaceous times, probably about 100 000 000 years ago, and became a place of radiation of marsupials and a few monotremes and not of placental mammals. With the exception of dingos and rodents (placental mammals that reached Australia more recently) the fossil mammalian remains are nearly all of marsupials.

The Americas

Mammalian remains have been found in many caves in North and South America. Most are of upper Pleistocene age, though some earlier sites are also known, including Port Kennedy Cave, Pennsylvania, and Conard Fissure, Arkansas, which are probably middle Pleistocene. A mammal of special interest from the former locality is the early sabre-toothed cat, *Smilodon gracilis*, an ancestor of the upper Pleistocene sabre-tooths found in some later cave deposits.

Mammals of late Pleistocene age of which remains have been commonly found in North American caves, include ground sloths, porcupine, bears, sabre-toothed cat, mammoth, tapir, peccary, camel, reindeer and bison. This assemblage, which includes forms with South American affinities (porcupine, ground sloths) and Eurasian affinities (mammoth, reindeer, bison) vividly illustrates the mixed origin of the North American Pleistocene mammalian fauna. This mixing will be further discussed in chapter 12. Some of the Florida bone caves have been flooded to a depth of 25 metres by subsequent rise of sea-level and can be excavated for their mastodon and other remains only with the aid of diving equipment.

One of the best known South American caves is that of Ultima Esperanza in Patagonia. So favourable were the conditions for preservation (probably the combined effects of aridity and mineral salts) that large pieces of 13 000-year-old skin with hair of the extinct Pleistocene ground sloth *Mylodon* still survived in the deposits of the floor of the cave. This site is further discussed in chapter 12.

Excavating caves

When a palaeontologist or archaeologist sets about excavating a cave, how does he do this and what sort of information does he seek? A cave may contain a record of past events and the reconstruction of this is usually his main objective. Excavation, however, is destructive and once the deposits have been removed from a cave the relationship of the various objects found in them is lost and cannot be restored. In a well-conducted excavation a team of specialists work together, so that no line of evidence is overlooked. Above all, special care has to be taken to record any stratification or layering in a cave, since the deposits there may have accumulated over a long period of time and may provide evidence of the evolution of man's industries or of past changes of climate, fauna and flora.

At a complex cave site it is probable that specialists would be needed to study separately human artifacts; human skeletal remains; bones of other large mammals; bones of small mammals such as rodents, shrews and bats (which can best be recovered by sieving); shells of molluscs; plant remains (including pollen); and sediments of the cave floor. In addition, samples of bone or charcoal, collected in the cave, would be submitted to a laboratory for carbon^{14} dating, likewise samples of travertine for uranium series dating.

Whenever possible part of the deposit in a cave should be left intact, partly as a permanent record and partly as material for future investigation.

Chapter 8 ***What the cave men saw***

Polychrome painting, in red, brown and black; mostly of bison, but also including three boar, two red deer hinds and a horse, on the ceiling of the Great Hall of the Cave of Altamira, Spain. Middle Magdalenian. Length of group 14 metres. (After Breuil).

The palaeontologist, when trying to reconstruct the physical appearance of extinct Quaternary mammals, is often limited in the amount of information available to him. Sometimes, as in the case of the carcasses of mammoths, woolly rhinoceros and bison from the frozen ground of the Arctic, even the skin and hair have been preserved, providing valuable information; though more frequently he has only skeletal material from which to work. If asked by an artist who is painting a reconstruction for him whether a fossil tiger should be shown with stripes or a leopard with spots he faces a dilemma; since he has no means of knowing and whatever is shown can only be the result of carefully considered guesswork.

Fortunately for the palaeontologist, there are a few instances of extinct mammals which attracted the attention of contemporary artists; and some of these early pictures still survive for reference at the present day. Most notable is the upper Palaeolithic cave art of Western Europe which includes, among the diverse mammals depicted, several hundred paintings and engravings of woolly mammoths – providing independent witness from a considerable number of people who had actually seen these animals!

The practice of making such depictions is not peculiar to the upper Palaeolithic of Europe, but ranges over a long period of time throughout most continents of the world. It can still be seen in action in some parts of Africa at the present day, for example by the Masai tribesmen in the rockshelters of Tanzania. Other important instances of rock art include the Bushmen paintings of people and other animals in South Africa (the latest of which were made after the arrival of Europeans; the date of the earliest is uncertain but may be several thousand years); the rockshelter paintings and engravings of the Sahara Desert, especially those on the Plateau of Tassili n'Ajjer (earlier than 4700 years ago – the date of the end of the Neolithic humid period); the American Indian pictographs on cliffs in southwestern North America (some of them probably only a few hundred years old); and the Australian Aboriginal rockshelter paintings, many of them quite late, though it has been suggested that the earliest could be of Pleistocene age.

Each of these picture series has provided valuable historical information about the past faunas of the areas concerned. In our study of the mammals of the Ice Age one of these groups of 'art galleries' – the upper Palaeolithic cave art and sculptures of Western Europe – is of special importance and will be considered here in greater detail.

Remarkably, the principal sites of European Palaeolithic art are restricted to an area only about five hundred kilometres square, with especially important centres at Les Eyzies in the Dordogne region of France (including the caves of Les Combarelles, Font de Gaume and Lascaux), the Pyrenees (Niaux and Trois Frères) and along the north coast of Spain (including Altamira). There are also other more widely scattered sites with Palaeolithic works of art (many of them open sites at which only small sculptures have been found) in eastern France, Germany, Switzerland, Italy (including Sicily), Czechoslovakia and Asia. The most outlying examples of cave art are the engravings discovered in Addaura Cave, Sicily in 1953, and the mammoth paintings discovered in 1959 in Kapovaya Cave, in the southern Ural mountains, Russia and the engraved mammoth tusk found at the Berelekh site in northern Siberia (fig. 8.3d).

These upper Palaeolithic artists employed several media for their work. On the walls and ceilings of caves they made paintings and engravings, some in a single colour (monochrome); others polychrome. Sometimes pictures, such as the salmon in the cave of Bédeilhac, France, were drawn in the mud on the floors of caves. Artists also made three-dimensional clay models on the floors of caves, the finest examples being the bison group in the Tuc d'Audoubert and the headless clay bear in the cave of Montespan. Their remarkable preservation has been possible only because of their sheltered underground situation. Such representations are known by the archaeologists as 'parietal art', an Anglicization of the French *art pariétal,* art attached to any permanent surface, such as a cave wall.

In addition, there is the portable art (*art mobilier*). It includes engravings on pebbles, bones, tusks and antlers and the rare baked clay models of rhinoceros, lions and other animals from the site of Dolni Vestonice in Czechoslovakia.

Mammals are not the only subjects depicted in European Palaeolithic art. Less frequently, birds, fish, snakes, insects and obscure signs and patterns have been found, the meaning of which is difficult to decipher. Sometimes man portrayed his own species and there are instances (for example in the Cave of Gargas, France) where he stencilled around his hands on the walls of caves. Many of these works are items of exquisite craftsmanship and beauty and they have attracted the attention of present-day artists no less than that of archaeologists.

How old are these various works of art and what was the artist's purpose in making them? The dating of the *art mobilier* is more straightforward than that of the cave paintings since such objects are usually found as the result of excavation of stratified deposits; and their association with charcoal (which can be dated from its carbon ^{14} content) and other datable finds can be demonstrated. Dating of the cave paintings is more difficult, though sometimes it has been possible to infer that they are of the same age as datable finds of artists equipment on the cave floor, or to demonstrate that they are buried behind and thus older than datable floor deposits.

All the Western European cave art is not of the same age but spans a time of at least 20 000 years, from before 30 000 years ago until after the end of the Palaeolithic. This was the time of Cro-Magnon man who had recently superceded Neanderthal man in Europe. His main associated archaeological industries were the Aurignacian, the Solutrean and the Magdalenian, all included in the upper Palaeolithic. There are also some paintings of Mesolithic age. Various schemes, evolved gradually over the years, have been employed for classifying the upper Palaeolithic of Europe. A simplified chronology is shown in Table 3.

The earliest works of art are all *art mobilier.* They include baked clay representations from the previously mentioned open site of Dolni Vestonice, believed to be Aurignacian, about 27 000 years old; and a 12 cm-long carved ivory mammoth, of similar age from Predmosti, both sites in Czechoslovakia. Among the earliest examples of dated paintwork are some blocks with engraved animals, symbols and traces of paint found stratified with Aurignacian deposits in the cave of La Ferrassie, France. Subsequently monochrome wall paintings preceded those in polychrome, which flowered during the Magdalenian. The most magnificent of the European cave paintings are of Solutrean and Magdalenian age. They include those in Lascaux (carbon date of main occupation level 17 000 years) and in Altamira (15 500 years). At both sites paintings apparently continued to be produced over a long period of time. The Berelekh mammoth engraving is only 10–11 000 years old.

The motive behind the manufacture of these works of art has been a source of much discussion among archaeologists, but it is improbable that all the examples were made for the same purpose. Among the many explanations that have been suggested for the various works of art are that they were documentary records of what was seen; artistic creations; were concerned with hunting (for example the speared clay bear in Montespan); were concerned with fertility of the artists' prey animals (the pregnant mares in Lascaux); or with religion or magic (the 'Sorcerer' in Trois Frères); or with man's own fertility.

Palaeolithic art as a source of palaeontological information

Let us now look more closely at some of these Palaeolithic depictions of Pleistocene mammals and see what they tell us about the life appearance or even habits of the animals concerned. Of special importance to us are the species that have since become extinct, such as the woolly mammoth; though much of interest is also to be learned about those which survive (if not

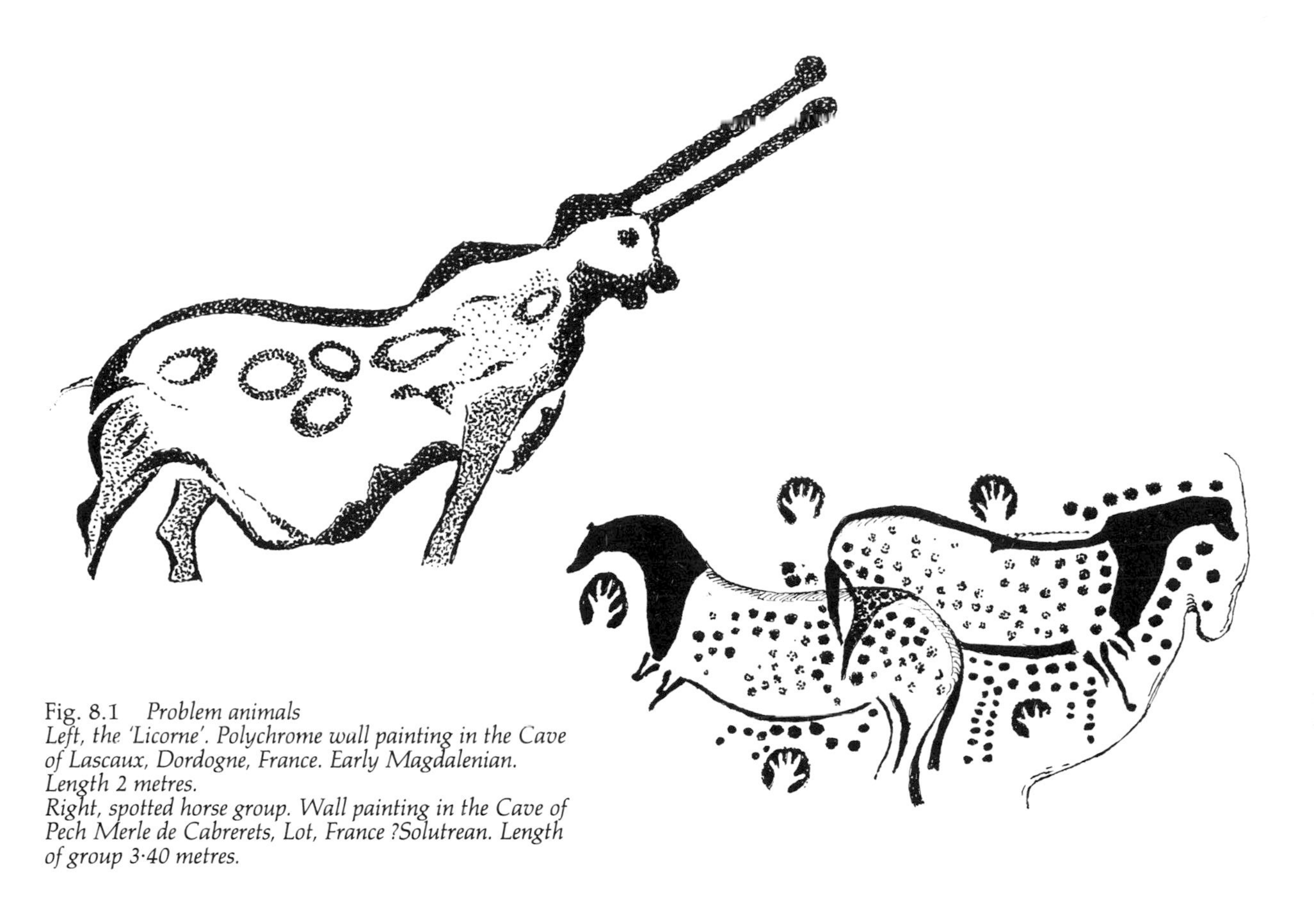

Fig. 8.1 *Problem animals*
Left, the 'Licorne'. Polychrome wall painting in the Cave of Lascaux, Dordogne, France. Early Magdalenian. Length 2 metres.
Right, spotted horse group. Wall painting in the Cave of Pech Merle de Cabrerets, Lot, France ?Solutrean. Length of group 3·40 metres.

in Europe, then in other parts of the world) at the present day.

Firstly, we will consider some of the problems of isolating palaeontologically significant information from these Palaeolithic works of art. If we believe that all the animals depicted really appeared as they are shown, then there were some very bizarre creatures living in Europe during the Pleistocene. Other portrayals, on the other hand, can be accepted as being anatomically very accurate. One of the commonest problems affecting parietal art lies in later paintings and engravings having been emplaced on top of earlier ones or allowed to overlap with them. In such instances it may be difficult to determine whether a particular pair of lines represents the horns or tusks of one animal or the back legs of the one adjacent to it.

Sometimes details of an animal are obscured by the artist having incorporated some features of the cave wall into his work. Some of the bison in the Cave of Altamira are painted on bulges of rock on the ceiling; and one of the horses in the Niaux has an eye which is a natural circle of calcite on the cave wall. The cave artists are likely to have portrayed only those animals which interested them, and this may explain the preponderance of horses, bison, ibex, all food sources; although sometimes carnivores are shown (as, for example, in Trois Frères, where there are some fine examples of lions in addition to the more edible mammalian species). The absence of paintings of a particular species does not necessarily mean it did not exist in the area. Some of the cave paintings and engravings (for example those in Niaux and Rouffignac) occur so far underground that there is no possibility that the artist was able to paint direct from life. If he was relying on his memory of what he had seen outside, can we be sure that this was reliable; and could he accidentally have combined the characters of more than one species in the same drawing? Are some of the paintings the work of children?

One wall painting that has generated an especially great amount of discussion is the so called 'Licorne' in Lascaux (fig. 8.1), variously interpreted as a chiru (a tibetan antelope, unknown in Europe from fossil evidence) and a mythical animal. The horns suggest a male animal (there are no horns in the female chiru; those of the female ibex are quite small), yet the

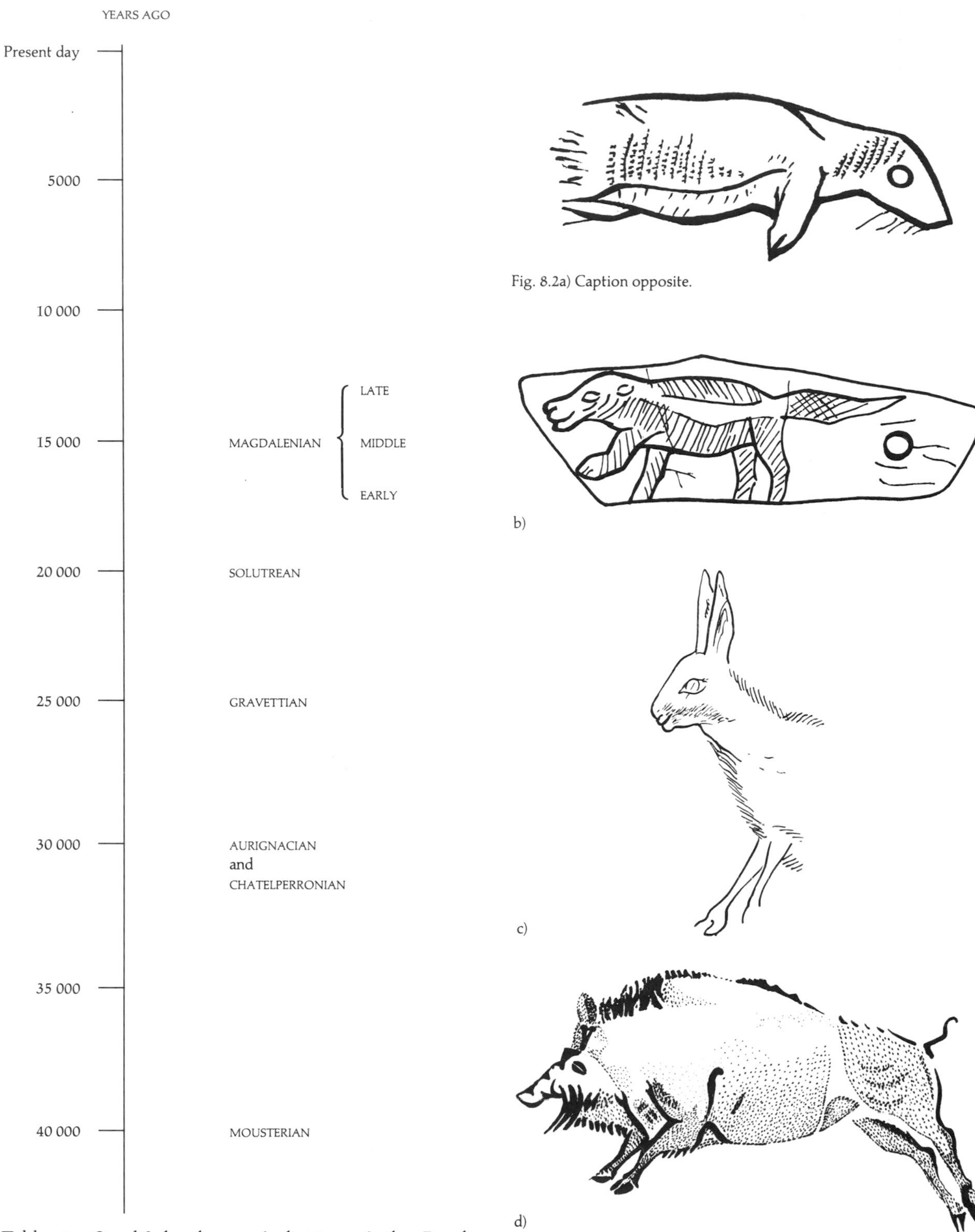

Fig. 8.2a) Caption opposite.

b)

c)

d)

Table 3. *Simplified scheme of division of the French middle and upper Palaeolithic cultural sequences. The actual dates of some of the industries overlap.*

Fig. 8.2 *Cave men used a variety of media for their reconstructions which not only depicted large 'everyday' mammals shown elsewhere in this chapter but also some smaller and less frequently seen species such as (a) a seal engraved on bone (La Vache, France), (b) a wolverine on bone (Les Eyzies, France), (c) a hare engraved on stone (Cave of Isturitz, France), (d) a wild boar painted in polychrome on a cave wall (Altamira, Spain), and above, a member of the weasel family painted in monochrome on a cave wall (Niaux, France). (Photo: J. Vertut).*

creature is clearly pregnant. Furthermore why has it no ears? The picture is at once suspect. The spotted bison in the Cave of Marsoulas and the spotted horses in the Cave of Pech-Merle de Cabrerets, France (fig. 8.1) look nearly as improbable. On close inspection of the horses it is seen that not all the spots are on the bodies of these animals; some of them are situated around them; and in other parts of the cave similar spots occur without any horses. Spotted bison and horses were not often portrayed by other Palaeolithic artists and, delightful though these paintings are to the eye, it seems better not to base any scientific reconstructions on them.

When we have eliminated paintings such as these, the palaeontological value of many of the other depictions at once becomes apparent. From site after site we see paintings, engravings, sometimes even clay models and carvings of mammoths, bison, horses and other animals – the work of artists totally unknown to one another, living in different places, even at slightly different times – all of which show the same stance and the same anatomical details. The similarity of these portraits can be no coincidence and there is good reason to believe that they provide between them a valuable source of palaeontological information.

Since our main concern here is to find out what these Pleistocene mammals looked like, we will select for discussion in this chapter a few of the clearest depictions available, since it is these that reveal the most interesting anatomical details. They are mostly of late Solutrean and Magdalenian rather than earlier age.

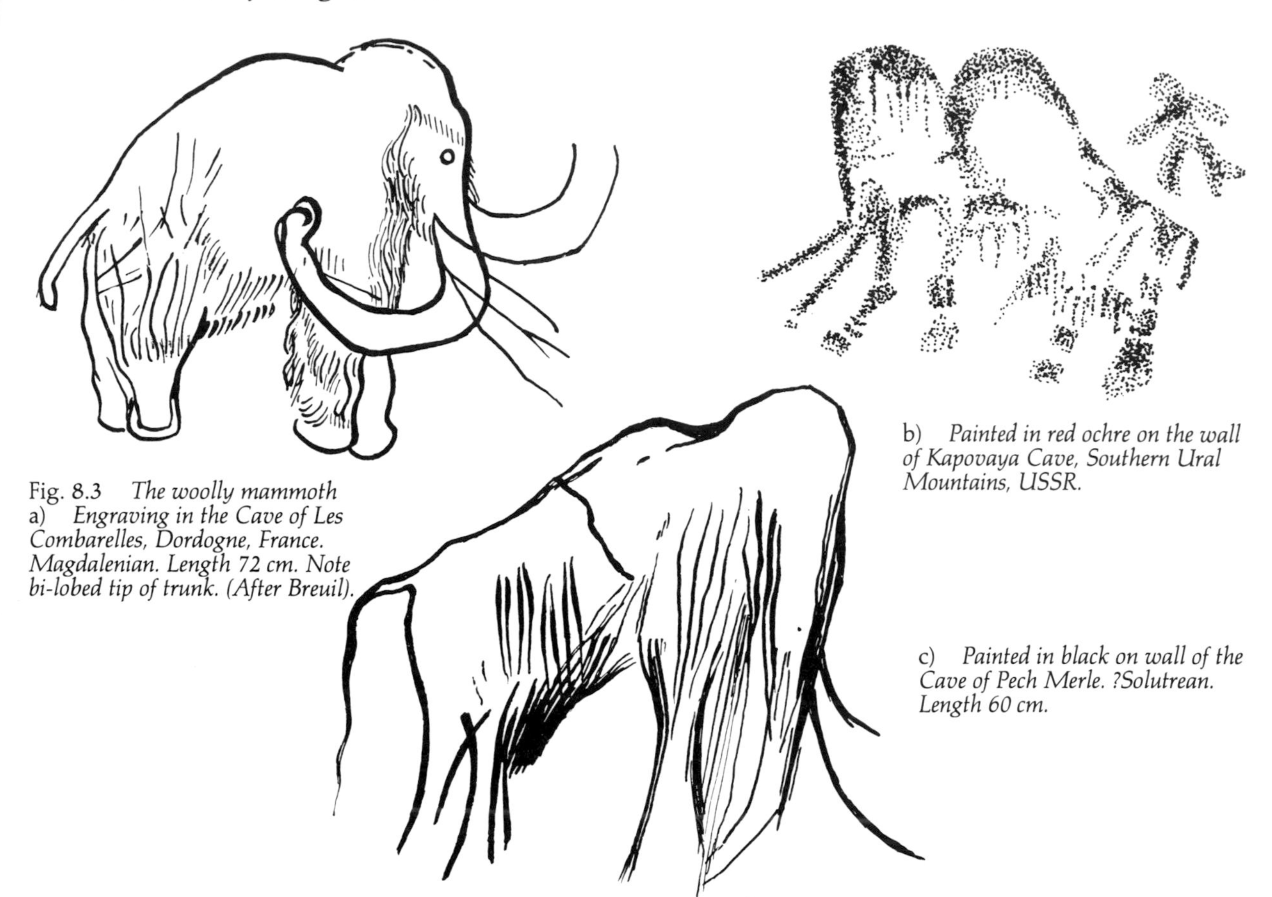

Fig. 8.3 *The woolly mammoth*
a) *Engraving in the Cave of Les Combarelles, Dordogne, France. Magdalenian. Length 72 cm. Note bi-lobed tip of trunk. (After Breuil).*

b) *Painted in red ochre on the wall of Kapovaya Cave, Southern Ural Mountains, USSR.*

c) *Painted in black on wall of the Cave of Pech Merle. ?Solutrean. Length 60 cm.*

The woolly mammoth

A clear picture of the appearance of the woolly mammoth is provided by the hundreds of paintings, engravings and occasional carvings of this animal found over an area extending from northern Spain to northern Siberia. Although providing less anatomical detail than the Siberian and Alaskan frozen carcasses, they provide a much better idea of the general stance of the mammoth than these remains, most of which are incomplete, shrunken and distorted.

Many characters can be seen repeatedly and there can be little doubt that they provide a good record of what the Palaeolithic artists saw. The principal characters shown are a high domed head, separated from the humped shoulders by a deeply depressed neck, and a steeply sloping back. Such an outline is essentially different from that of either living elephant. There is great variation in the shapes of the tusks. Some of the mammoths are tuskless, whereas others (presumably male animals) have strongly curved tusks of great size; a character repeatedly displayed among fossil remains. Long hair is commonly indicated, apparently terminating short of the ground as a fringe. The feet are often visible beneath this. The facial portraits in the caves of Rouffignac and at Arcy-sur-Cure show a hairy trunk, especially along its posterior border. The eyes were situated in a forward position. There is seldom any indication of ears, suggesting that these were small; and there is one portrayal (fig. 8.3a), showing what appear to be two grasping 'lips' on the tip of the trunk, unlike those on living elephants which are of unequal size. Both these features have been independently confirmed from a study of the Siberian frozen remains. There is no indication of the colour of the woolly mammoth from the cave paintings, since there are no polychrome examples. Sometimes the hair of the frozen carcasses from the Arctic is found to be of ginger colour; though whether this is original or whether the colour has changed from black over the years is still uncertain.

It has been suggested that some depictions of elephants in caves in northern Spain represent the more southerly straight-tusked elephant, although there is no evidence from fossil remains to indicate that this animal survived as late in Europe as the coming of the cave artists. Another interpretation is that they are juvenile mammoths.

Fig. 8.3d) *Engraving of a mammoth on a mammoth tusk from Berelekh, Siberia (length of legs exaggerated by the artist). Radiocarbon date on tusk c. 12 000 years. Height about 20 cm.*

e) *Taxidermist's reconstruction of a woolly mammoth, based on cave paintings and engravings and on palaeontological evidence. (Model: A. Hayward British Museum (Natural History)). Length 60 cm.*

f) *Engraving of a mammoth (the 'Patriarch') in the Cave of Rouffignac, Dordogne, France. Middle–late Magdalenian. Length 71 cm. (Photo: J. Vertut).*

The woolly rhinoceros

Although depictions of the woolly rhinoceros are less numerous than those of the woolly mammoth, the appearance of the animal was nevertheless very clearly recorded by the Palaeolithic artists. Its most striking features were a remarkable shoulder hump and, when at rest, a steeply sloping neck and a downward inclined head. At first sight this posture seems unlikely, but it was recorded frequently by different artists and the downward inclination of the head is also substantiated by skeletal evidence and is in keeping with the inferred habit of this animal as a grazer. Some of the illustrations show hairiness of greater or lesser magnitude, especially along the lower jaw, around the back of the head and along the belly. The two horns, of which the anterior one was generally the longer, varied greatly in shape and direction, as in some living rhinos.

Fig. 8.4 *The woolly rhinoceros*
a) *Red line drawing in the Cave of Font de Gaume, Dordogne, France. ?Magdalenian. Length 70 cm. (After Breuil).*

b) *Engraving on the wall of the Cave of Les Combarelles, Dordogne, France. Magdalenian. Length about 40 cm. (Photo: Monuments Historiques pour l'Aquitaine (Ministère de la Culture)).*

Horses and asses

With the possible exception of the bison no animal appears so frequently in Palaeolithic art as the horse, giving us not only a fine record of proportions and posture but also very detailed information – unmatched in any other species portrayed – about the appearance of the pelage.

Present-day horses can be divided into four main groups, with the domestic horse and the Przewalski or Mongolian wild horse in one, while the other three comprise the Asiatic wild asses, the African asses (including the donkey) and the zebras.

It is tempting to try to identify the Palaeolithic depictions as examples of existing subspecies and races, but the differences that age, health and season make to individual appearance, not to mention the possibility of extinct forms, makes this a dangerous approach. We will restrict ourselves therefore to the features which were depicted frequently and appear to be characteristic; without suggesting that identical forms survive anywhere at the present day, since this is surely very improbable.

Although most of the pictures clearly represent true horses there also exist a few that must be asses of some sort, notably two engravings in the Cave of Trois Frères (fig. 8.5a) and a painting in Lascaux. Of these, the finest depictions are those in Trois Frères. The long ears, long necks and the elevated position of the heads of both animals are typical asinine characters. In true horses, the neck is much shorter and the head is not usually held so high.

Most of the hundreds of other equine portraits represent true horses. So great is the variety of artistic style that it is difficult to ascertain how many races are involved; although, as annual changes of pelage are

Fig. 8.5 *Horses and asses*
a) *Engraving of an ass in the Cave of Trois Frères, Ariège, France. Middle Magdalenian. Length 40 cm. (Photo: J. Vertut).*

Fig. 8.5b) *Bearded horse painted in black pigment on the wall of the cave of Niaux, Ariège, France. Middle Magdalenian. Length 1·70 metres. (Photo: J. Vertut).*

probably sufficient to account for the variation in appearance, there seems to be no good foundation for claiming more than one. Wherever sufficient detail is portrayed, most of the depictions show an animal very similar to the living Przewalski horse with a mane of short stiff upwardly directed hairs which terminates abruptly at the shoulder (fig. 8.5b). Quite commonly, hair is emphasized along the margin of the lower jaw (this would indicate an animal in its winter coat) and above the hoofs. There is sometimes also an indication of one or more stripes across the neck and of a lighter coloured underbelly. In one of the Lascaux paintings the mane is portrayed in black; the back and flanks in yellow ochre; and the underbelly has been left white – the unpainted wall of the cave. If allowance is made for the limited range of available pigments, this probably gives a good idea of the appearance of the animal in life.

It was not until 1968 that the best preserved, the most beautiful and most detailed of all the horse paintings were to come to light. Among the pictures of bison, ibex, deer and bears which adorned the walls of the newly discovered Cave of Ekain in northern Spain (fig. 8.5c) were no less than 26 horses; thirteen of them showing entire animals, ten in monochrome, three in polychrome; most of them with detailed indications of the appearance of the pelage. Some of the characters shown (for example, single and multiple shoulder stripes) were already familiar from previously known paintings, but there were also other characters – most notably leg stripes – which were new. Similar striping still occurs in some Przewalski and domestic horses at the present day and has been a source of special interest among those concerned with studies of the origin of domestic breeds, including Darwin in 1868.

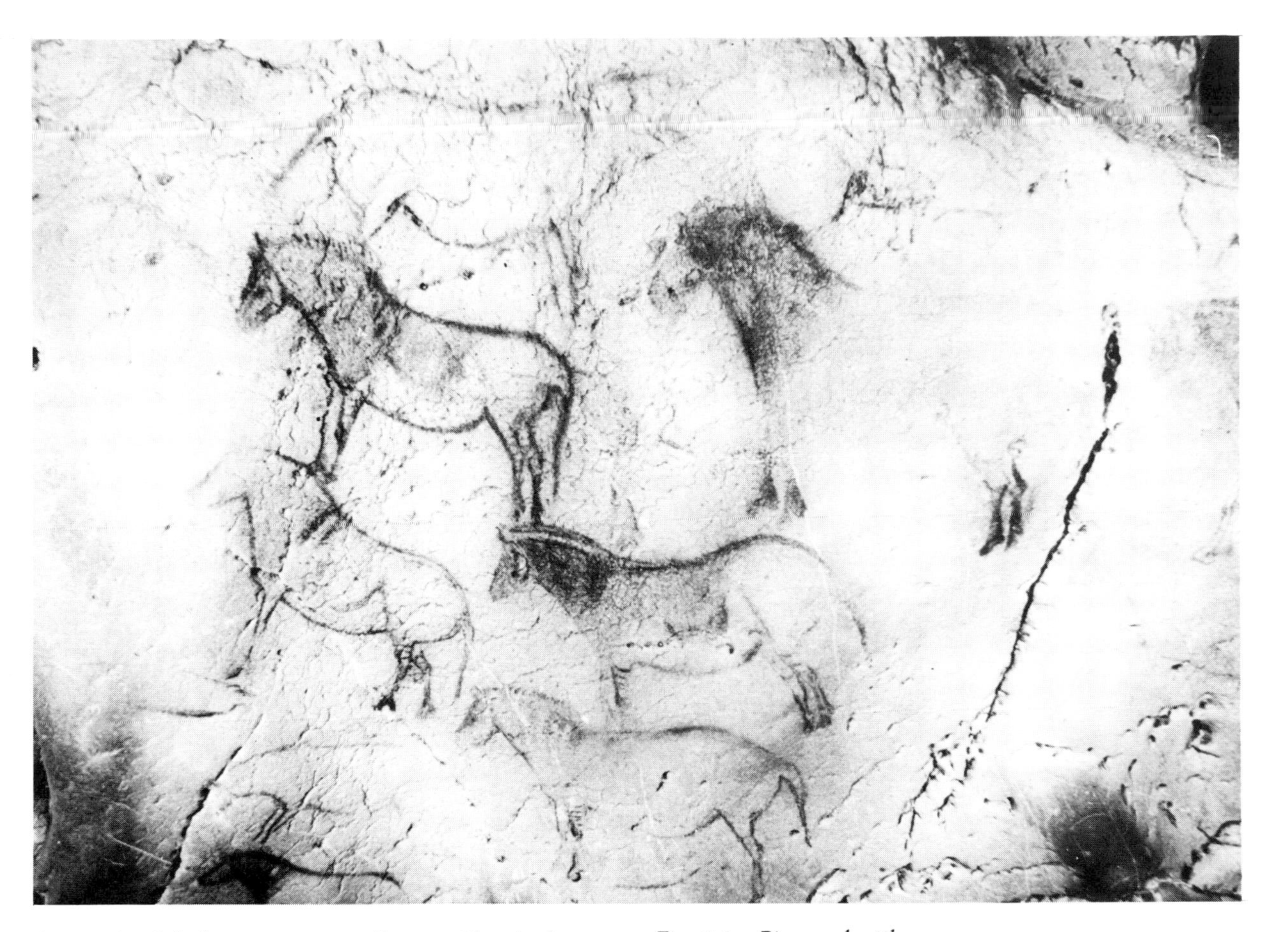

Fig. 8.5c) *Polychrome paintings of horses with striped shoulders and striped legs in the Cave of Ekain, Northern Spain. Magdalenian-Mesolithic. Length of lowermost horse 80 cm. (Photo: J. Vertut).*

Fig. 8.6 *Bison and cattle*
a) *Bison (one of a group of two) modelled in clay against a limestone block in the Cave of Tuc d'Audoubert, Ariège, France. Magdalenian. Length 60 cm. (Photo: J. Vertut).*

Bison and cattle

Many hundreds of depictions of upper Pleistocene bison are known, the uniformity of most of which can leave no doubt that they represent accurate portraits of the animals seen by the artists. Characters repeatedly portrayed (fig. 8.6a) include a beard of coarse, usually forward pointing, pendant hairs which terminates abruptly just behind the mouth; and upward or forward pointing horns, which are relatively small, unlike those of some of the earlier Pleistocene bison, which were very large. Only one form of bison appears to be represented and this is indistinguishable from the present-day European species, which still survives in small numbers in Eastern Europe (fig. 8.6c).

Although less commonly depicted by the Palaeolithic artists than the bison, sufficient pictures of

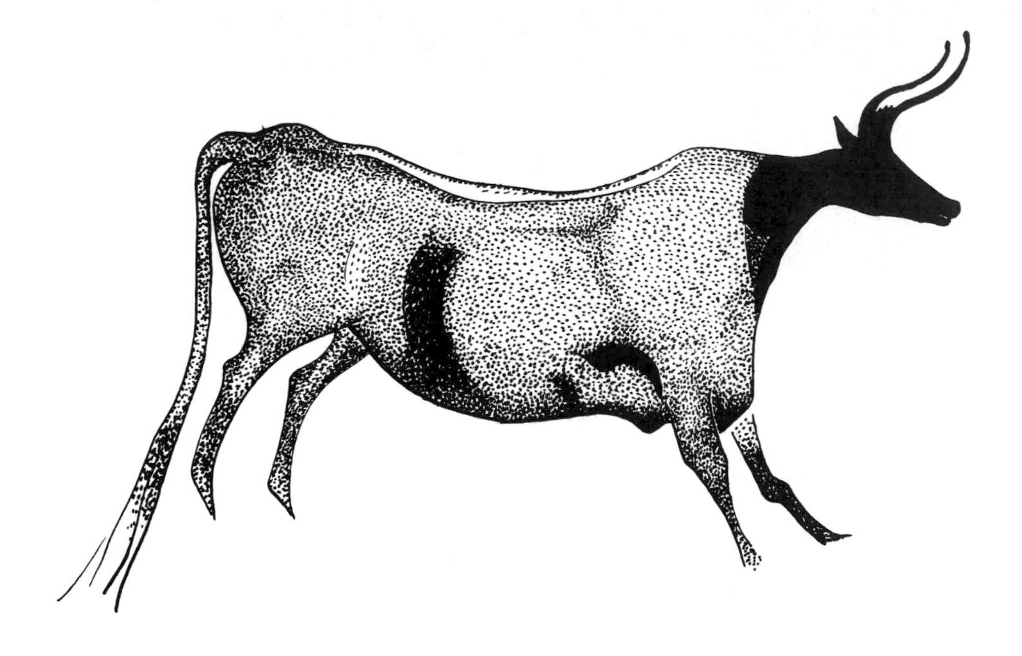

Fig. 8.6b) *Aurochs cow. Polychrome painting (dark head, red body, white line along back) in the Cave of Lascaux, Dordogne, France. Early Magdalenian. Length 3 metres.*

Palaeolithic cattle (giant oxen or aurochs) do nevertheless exist to give a good idea of the appearance of these animals, which survived in Europe (where they were also portrayed by medieval artists) until the sixteenth century. Most magnificent of the Palaeolithic examples are the polychrome bulls and cows in the Cave of Lascaux, which also show some interesting patterns of colouring. Zeuner, writing in 1963, commented on the white stripe on the shoulder of both a bull and a cow in this cave, and on the dark head of one of the cows (fig. 8.6b), a type of coloration still seen in some domestic breeds. Sexual dimorphism is well shown, the bulls being more massive in both body and face than the elegant cows. Although the heavy build of these beasts is clearly shown, there is also a lively indication of animals more supple and active than their domesticated descendents of the present day. The Lascaux depictions are nevertheless surprisingly lacking in some details and contain an element of caricature. The horns, (so well known from fossil remains), for example, are very poorly shown. The engraving from Teyjat (fig. 8.6d) in which the characteristic large size and forward direction of the horns are clearly recorded, apparently provides a truer to life record of the appearance and stance of the aurochs than the Lascaux paintings.

Fig. 8.6c) *Present-day European bison bull. Oka River Bison Reserve, USSR. (Photo: A. J. Sutcliffe).*

Fig. 8.6d) *Head of cow giant ox, engraved in the Cave of La Mairie, Teyjat, Dordogne, France. Late Magdalenian. Length of whole cow 50 cm.*

Fig. 8.7 *Ibex*
a) *Black painted ibex of Alpine type in the Cave of Niaux, Ariège, France. Middle Magdalenian. Length 27 cm. (Photo: J. Vertut).*

b) *Present-day Alpine ibex, Pontresina, Switzerland. (Photo: A. J. Sutcliffe).*

Ibex

Depictions of ibex by European Palaeolithic artists are quite common and are potentially of special zoological interest, since they provide information about an animal that was formerly continuously distributed across Western Europe but which has since been geographically separated into two isolated populations which are now regarded as separate subspecies. Can we learn anything about the circumstances or rate of subspeciation from this evidence?

The two living European forms are the Alpine ibex, *Capra ibex ibex*, now restricted to the Alps and Tatra Mountains; and the Spanish ibex, *C. i. pyrenaica*, probably already extinct in the Pyrenees, although still surviving in other high Spanish mountains. They are very different in appearance, especially the male animals, which have larger horns than the females. In the male Alpine ibex the horns are typically curved in only a single plane (so that they could be laid flat on a table) and have thick transverse ridges on their anterior surface; whereas those of the Spanish males are lyre-shaped and are usually without prominent ridges. In some of the better drawn Palaeolithic representations it is possible to make out animals which distinctly seem to show the Alpine or Spanish condition. Unfortunately, these overlap geographically, making it difficult to draw any conclusions from them and their significance must remain uncertain. Most cannot be accurately dated and a study of fossil remains is likely to give a more detailed record of ibex palaeogeography. Of one matter we are nevertheless sure. About 18 000 years ago the ice sheets of the world were far more extensive than at present and the ibex populations of Spain and France were probably separated for several thousand years by a continuous barrier of ice extending all the way along the Pyrenees from the Mediterranean Sea to the Atlantic Ocean. Part of the record of how the European ibex fared at this time and what happened subsequently is to be sought in the record of the Palaeolithic artists.

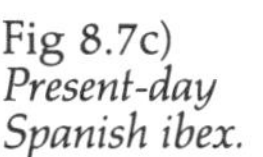

Fig 8.7c)
Present-day Spanish ibex.

Fig. 8.8
a) *Present-day bull musk ox. Ellesmere Island, Canada, 79°N. (Photo: A. J. Sutcliffe).*

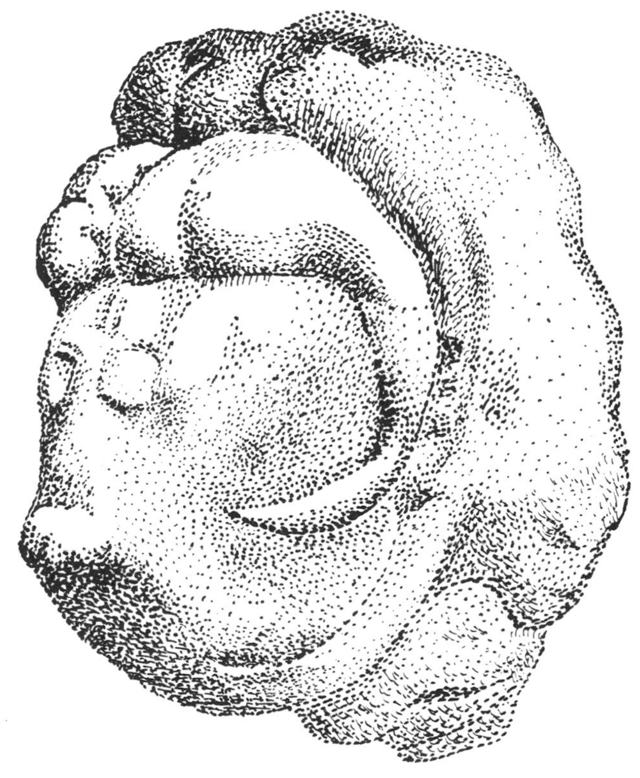

b) *Head of a bull musk ox carved on a block of limestone from the Cave of Laugerie Haute, Dordogne, France. Height 27 cm.*

Musk ox

Although not commonly depicted in cave art, the musk ox is geographically perhaps the most interesting of all the species represented. At present an inhabitant of the high arctic, where it can survive north of 80° latitude, the former occurrence of this tundra animal in central France, 30° south of its present range (fig. 8.8), strikingly illustrates the magnitude of the vegetational and faunal displacements that must have accompanied the last major glacial advance in Europe.

Fig. 8.9 *Deer*
a) *Bull reindeer engraved on a reindeer antler. Kesslerloch Cave, Germany. Magdalenian. Length 6·2 cm.*

b) *Head and back of a red deer stag, painted in sepia on the wall of the Cave of Lascaux, Dordogne, France. Early Magdalenian. Height 1·4 metres. (Photo: J. Vertut).*

Deer

At least four species of deer were depicted by the Palaeolithic artists – reindeer, red deer, giant deer and fallow deer. The reindeer, today an inhabitant of northern regions, provides another instance of a mammal recorded at a time when colder climatic conditions caused a southward displacement of its range. All the characters portrayed suggest an animal that was indistinguishable from the living European reindeer. In the few instances in which the tones of the pelage were indicated, the back is clearly darker than the belly and there is a dark horizontal stripe on the flanks, a colouring common among present-day reindeer.

The red deer was also commonly depicted. Its antlers are of a special interest in our studies since their individual size and structure provide a general indication of the nutritional well being of the populations concerned. At the present day there is considerable variation in the complexity of antlers of stags of similar ages in different geographical regions. Scottish red deer, for example (driven by man onto marginal habitat), commonly have one less point (tine) near the base of each antler than their better fed relatives further south and on the continent of Europe. A red deer with twelve tines is sufficiently unusual to be known by hunters as a 'royal'. Many of the Palaeolithic depictions show more tines than this, for example 15 on the two antlers of a deer engraved on a piece of antler from the Cave of Lorthet, Haute Pyrénées; 18 on a painted head in Lascaux (8.9b). Even allowing for some artistic licence, (in the Lascaux example both antlers have identical configurations and were almost certainly drawn from a single specimen or from memory) the European Palaeolithic red deer was apparently an animal of a regal dimensions; a conclusion further endorsed by the evidence of fossil remains.

Since the giant deer (the so called 'Irish elk') is now totally extinct, the few possible representations in Palaeolithic art are of special interest. The best example is that in the Cave of Cougnac, France, which shows large palmate (flattened) antlers, a pointed nose and large shoulder hump. The animal concerned cannot be a fallow deer, which is of more slender build and without a shoulder hump (a character of animals with heavier heads), and is unlikely to be an elk (American: moose), since the present-day bull elk is characterized by a very bulbous nose and a beard, neither of which is shown in the painting. The animal is therefore probably a giant deer.

Dipictions of the fallow deer are uncommon, partly because this is a southern species that did not range as far north as France at the time when the artists were painting mammoths and reindeer there. A head with palmate antlers engraved on the wall of the Cave of Addaura in Sicily can reliably be identified as a fallow deer.

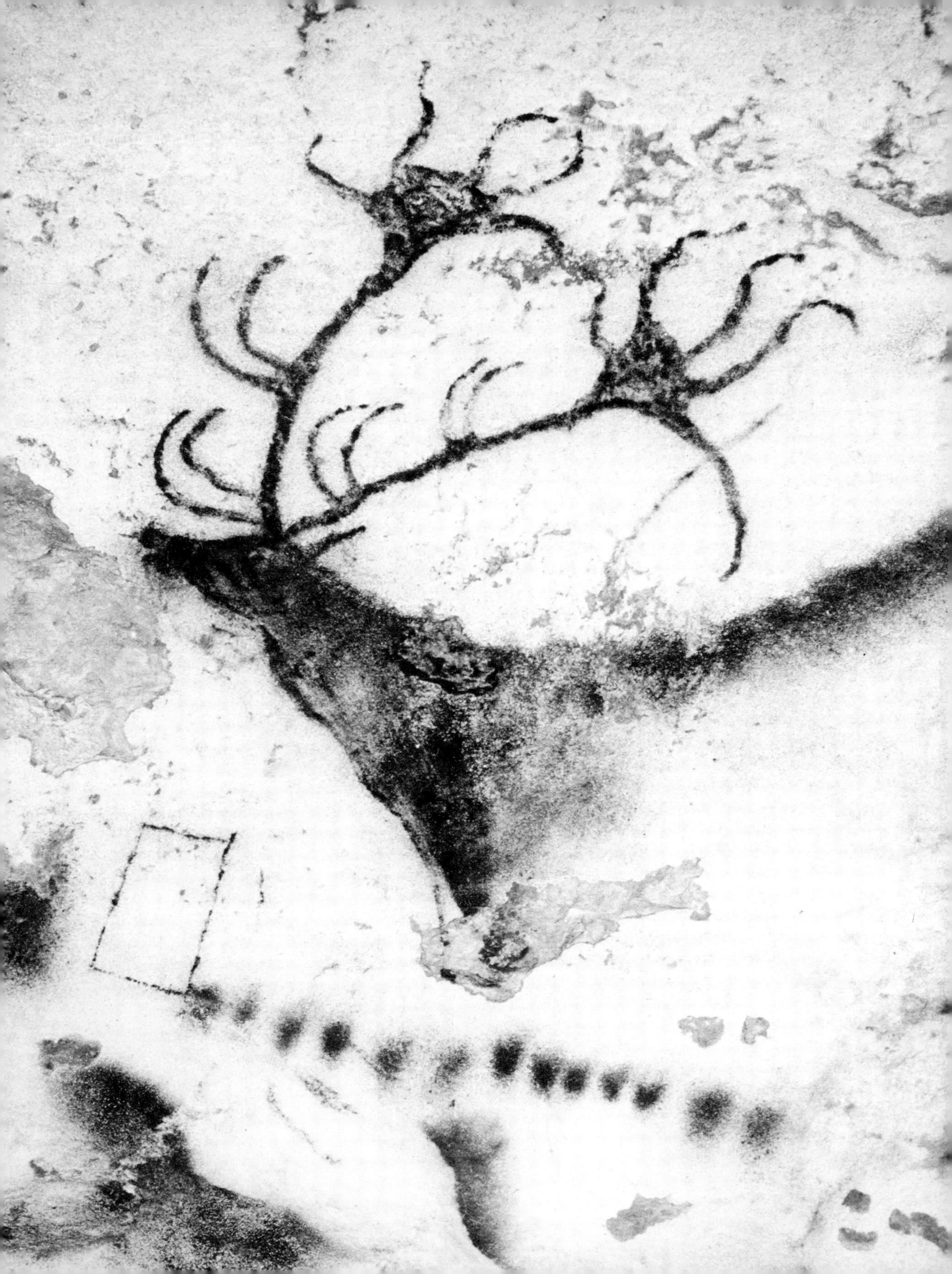

'Cave lion'

Although the cave lion was not often recorded by the Palaeolithic artists, possibly because it was not a source of food to them, the few known representations are of special interest for the remarkable details portrayed. The name 'cave lion' has been widely applied in both scientific and popular writings to a lion-sized member of the cat family, the fossil remains of which have commonly been found in the caves of Europe. It was still living at the time of the Palaeolithic artists and became extinct subsequently. Palaeontologists making anatomical studies of its remains have nevertheless not been unanimously convinced that this was in fact a lion. There have been suggestions that it was a tiger; or that possibly more than one large feline is involved. What do the cave paintings tell us about this matter? The finest representations are the magnificent engraved frieze from the Cave of La Vache, France (fig. 8.10a) and the four engraved and painted frontal views in the Cave of Trois Frères. All the diagnostic characters shown are lion-like rather than tiger-like, especially the tufted tail on the left hand animal from La Vache. As on a domestic cat, the end of the tail of a tiger usually ends without a terminal tuft.

We must note, however, that so far no manes have been recognized in any of the pictures. Were all the animals portrayed lionesses, or was the 'cave lion' without a mane?

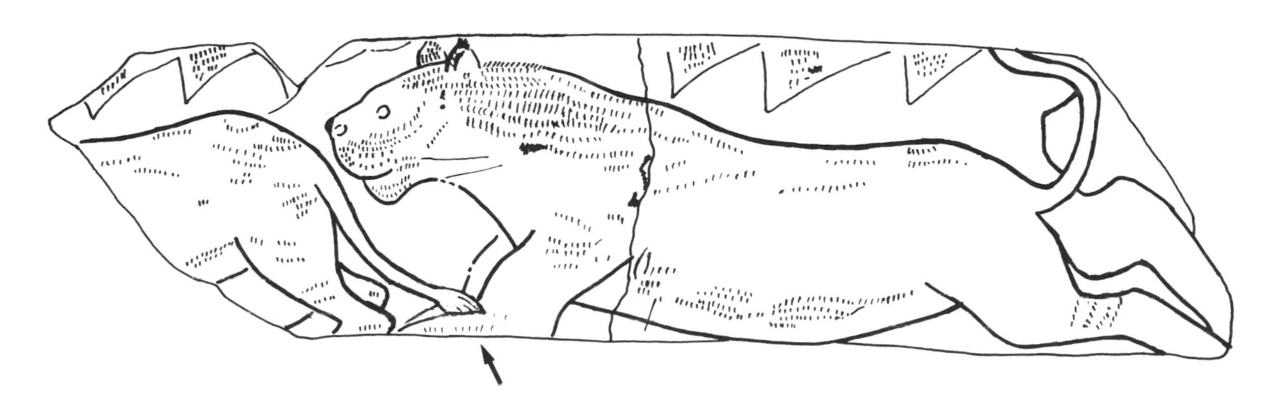

Fig. 8.10 *'Cave lion'*
a) *'Frieze of lions'. Engraving on a rib from the Cave of La Vache, Ariège. France. Late Magdalenian, radiocarbon date* c. *10 900 years old. Length of central lion 10 cm. Note tuft on end of tail (arrowed).*

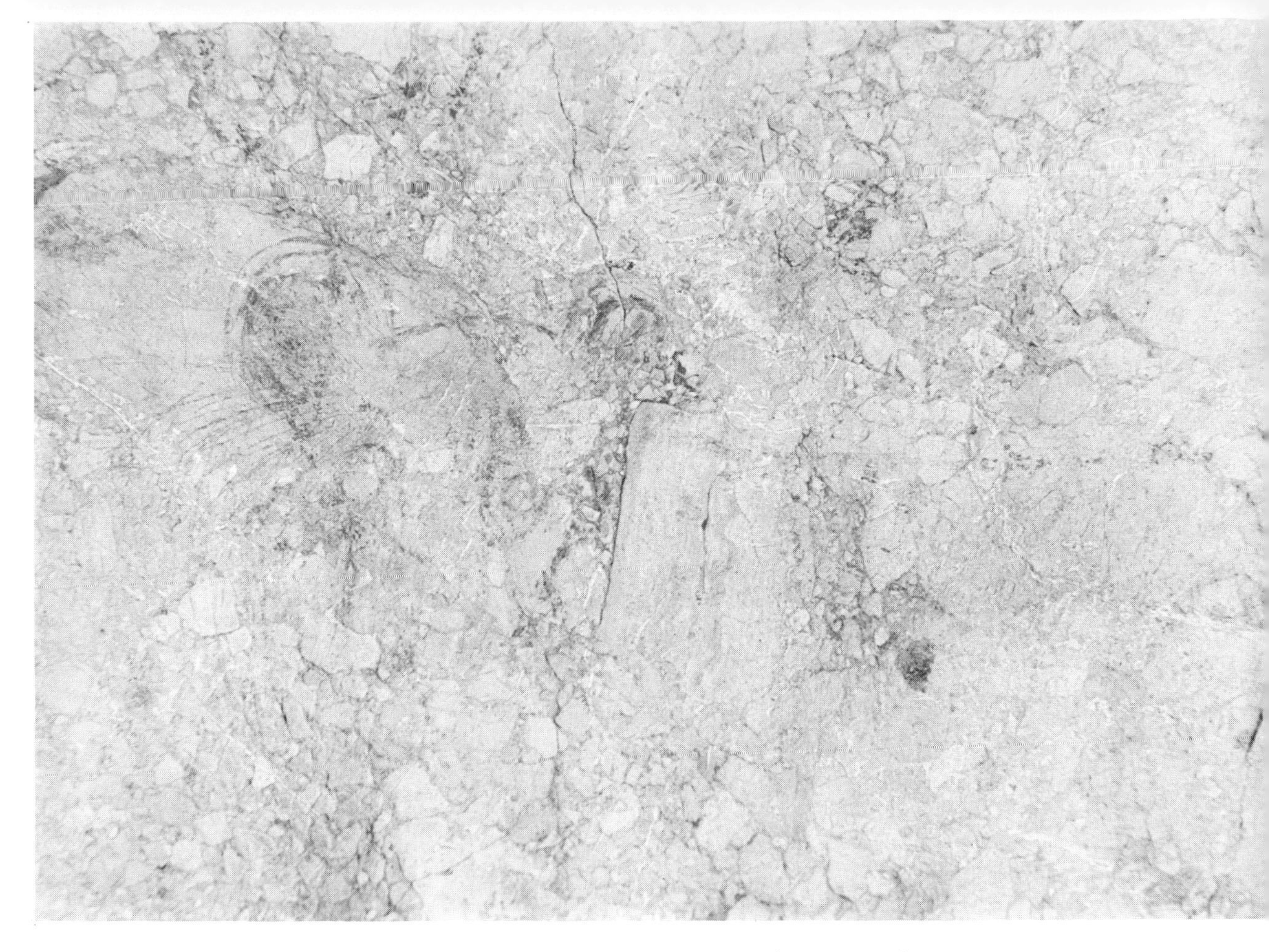

Fig. 8.10b) *Lion, possibly a cub (compare with (c)) with its head shown in frontal view. Painted in black on the wall of the Cave of Trois Frères, Ariège, France. Middle Magdalenian. Height of head 40 cm. (Photo: J. Vertut).*

Fig. 8.10c) *Line drawing interpretation of (b).*
d) African lioness and cubs. Serengeti National Park, Tanzania. (Photo: A. J. Sutcliffe).

Fig. 8.11 *Bears*
a) *Engraved bear (either a brown bear or a cave bear) on the wall of the Cave of Les Combarelles, Dordogne, France.*

b) *Line drawing interpretation of (a). (After Breuil).*

Bears

Unlike other carnivores, the bear was commonly portrayed by the Palaeolithic artists. Over a hundred pictures are known. The identity of some of the animals is questionable, but the best are extremely fine and provide a good idea of the appearance of the bears concerned.

Three species of bear could have been recorded – the brown bear, still found in the Pyrenees today, the cave bear and the polar bear (though its fossil remains are as yet unknown further south than London). The main feature distinguishing the first two species is the slope of the forehead, which is sharply stepped in the cave bear. It is probable that most depictions are of the brown bear, but the foreheads vary from flat to stepped so that some may also represent cave bears.

Other mammals portrayed by the Palaeolithic artists include saiga antelope, chamois, wild boar, spotted hyaena, wolf, wolverine, stoat or weasel, hare and seal (fig. 8.2, p. 86).

Fig. 8.11c) *Present-day European brown bear. (Photo: G. Kinns).*

Palaeolithic art as an indication of the former distribution of Pleistocene mammals

We have seen how much can be learned about the appearance of certain Pleistocene mammals from Palaeolithic works of art. Since not all the depictions are of the same age it could be hoped that this source of information might also provide us with details of temporal changes of mammalian faunas. Certainly, the combinations of animal species differ quite substantially from cave to cave. In the Cave of Rouffignac, for example, there are many mammoths and some woolly rhinos, whereas in Lascaux one of the commonest animals shown is the giant ox (absent from Rouffignac); there are no mammoths; and the only portrayal of a rhinoceros is in a separate part of the cave and possibly of different age from the rest of the paintings. Although detailed studies of the composition of the mammalian species represented at the various sites have been undertaken by archaeologists, from the palaeontological viewpoint the conclusions are of rather limited stratigraphic value. In terms of inter-glacial–glacial cycles the length of time covered is relatively short, so that massive faunal movements would not be expected; there is the likelihood that at some sites occupation and artistic activity continued over a period of time so that all the animals depicted need not have lived contemporaneously; and there is the additional problem of working with pictures on the walls of caves, which cannot be as accurately dated as stratified objects buried under the floors of caves. Furthermore we know that the artists were selective in the animals they portrayed, so that lack of depictions does not necessarily mean that particular animals were absent from a region. For studies of changing mammalian faunas it is better to refer to fossil remains (which are likely to be more representative of a total fauna and which are more readily dated) from stratified sites.

The Palaeolithic works of art do nevertheless broadly confirm what is already known from the palaeontological evidence. The best of the depictions probably span the period of time from maximum of the last glacial advance, about 18 000 years ago until the final disappearance of this ice at the end of the Pleistocene, about 10 000 years ago. The southward displacement of the mammalian fauna which accompanied this event, is readily apparent. The musk ox (today an animal of the northern arctic) reached central France. Woolly mammoth, woolly rhinoceros, reindeer and wolverine are other northern species, which spread at least as far south as the Pyrenees. Further south these cold faunal elements disappear. The red deer gradually replaces the reindeer, and the wild ass, wild boar (a forest animal) and fallow deer also appear. These differences can be explained on geographical grounds and it is not necessary to postulate any spectacular changes of climate to account for them. It is known from geological evidence that, during the preceding interglacial stage (about 120 000 years ago) many warm-climate mammals, including the hippopotamus, straight-tusked elephant, narrow-nosed rhinoceros and fallow deer spread northwards, at least as far as Yorkshire in England (chapter 10). It is not surprising that they were not recorded by the Palaeolithic artists, who did not arrive until much later. There are no depictions of the hippopotamus; and the supposed narrow-nosed rhinoceros and straight-tusked elephants are of doubtful identity. All three animals were probably already extinct in Europe before the advent of Palaeolithic art.

Fig. 9.8, pp. 106–107 *Artist's reconstruction of a scene in northern Siberia during the late Pleistocene, about 10–11 000 years ago, based on geographical data, on the evidence of frozen carcasses preserved in the permafrost, on plant remains and pollen from the contents of their alimentary canals and on anatomical details recorded by the French Palaeolithic cave artists (see fig. 8.3 p. 88). The landscape, such as might have been observed in many places over large areas of mammoth territory, is mainly imaginary but also includes some details based on observations along the mid part of the River Lena.*

Standing prominently are four woolly mammoths: on the right a large male, height at the shoulders about 3 metres; on the left two females, which are slightly smaller, and a half-grown calf. Note the long hair, small ears, relatively slender tusks on the females and bi-lobed trunk tips, interpreted by Flerov as an adaptation for collecting herbaceous food (present-day elephants, in contrast, prefer a mainly arboreal diet). Also shown are wild horses, height at shoulder about 1·35 metres, and a wolverine feeding on a carcass. Other animals represented by frozen remains, not illustrated, include the woolly rhinoceros, bison, ground squirrel and ptarmigan.

The scene is set in spring. After a very cold winter with little precipitation, the frozen ground swept clear of snow over large areas by the wind, a few snow patches still survive on a sheltered slope. The last small icebergs float down the river which, only a short time previously, had been frozen from bank to bank.

The first flowers have already appeared. Three plant communities are shown. In the distant gullies are small woods of larch, Larix dahurica, *and occasional pines,* Pinus pumila. *The male mammoth stands beside a local wet patch with sedges,* Carex *sp. and* Kobresia *sp., and the seed heads of cotton grass,* Eriophorum vaginatum, *preserved under a snow bank from the previous autumn. All the rest of the area is much drier and carries a meadow flora with the grass* Poa *sp.; wormwood,* Artemisia *sp.; dwarf birch,* Betula nana; *mountain avens,* Dryas punctata; *moss campion,* Silene acaulis; *poppy,* Papaver lapponicum, *and many other herbs. This is the so-called 'mammoth steppe', source of abundant nourishment for the mammals which inhabited it. The disappearance of this highly nutritious biome at the end of the Ice Age about 10 000 years ago, as the result of increased precipitation, longer annual snow cover and the development of boggy tundra, is one of several possible reasons which have been proposed for the mammoths' final extinction.*

Painted by Peter Snowball under the direction of the author, 1985.

Chapter 9 **Frozen mammoths and other beasts**

Fascinating though it is to speculate about past mammalian life from a study of fossil bones and teeth, such remains cannot create so dramatic an impression as carcasses of extinct mammals, complete with skin and hair and with food still preserved in mouths and stomachs. Foremost among such discoveries, both for their popular appeal and for the scientific information that they have provided, are the frozen carcasses of mammoths and other animals from the permafrost of Siberia and Alaska.

At the present day, permafrost forms a broad belt across the arctic zone of the northern hemisphere from Siberia to Alaska, northern Canada and Greenland. Much of this region has been continuously frozen since Pleistocene times, in places to depths of as much as 1400 metres. As we have already seen, even at the height of the Last Glaciation, northeastern Siberia and northern and central Alaska, being areas of little precipitation, supported only relatively small valley glaciers in major highland areas. Most of the region where the carcasses have been found (unlike the rest of North America and Europe, which carried vast ice sheets), though deeply refrigerated, was not glaciated. With the sea-level lower than at the present day the New Siberian Islands, Siberia and Alaska became joined together, allowing unrestricted movement of animals between all these areas (fig. 9.1).

Only where the ground has remained continuously frozen ever since Pleistocene times has it been possible for carcasses to remain preserved by freezing. Mammoths are not the only animals to which this happened. Others include the woolly rhinoceros, horse, bison, musk, ox, reindeer, wolverine, ground squirrel and ptarmigan.

Today, the northern part of the terrain where the frozen carcasses are found is almost uninhibited bleak

Fig. 9.1 *Map showing where some of the more important frozen mammalian carcasses and parts of carcasses have been found in Siberia, Alaska and Canada. Mammoths: (1) Yuribei, 1979, (2) Taimyr, 1948, (3) Khatanga, 1977, (4) Adams, 1799, (10) Berelekh, 1970, 73, 77, (11) Terektyakh, 1971, (13) Dima, 1977, (14) Shandrin, 1971, (15) Berezovka, 1900; Bison: (12) Mylakhchyn, 1971, (16) Fairbanks, 1951, 79; woolly rhinoceros: (5) Churapcha, 1972; ground squirrels: (7) Dirin–Yuryakh, 1946, (17) Glacier Creek, 1981; narrow-skulled vole: (6) Dirin–Yuryakh, 1946; horse: (8) Selerikan, 1968; wolverine (9) Berelekh,* c. *1970.*

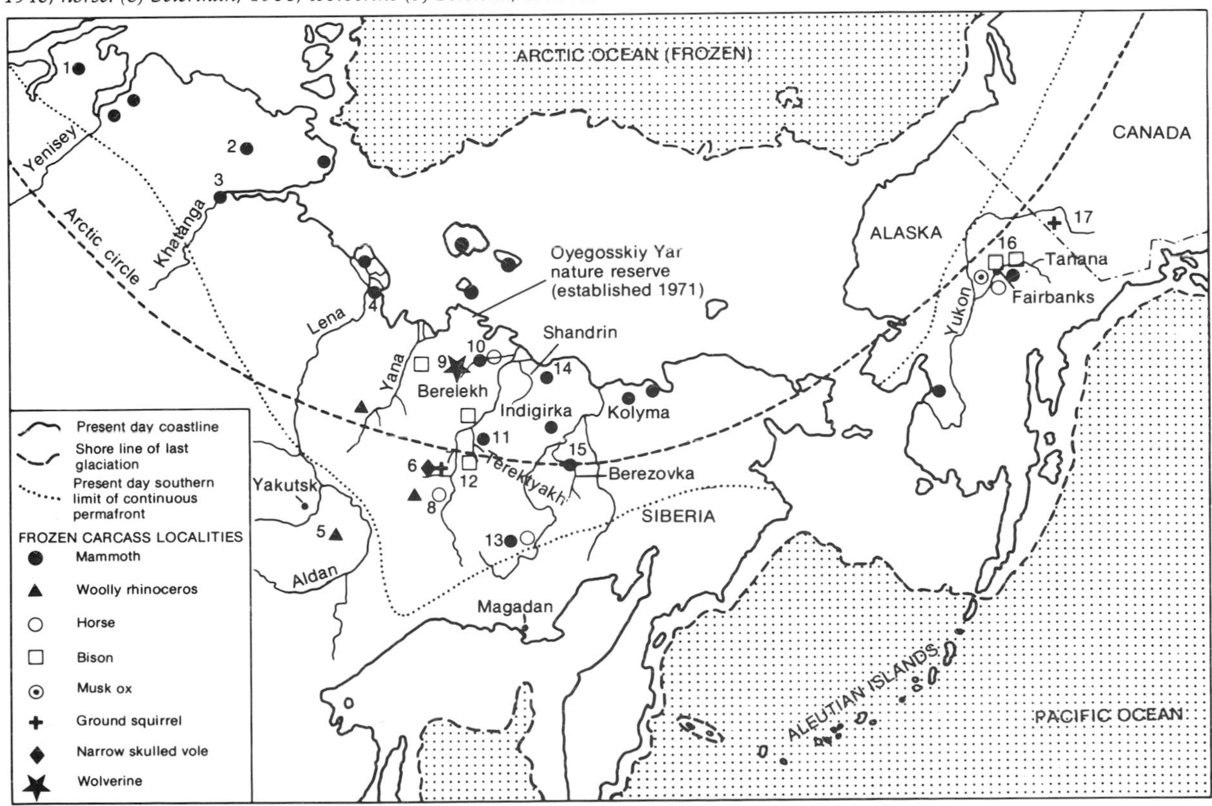

open tundra, with continuous daylight during the summer; but with extensive snow cover, low temperatures and continuous darkness during the winter. Further south, scattered trees, stunted and sprawling on the ground, their roots unable to penetrate below its seasonally thawed surface, provide a transition to woods of dwarf birch, larch and pines.

So abundant are mammoth tusks in northern Siberia and so well are some of them preserved that fossil ivory has long been an important item of commerce. Not only have the inhabitants used it themselves for making combs, knives, rings, caskets and other household equipment but it has also been extensively traded to other countries. Some fossil ivory may have reached China as early as the fourth century BC. It was already known in London soon after the beginning of the seventeenth century. Whilst no exact figures are known for the amount collected in Siberia the total quantity is astonishing and it has recently been estimated that 550 000 tonnes of mammoth tusks still lie buried off shore along the 1000 km of coast between the Rivers Yana and Kolyma.

Early ideas about mammoths

From early times ivory hunters came across mammoth remains that had not entirely decomposed, and these remains, which were considered to be a cause of bad luck, were variously explained. Isbrants Ides, writing in 1706 of an expedition that took place about 1693 reported of the animals concerned:

> *'The Heathens of Jakuti, Tungusi, and Ostiacki say, That they continually, or at least by reason of the very hard Frosts, mostly live under Ground, where they go backwards and forwards; to confirm which, they tell us, That they have often seen the Earth heaved up, when one of these Beasts was on the March, and after he was past the Place, sink in, and thereby make a deep Pit. They further believe, that if this Animal comes so near the Surface of the frozen Earth, as to smell, or discern the Air, he immediately dies, which they say is the Reason that several of them are found dead on the high Banks of the River, where they unawares come out of the Ground. This is the Opinion of the Infidels concerning these Beasts, which are never seen. But the old Siberian Russians affirm, that the Mammoth is very like the Elephant, with this only Difference, that the Teeth of the former are firmer, and not so straight as those of the latter. They also are of Opinion, that there were Elephants in this Country before the Deluge, when this Climate was warmer, and that their drowned bodies floating on the Surface of the Water of that Flood, were at last washed and forced into subterranean Cavities: But that after this Noachian Deluge, the Air, which was before warm, was changed to cold, and that these Bones have lain frozen in the Earth ever since, and so are preserved from Putrefaction till they thaw and come to Light, which is no very unreasonable Conjecture; though it is not absolutely necessary that this Climate should have been warmer before the Flood, since the Carcasses of drowned Elephants were very likely to float from other Places several hundred Miles distant, to this Country, in the great Deluge which covered the Surface of the whole Earth'.*

The first scientific studies

Being so familiar in Russia, the mammoth remains soon attracted the attention of scientists. In 1799, the botanist Adams, who was on an expedition to Siberia, heard of a frozen carcass on the banks of the River Lena, and attempted to salvage it. By the time he arrived most of the soft parts had disappeared, but he was able to recover most of the skeleton, two feet, the skin covering the head, an ear and a great quantity of hair. The remains were taken to St Petersburg (now Leningrad) in 1806.

During the nineteenth century the Russian Academy of Sciences established a system of monetary rewards in order to promote further discoveries of this kind. The earliest advertisement for such rewards was issued in 1860, with the recommendation that it should reach the farthest branches of the administrative network of Siberia and nomads to whom its contents should be explained by tax collectors, priests and merchants. The amount offered ranged from 100 silver roubles for a complete skeleton of a large antediluvian animal, not removed from the ground, to 300 silver roubles for a carcass with at least part of the flesh and skin undecomposed. Similar advertisements have continued to be issued from time to time up to the present day (fig. 9.2).

The Beresovka mammoth

Although the reward system led to the reporting of a number of imperfect carcasses, the Adams mammoth remained the most complete that had been collected until 1901, when news of a fresh discovery on the banks of the River Beresovka, a tributary of the Kolyma, reached St Petersburg. The Academy of Sciences quickly arranged an expedition, which set

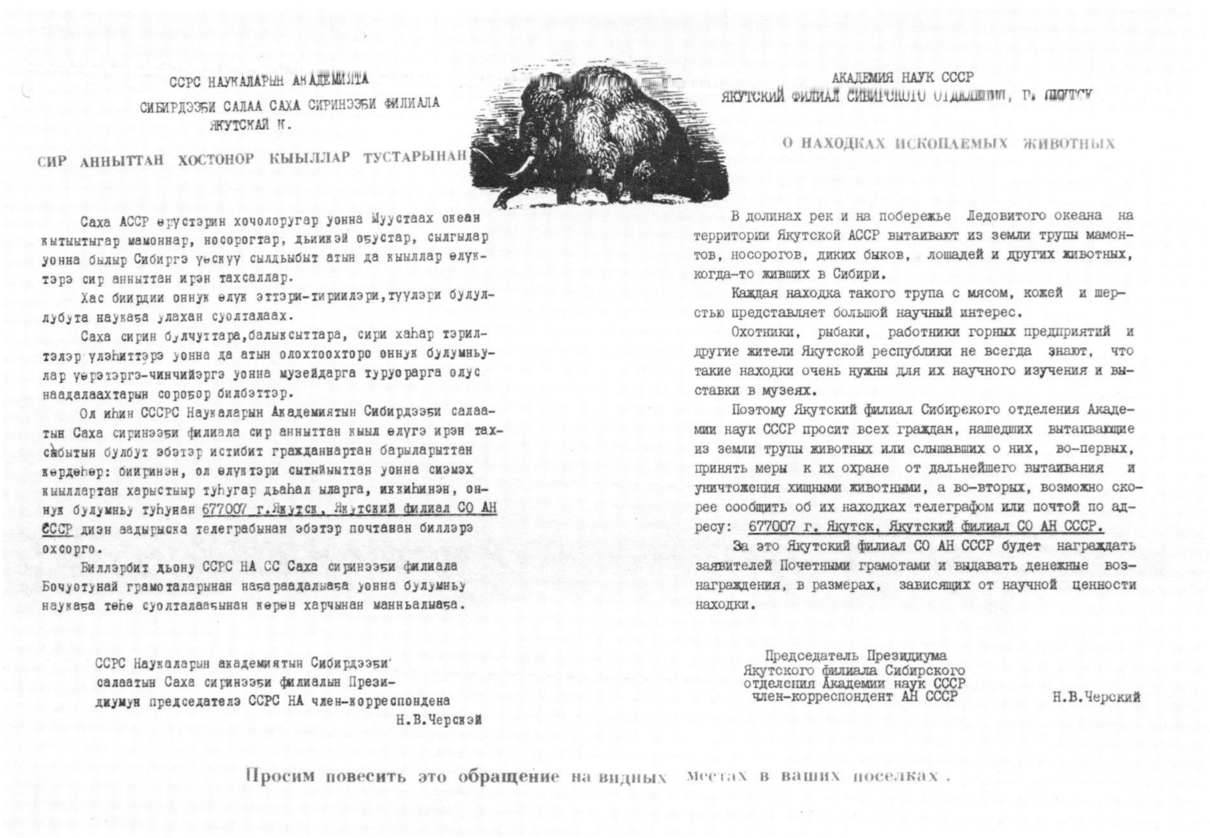

ССРС НАУКАЛАРЫН АКАДЕМИЯТА
СИБИРДЭЭҔИ САЛАА САХА СИРИНЭЭҔИ ФИЛИАЛА
ЯКУТСКАЙ К.

СИР АННЫТТАН ХОСТОНОР КЫЫЛЛАР ТУСТАРЫНАН

Саха АССР өрүстэрин хочолоругар уонна Муустаах океан кытыытыгар мамоннар, носорогтар, дьиикэй оҕустар, сылгылар уонна былыр Сибиргэ үөскүү сылдьыбыт атын да кыыллар өлүктэрэ сир анныттан ирэн тахсаллар.

Хас биирдии оннук өлүк эттэри-тириилэри, түүлэри булуллубута наукаҕа улахан суолталаах.

Саха сирин булчуттара, балыксыттара, сири хаһар тэриллэр үлэһиттэрэ уонна да атын олохтоохторо оннук булумньулар үөрэтэргэ-чинчийэргэ уонна музейдарга туруораргa олус наадалаахтарын сороҕор билбэттэр.

Ол иһин СССРС Наукаларын Академиятын Сибирдээҕи салаатын Саха сиринээҕи филиала сир анныттан кыыл өлүгэ ирэн тахсыбытын булбут эбэтэр истибит гражданнартан бырыларыттан көрдөһөр: бииринэн, ол өлүктэри сытыйыыттан уонна сиэмэх кыыллартан харыстыыр туһугар дьаһал ыларга, иккиһинэн, оннук булумньу туһунан 677007 г.Якутск, Якутский филиал СО АН СССР диэн аадырыска телеграбынан эбэтэр почтанан биллэрэ охсоргo.

Биллэрбит дьону ССРС НА СС Саха сиринээҕи филиала Бочуотунай грамоталарынан наҕараадалыаҕа уонна булумньу наукаҕа төһө суолталааҕынан көрөн харчынан манньалыаҕа.

ССРС Наукаларын академиятын Сибирдээҕи
салаатын Саха сиринээҕи филиалын Президиумун председателэ ССРС НА член-корреспондена
Н.В.Черскэй

АКАДЕМИЯ НАУК СССР
ЯКУТСКИЙ ФИЛИАЛ СИБИРСКОГО ОТДЕЛЕНИЯ, г. ЯКУТСК

О НАХОДКАХ ИСКОПАЕМЫХ ЖИВОТНЫХ

В долинах рек и на побережье Ледовитого океана на территории Якутской АССР вытаивают из земли трупы мамонтов, носорогов, диких быков, лошадей и других животных, когда-то живших в Сибири.

Каждая находка такого трупа с мясом, кожей и шерстью представляет большой научный интерес.

Охотники, рыбаки, работники горных предприятий и другие жители Якутской республики не всегда знают, что такие находки очень нужны для их научного изучения и выставки в музеях.

Поэтому Якутский филиал Сибирского отделения Академии наук СССР просит всех граждан, нашедших вытаивающие из земли трупы животных или слышавших о них, во-первых, принять меры к их охране от дальнейшего вытаивания и уничтожения хищными животными, а во-вторых, возможно скорее сообщить об их находках телеграфом или почтой по адресу: 677007 г. Якутск, Якутский филиал СО АН СССР.

За это Якутский филиал СО АН СССР будет награждать заявителей Почетными грамотами и выдавать денежные вознаграждения, в размерах, зависящих от научной ценности находки.

Председатель Президиума
Якутского филиала Сибирского
отделения Академии наук СССР
член-корреспондент АН СССР
Н.В.Черский

Просим повесить это обращение на видных местах в ваших поселках.

Fig. 9.2 *Advertisement, as currently distributed to places of habitation in Northern Siberia, requesting that discoveries of frozen remains of mammoths, rhinoceros, 'wild bull', horses and other animals are reported to the Yakutian Branch of the USSR Academy of Sciences. A reward is offered for such notification. The language on the left is Yakutian, that on the right Russian.*

out at the beginning of May, to investigate the new find. Among those taking part was E. W. Pfizenmayer, whose subsequently published book provides a vivid account of the excitements and hardships of the expedition.

The carcass was found to be lying in a vast landslide where thawing had caused part of the frozen ground along the river bank to collapse. It was emplaced in a sitting position, its hind legs beneath it, its front legs outstretched as though the animal had been struggling to get up when it died (fig. 9.3). A fractured foreleg and evidence of extensive bleeding suggested that it had been involved in a fall.

At the time of the discovery of the carcass in 1900 most of the soft parts, including the head and trunk, were probably still preserved, but the tusks were removed by the discoverer and sold in Kolymsk and by 1901 a large part of the head and back had already been eaten by wolves, exposing the underlying skull. The carcass was decomposing and the stench was des-

Fig. 9.3 *The Beresovka mammoth (more than 39 000 years old) in situ at the time of its excavation in 1901. (Photo: USSR Academy of Sciences).*

Fig. 9.4 *The front leg of the Beresovka mammoth, after Museum preparation, showing the dried muscles. (Photo: USSR Academy of Sciences).*

cribed as like the smell of a badly kept stable blended with that of offal. The lower part of the body and legs were nevertheless still frozen and in a magnificent state of preservation (fig. 9.4).

The animal was found to have a layer of fat beneath its skin, a covering of short hair and an outer layer of coarser hair up to 50 cm long (much of which had fallen out and was recovered from the surrounding sediment), all suggesting that, like present-day reindeer and musk ox, it was well adapted for withstanding cold conditions.

The hair was reddish-brown when first exposed, but became lighter as it dried. It must not necessarily be assumed, however, that the mammoth was reddish-brown in life. Other frozen Siberian mammoths have also been found to have ginger hair, others hair of dark grey colour. Post-mortem colour changes from black to ginger have commonly been encountered by anthropologists studying the hair of human mummies; and that of the mammoths may sometimes have been affected by similar changes. The contents of the stomach were preserved and there was still food in the mouth, including grasses and seeds. It is now thought that the mammoth died about June or July.

Excavation took six weeks to complete. The excavators constructed a wooden hut over the carcass, heated the inside in order to thaw the remains so that they could be cut into manageable pieces and then allowed them to freeze once more in the open, ready for transport to St Petersburg. Excavation complete, the party started the return journey, with the remains carried on sledges drawn by dogs and later by reindeer and then horses.

In Kolymsk, the left tusk, which had been removed from the skull the previous year, was recovered and added to the other remains. It is reported that a reward of 1000 gold roubles was paid to the Cossak, Yavlovski, who had first reported the carcass; but soon afterwards he lost the entire amount and fell into debt while gambling, a misfortune which he blamed on his involvement with the mammoth, and he determined never to have anything to do with such remains again!

At Irkustsk, a refrigerating car was attached to the mail train to carry the carcass to St Petersburg. It finally reached its destination early in 1902, nearly a year after the expedition had first set out. At the museum the skeleton and reconstructed skin were mounted separately for display.

Other twentieth century mammoth finds

The development of better communications and easier exploration during the present century has led to a whole series of further discoveries of frozen remains of mammoths and other animals in various states of completeness. In 1908, another mammoth carcass was found beside the River Sanga-Yurjach with most of the trunk (missing in the case of the Beresovka mammoth) still preserved for study. More recent finds investigated by the Soviet Academy of Sciences include a carcass of a horse beside the River Selerikan in 1968, parts of carcasses of mammoths on the Taimyr Peninsula in 1948 and in 1977, another on the bank of the River Terektyakh in 1971 and another beside the River Shandrin in 1972 (fig. 9.5).

Although no wool, muscle or ligaments of the Shandrin carcass had survived, the internal organs, protected by the rib cage, were almost intact. Palaeontologists recovered not only the contents of the stomach of the mammoth but also 12 large larvae of a parasitic warble fly of the genus *Cobboldia*, today known only in the Indian elephant and restricted in distribution to Burma and India.

In the early 1970s special attention was directed to a massive accumulation of mammoth bones that was being washed out of the banks of the River Berelekh, a tributary of the Indigirka. Although no complete carcasses have yet been found at this locality, investigations by the Soviet Academy of Sciences led to the recovery of over 8000 mammoth bones, weighing approximately ten tonnes and representing about 140 individuals, and also hair and some frozen soft parts; and the frozen carcasses of a wolverine and of a ptarmigan. It was concluded that the remains had accumulated over a long period of time by natural processes of concentration in an old river bed (fig. 9.6).

In 1971 the mammoth-rich area known as the Oyegosskiy Yar was declared a nature reserve and the collecting of tusks there is now forbidden.

Fig. 9.5 *Excavation of the Shandrin mammoth (about 42 000 years old), in 1972. A jet of water, drawn from the nearby river by a petrol driven pump, is being directed on to the remains in order to melt the enveloping permafrost. (Photo: Yakutian Branch of the USSR Academy of Sciences).*

Fig. 9.6 *Leg of a mammoth from the Berelekh River. About 12 000 years old. (Photo: Novosti Press Agency).*

Dima

The most spectacular discovery of all was yet to come. On 23 June, 1977, gold miners who were thawing the permafrost with jets of water near the River Kirgiliakh, a tributary of the River Kolyma, uncovered a complete frozen carcass of a baby mammoth, which measured only 115 cm long and 104 cm high and was later estimated to be only 6–7 months old at the time of its death (fig. 1.1, p. 10). The find was at once reported to the scientific authorities who flew it to Leningrad, where its study was co-ordinated by Professor Nikolai Vereshchagin. News of the discovery appeared in newspapers throughout the world. The ensuing sensation was two-fold. Everywhere the early demise of this baby animal (now to be known as Dima)

Fig. 9.7 *In 1979 the carcass of the baby mammoth 'Dima', discovered in 1977 (compare with fig. 1.1) was displayed after laboratory treatment, as part of the Soviet Exhibition at Earl's Court, London. Here it is being unpacked from its special travelling case. On the right is Dr Andrei Kapitsa, who travelled with the carcass, with 'Dima's' British police guards. (Photo:* The Observer).

appealed to all. Much more important, however, was the opportunity which the new discovery provided for scientific study, using all the new techniques that had been developed since the discovery of the Beresovka mammoth three-quarters of a century earlier. In Leningrad, after thawing and chemical treatment, it was placed on public display in the museum, where it attracted a long queue of fascinated visitors. Although in most respects not unlike a baby Indian or African elephant it showed some features that were totally different, most notably the long hair (again ginger in this instance) and the very small ears, a character also portrayed by Western European upper Palaeolithic cave artists.

Soon afterwards Dima was sent on tour overseas, eventually reaching London in May, 1978, where, protected by an around-the-clock guard and heavily insured, it was shown as part of the Soviet exhibit at Earls' Court (fig. 9.7).

Every year now brings additional discoveries from the permafrost. In 1977, the skull and a well-preserved foot of a mammoth were found on the Taimyr Peninsula, and in 1979, the remains of a 12-year-old female mammoth with well-preserved stomach contents were found beside the River Yuribei, the most westerly frozen mammoth carcass yet discovered.

Frozen remains of mammoths and other animals are not restricted to Siberia. In 1948, the head and one forelimb of another baby mammoth had been found in permafrost being washed for gold near Fairbanks in Alaska, and less well-preserved remains have been found in Erscholtz Bay. In 1951, a complete carcass of a bison was found in another gold working near Fairbanks (fig. 4.8, p. 41), yet another in 1979.

Interpreting the remains

With such a wealth of frozen remains now known to science, let us consider some of the problems that scientists have tried to solve. How were the remains preserved? What did the mammoth look like and what were its habits? How long ago did it live, and when and why did it become extinct?

The woolly mammoth was not restricted in its distribution to areas where permafrost occurs today. It spread over a broad belt from Western Europe through nothern Asia to North America. Outside the permafrost area only skeletal remains are known.

It is not necessary to postulate any dramatic catastrophe to account for the permafrost carcasses. There was no sudden oncoming of cold, so that individual mammoths were frozen as they grazed, with food still in their mouths. It is not necessary to postulate that the remains had been carried northwards, from more temperate regions, by diluvial flood waters. Some carcasses did find their way into rivers, but the field evidence suggests that breeding populations of mammoths were able to spread as far north as Taimyr Peninsula (76°N) and the New Siberian Islands (then connected to the Siberian mainland). Most of the individual animals apparently lived, at least seasonally, not far from their place of burial.

Many suggestions have been made to account for the quick burial and refrigeration of the carcasses. Probably all the mammoths did not die or become buried by the same process. The Shandrin mammoth was an old animal that had perhaps died a natural death, for it lay on its stomach with its legs in front, a position characteristic of dying elephants. Other mammoths may have died by falling through ice or into permafrost collapse pits. Pfizenmayer thought that the Beresovka mammoth had fallen down a steep river bank and had been buried by land slides and snow drifts, though later investigations have suggested that it was buried in river deposits. The process of gelifluction, as described on p. 17 may have caused the entrapment of some of the mammoths, and was probably the cause of burial and freezing of most of them.

Freezing of carcasses must nevertheless be regarded as abnormal, even in Siberia and Alaska. The enormous quantity of bones, lacking soft parts, that are found there shows that most carcasses decomposed before they ever became buried. Even some of those that were preserved display evidence of damage by carnivores; or contain the puparia of blow-flies; or show other signs of partial pre-burial decomposition. Preservation by freezing can continue only so long as the remains stay frozen. On thawing the process of decomposition is at once resumed. Vereshchagin records that his dogs tried to get at the remains of the Terektyakh mammoth, even though 'it smelt like old rubbish'. Preservation of some carcasses may have been enhanced by freeze-drying before burial, in the cold dry climate of the Arctic.

Appearance

The appearance of the woolly mammoth is now well-known. A member of the elephant family, it had the general appearance of a present-day elephant, although it differed in many details. The Siberian race was relatively small, with a shoulder height of only about 3 m in males, 2·5 m in females (less than the same measurement in large Recent Indian and African elephants). In fully grown male animals the immense tusks were wildly twisted, their tips sometimes pointing together; the head was high-domed and the shoulders humped, giving the neck a notched appearance; and its ears were very small. The greatest peculiarity of the mammoth was its long coat, consisting of an inner covering of short hair and an outer layer of long coarser hair. As we have already seen, most of these features are also well shown in European upper Palaeolithic cave paintings. The hair and fat found on the frozen carcasses are interpreted as protection against cold conditions, a conclusion supported by the associated field evidence. The mammoth apparently changed its hair, as occurs today in many arctic mammals, at the beginning of summer.

Muscle tissue and blood

Biochemical geneticists in the United States were recently given the opportunity to study dried and frozen samples of Dima's muscle tissue and blood supplied by members of the mammoth Research Committee of the Soviet Academy of Sciences. They found that the state of preservation of the tissue cells was remarkable. Scanning and transmission electron microscopy revealed almost perfectly shaped red and white blood cells and clearly striated muscle fibres.

Unfortunately, many of the original muscle proteins had undergone extensive post-mortem modification through external leaching by soil water and the natural processes of internal denaturation. However, one of the important globular protein group, albumin, was found to have survived the time-span of about 40 000 years and produced weak, but recognizable, indirect immunological reactions. These were clear enough to demonstrate a close qualitative relationship between mammoth albumin and albumins obtained from Indian and African elephants.

Molecular biologists have used albumin extensively in their attempts to reconstruct evolutionary relationships between the proteins of living organisms. The detection of immunologically active albumin in an animal of such antiquity therefore opens up important possibilities for future research.

EARLIER AGE GROUP		LATER AGE GROUP	
SIBERIA			
Adams (Lena River) mammoth, 1799	36 000–37 000 years	Taimyr Peninsula mammoth, 1948	11 500 years
Beresovka mammoth, 1900	more than 39 000 years	River Berelekh mammoth remains, 1970	12 000 years
Shandrin mammoth, 1971	42 000 years	Yuribei mammoth, 1979	9700 years
River Indigirka woolly rhinoceros	38 000 years		
Selerikan horse, 1968	35 000–40 000 years		
'Dima' 1977	40 000 years		
Khatanga mammoth, 1977	more than 50 000 years		
ALASKA			
Fairbanks, mammoth hair	32 000–34 000 years	Fairbanks, mammoth	15 400 years
Fairbanks bison, 1951 (fig. 4.8, p. 41)	31 000 years	Fairbanks, another bison	12 000 years
Fairbanks bison, 1979	36 000 years	Fairbanks, hoof of horse, 1981	17 200 years
		Fairbanks, musk ox	17 000 years

Table 4. *Dates of discovery and approximate radiocarbon dates of some of the most important frozen mammalian carcasses from Siberia and Alaska.*

How old are the frozen remains?

The absolute age in years of the frozen carcasses was for a long time a subject of speculation. During recent years, with the availability of carbon^{14} dating, the exact age of many of them has become known, with surprising results. Their ages fall into two main groups, one ranging in age from about 45 000 years to 30 000 years and a smaller number of remains about 14–11 000 years old (Table 4).

Although skeletal remains lacking soft parts are known from the period 30–12 000 years ago, there is very little carcass material of this age. A tendon on a 22 000-years-old bone of a lion from Alaska is one of the rare examples. As we have already seen, this intervening period was a time of massive glacial advance, the ice sheets in the northern hemisphere expanding to their maximum extent about 18 000 years ago. There were minor, more temperate, periods from about 45–25 000 years ago and about 12–11 000 years ago. It was apparently during these ameliorations that most of the known carcasses became frozen. This appears to be a climate-related depositional phenomenon, related to the amount of available water (which reached its minimum at times of glacial advance) and does not reflect an absence of mammoths from the areas in question. Under cold arid conditions, with little moisture to supply mudflows, carcasses would have tended ultimately to rot on the surface with only the bones surviving for potential fossilization. Under moister conditions summer mudflows could rapidly have covered carcasses lying in their paths, which became permanently frozen when the permafrost level rose above them the following winter.

Mammoth habitat and diet

Most fascinating of all the problems related to the study of woolly mammoths is the nature of the habitat of these animals and the sort of food that was available to sustain their massive bulk. Did the mammoth really live in the bleak almost treeless snowy north, as has sometimes been portrayed by artists; and what was its diet? In theory, it should be simple to resolve both these problems from a study of the fossil leaves, fruit and pollen of plants in the sediments surrounding the carcasses and that preserved in the stomachs. In some instances, for example the Yuribei mammoth and Dima, well-preserved associated insect remains were also found. In practice, however, the operation is not as simple as this. The mammoths may have eaten selectively; plant assemblages may vary locally according to the terrain; the pollen production of trees occurs earlier in the year than that of grasses, so that the natural proportions of tree and grass pollen may vary according to the time of year; the mammoths may have been animals of broad ecological range which migrated southwards to forests in winter; and we have already seen that all the remains are not of the same age so that they may date from times of different climatic conditions.

Well-preserved stomach contents have been recovered from a number of frozen mammoths, among the most important being the Beresovka and Shandrin mammoths, both of the earlier age group. The stomach remains from both mammoths were separately examined for fruits, fragments of stems and leaves; and for pollen and spores. In both instances, it was found that herbs and herbaceous pollen and spores made up the bulk of the material, with parts of trees and arboreal pollen only sparsely represented. Macroscopic plant remains from the Beresovka mammoth included predominantly grasses and sedges, with rare buttercup, poppy and other herbs. In addition, fragments of bark of larch, birch and alder were beneath the carcass. Macroscopic remains from the Shandrin mammoth included abundant mosses and *Sphagnum* together with occasional woody fragments of bilberry and willow and needles of larch. Pollen and spore counts from both stomachs gave more extensive plant lists, although such evidence needs to be accepted with caution since some of the plants have wind-carried pollen and need not necessarily have been growing at the mammoths' feeding places. Although pine, birch and willow are represented by a few pollen grains from both mammoths, this line of study also points to a diet of predominantly herbs, though with a greater proportion of grasses and sedges in the Beresovka mammoth and more mosses in the Shandrin mammoth. Perhaps this difference is a seasonal one, the Beresovka mammoth having died during the summer; the Shandrin mammoth during the winter, when it might have had to accept food of lesser palatability. In general, mosses do not form an important item of diet among present-day mammals.

From this botanical evidence several ecological plant assemblages are recognizable, suggesting that the mammoths' habitat was very varied. There are plants characteristic of a dry steppe-like environment, others of slightly wet conditions, others of swampy conditions and a few arctic-alpine plants. The habitat of the mammoth was apparently most typically meadow-like with swampy areas and more sparsely growing trees; and its diet was predominantly of grasses and sedges with some other herbs and mosses and occasional parts of trees.

Most of the plants represented in the mammoth stomachs and in the sediment surrounding the carcasses are of species that still grow locally or within a few hundred kilometres of the finding places at the present-day. Slight changes of plant distribution, which could be of climatic significance, have nevertheless been claimed by some workers. From a study of the pollen in the stomach of the Beresovka mammoth, Tichomirov writing in 1960, observed a similarity to plants growing at the present day to the southwest of the fossil site and inferred a change to colder conditions since the death of the animal; and Solenovich, writing in 1977, found the closest present-day analogy to the Shandrin mammoth flora 320 km south of that locality. Both these mammoths belong to the earlier age group. Plant remains associated with the 1948 Taimyr mammoth (later age group) also have a slightly more southern distribution today. The dwarf birch, for instance, abundantly represented in the sediments at this fossil locality, no longer occurs there. The possibility that the wood had been carried northwards by rivers can be excluded since the local River Mamontovaya flows from west to east. Lozhkin, writing in 1977, on the other hand, concluded from a study of the plants associated with remains of a mammoth at the Berelekh site (younger age group) that the climate had then been colder, similar to that prevailing today 200 km further north, with the mean July temperature not rising above about 8–10°C.

When attempting to interpret such evidence it must not be assumed that changes of climate during and since the Pleistocene can be measured in terms of a simple displacement of the existing flora southwards at times of advance of the world's ice sheets and northwards at times of climatic amelioration. As we have already seen, recent studies have suggested that, at the end of the Pleistocene, there existed across three continents, from Europe to Canada, a cold steppe-like biome variously known as 'mammoth steppe' 'arctic steppe' and 'steppe tundra'. It was characterized by succulents, grasses and abundant *Artemisia* (wormwood) and became extinct, except for some local patches, at the beginning of the Holocene, about 9–8000 years ago.

Although the climate was very cold, the sea would have been much lower at this time, causing the whole of the Bering area and shores of the Arctic Ocean to become dry land with consequently greater aridity, lighter winter snow cover, deeper seasonal melting of the permafrost, a longer growing season for plants, and deeper root penetration. The consequence (a conclusion also borne out by evidence from fossil insects associated with some of the mammoths) was the development of widespread areas of cold steppe, which were highly productive and able to provide fodder for a greater concentration of herb-feeding animals than can survive there today.

The 'mammoth steppe' was not an additional biome between the tundra and taiga, which today merge gradually with one another. Rather the last two mentioned biomes appear to have been in parts a replacement of it. Among the mammals which coexisted on the 'mammoth steppe' were species which today would be regarded as both tundra forms (e.g. musk ox, reindeer and collared lemming) and steppe species (saiga antelope, horse and bison). There were also extinct species such as the woolly mammoth and woolly rhinoceros (though the last mentioned species did not cross the Bering Straits to the New World) and there were attendant carnivores such as the lion and wolverine.

Extinction

Those concerned with reasons for the extinction of the woolly mammoth have observed that the most recent dated remains are about 11–12 000 years old and that it is probable that extinction occurred soon after that time. Various causes have been suggested to account for the extinctions of this and other large mammals at the end of the Pleistocene, man and climatically caused change of habitat from 'mammoth steppe' to boggy tundra being those most generally favoured.

Whilst it is known that the woolly mammoth was hunted by upper Palaeolithic man in Europe and in more temperate parts of the USSR, (see chapter 14, fig. 14.2, p. 207) he appears to have had little influence on the mammoths of the Siberian permafrost areas. Until recently it was thought that he did not penetrate so far north, though recent discoveries on the River Berelekh (including stone artefacts and a piece of tusk with an engraving of a mammoth (fig. 8.3d)) show that man and mammoth were briefly contemporary, 12–10 000 years ago at this, the world's most northerly Palaeolithic site. The number of people reaching this area must nevertheless have been small and, in contrast to Europe, man can have played little part in the reduction of the mammoth population.

More likely some change of climatic or geographical conditions finally killed off the Siberian mammoths; and here the disappearance of the 'mammoth steppe' at the end of the Pleistocene has received especially close attention. Vereshchagin, writing in 1974, has pointed out that the final disappearance of these animals coincides with the end of the last glacial epoch, when important changes in climate, vegetation and landscape took place over much of the world. In the arctic, he suggested, post-glacial warming would have led to melting of arctic pack ice, which would have caused cyclones and winter snow storms. The replacement of grasses by sedge and moss communities would have led to a reduction of the pasturage needed to support the mammoths and other large herbivores.

Later, post-glacial rise of sea-level isolated the New Siberian Islands and greatly reduced the area of land previously frequented by the mammoths, but this did not occur until after their extinction and was not the cause of it.

Fig. 10.3, pp. 118–119 *Artist's reconstruction of an August scene on the Gower coast, South Wales, during the Last (Ipswichian) Interglacial, about 120 000 years ago, based on topographical data, on fossil mammalian remains from deposits in local 'raised' sea caves (see fig. 10.12, p. 142) and on palaeobotanical evidence from river and lake deposits, believed to be of equivalent age, in southern England. Ideas for the hyaena activity group are based on observations of living animals in the Ngorongoro Caldera, Tanzania.*

The area of sea shown is part of the Bristol Channel, with the county of Devon, 35 kilometres away, just visible in the distance. Sea-level during this interglacial episode is marginally higher than at the present day. There is forest vegetation, with the Montpellier maple, Acer monspessulanum *(which today does not occur further north than southern Europe), oak,* Quercus, *and many other common present-day British plant species. The forest is more open than its continuation further inland, because of the wind-swept location and trampling by large mammals.*

Five species of mammals are shown, all abundant in southern Britain during the Ipswichian, and a sixth is represented by a skull. These are the hippopotamus; straight-tusked elephant, Palaeoloxodon antiquus; *narrow-nosed rhinoceros,* Dicerorhinus hemitoechus; *red deer,* Cervus elaphus; *and spotted hyaena* Crocuta crocuta, *one of which has a skull of a fallow deer,* Dama dama. *The straight-tusked elephant was an animal of immense size, sometimes reaching 4 metres at the shoulder, substantially larger than any upper Pleistocene woolly mammoth. The herd of red deer shown is composed entirely of stags which, after shedding the velvet from their antlers in July, await the rut in September, when the strongest animals round up as many hinds as their status permits. The entrance of the small cave in which the chocolate brown baby hyaenas are sheltering is an assembly place for adult hyaenas, one of which is the mother. There is a scattering of splintered bone fragments and droppings near the lair, inside which would be more bones, carried there by the baby hyaenas.*

With the exception of the red deer and the spotted hyaena all these mammals probably disappeared from the British Isles soon after the above date, certainly by about 90 000 years ago. The elephant and rhinoceros of the succeeding Devensian cold stage were the woolly mammoth and woolly rhinoceros.

Fig. 10.4, pp. 122–123 *Artist's reconstruction of a scene on the Gower coast, South Wales, about 18 000 years ago—the same locality that is shown in Figure 10.3, but seen from a more distant view point.*

This was the time of greatest expansion of the Last Glaciation (Devensian) ice sheet, the 'front' of which lay only 10 kilometres away to the north east. Sea-level, as the result of water being transferred to the land to become ice, was about 100 metres lower than at the present day, reducing the width of the Bristol Channel and exposing additional land along its flanks. No glacial processes are involved in the creation of the landscape shown, which does nevertheless display many interesting periglacial features (see pp. 17–21). These include polygonal cracking of the ground, water-filled

continued on p. 121

Chapter 10 **The changing Ice Age mammalian faunas of the British Isles**

To the casual observer, looking at the globe or a map of the world, the British Isles looks a very small place indeed, unlikely to compare with the wastes of Siberia with their frozen mammoths or with the Rift Valley of East Africa with its important anthropological sites, as a finding place for remains of Pleistocene mammals.

But how mistaken any such inference would be! Britain's geographical situation in mid-latitudes – its climate drastically influenced by the ever changing position of the polar front and by a late lingering corridor of warmth in the North Atlantic at the end of oxygen isotope stage 5e – makes it, by chance, the focus of one of the most spectacular series of mammalian faunal changes that occurred anywhere in the world during the Pleistocene. So often elsewhere the palaeontologist is involved with the study of changes which, although they are of great significance to him because they indicate fluctuations of rainfall or evolutionary changes, are not spectacular to the layman. But the faunal changes of the British Isles, especially those which occurred towards the end of the Pleistocene (by which time most familiar present-day species of mammals had already evolved), are unsurpassed in their contrast.

Let us take up the story from the peak of Last Interglacial, *c.* 120 000 years ago (this was 5e of the marine oxygen isotope chronology, see p. 58). At this time the climate of the British Isles was appreciably warmer than today, supporting not only familiar present-day plants such as oak and hazel but also, locally, some southerly species, including the Montpellier maple, *Acer monspessulanum* and water chestnut, *Trapa*

Fig. 10.1 *Map showing area within which remains of hippopotamus dating from the Last, Ipswichian, Interglacial (about 120 000 years ago) have been found in the British Isles. Although abundant at this time in England and Wales, the hippopotamus apparently did not reach Scotland or Ireland.*

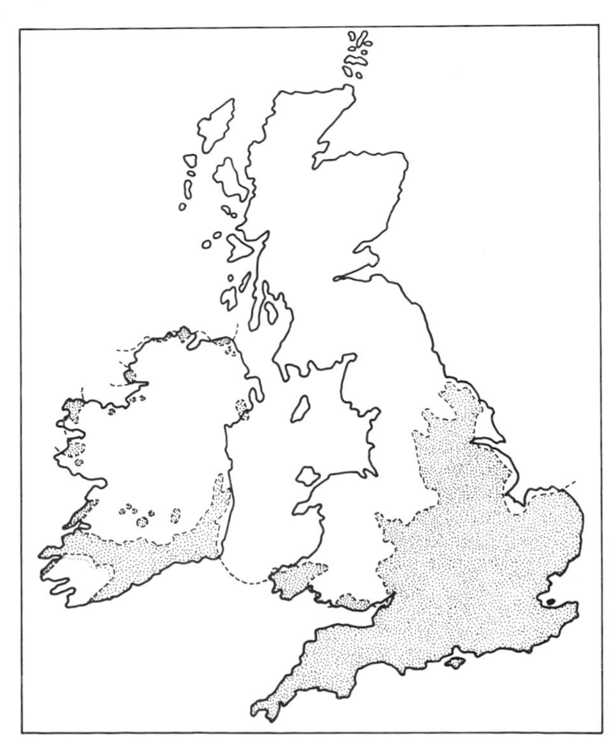

Fig. 10.2 *Probable maximum extent over the British Isles of the continental ice sheet of the last glacial advance about 20–17 000 years ago. As the ice subsequently retreated the mammals which had previously been restricted to the unglaciated southern region were able to spread northwards once more, following the advancing plant cover onto the terrain exposed by the shrinking glaciers.*

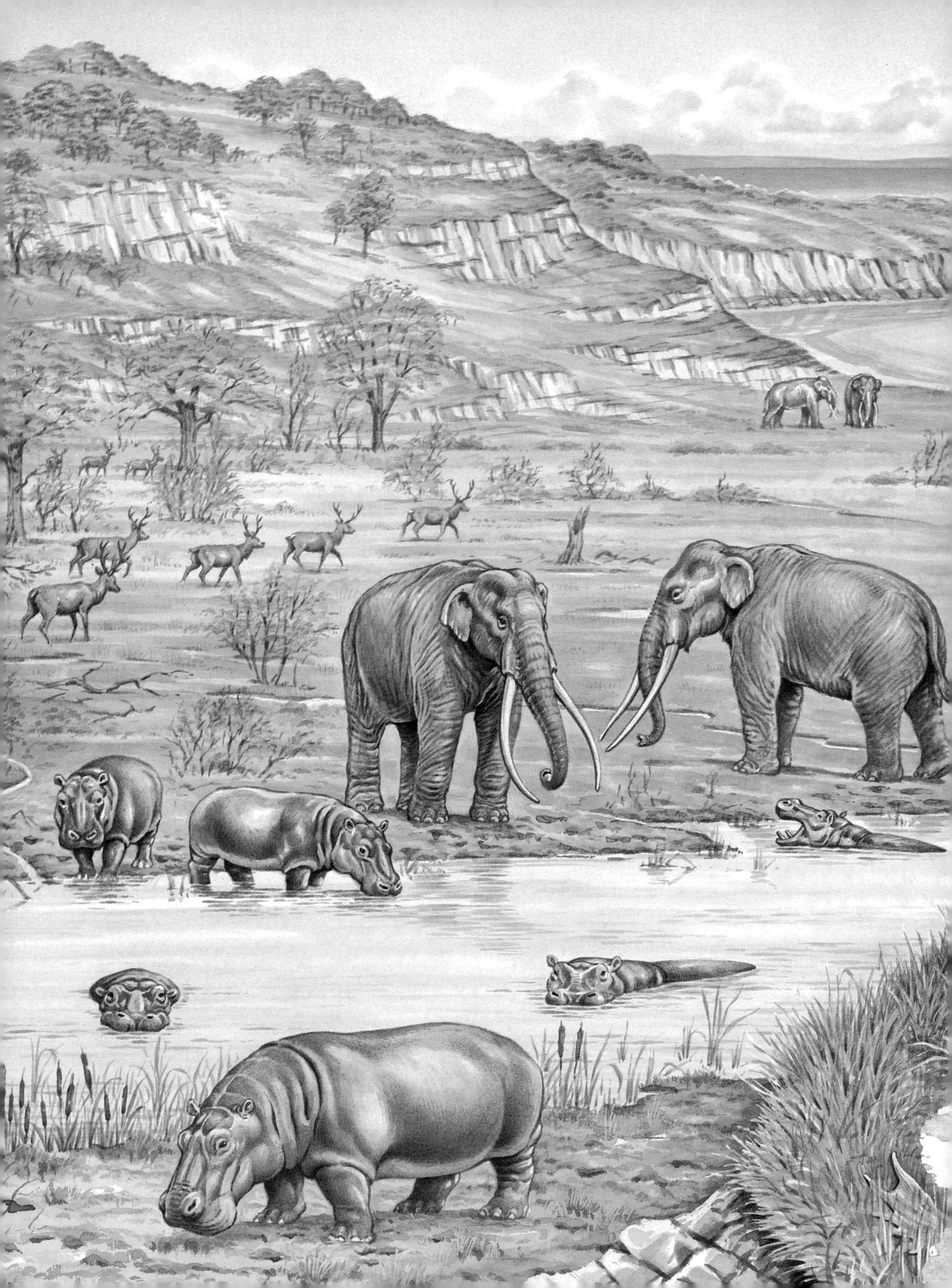

natans, which can no longer survive there. Finding conditions so favourable the hippopotamus (today an inhabitant of the equatorial regions) had been able to spread northwards throughout most of England and Wales, up to an altitude of 400 metres on the now bleak Yorkshire moors, where it was accompanied by the extinct straight-tusked elephant and narrow-nosed rhinoceros and also by red and fallow deer, lion, hyaena and wolf (fig. 10.1). Although the precise time of the disappearance of the hippopotamus from England is still uncertain, it seems unlikely that it survived very long after the above date. Certainly by about 70 000 years ago the series of genial and less genial phases of stage 5 gave way to conditions of deteriorating climate which, by 20 000 years ago, had brought glacial ice over Scotland, northern England and most of Wales and Ireland (fig. 10.2). Although the central and southern parts of England still remained unglaciated, tundra conditions extended over much of the area. Reindeer, musk ox and collared lemmings, mammalian species which today can survive at the most northerly limit of land in the world, only 700 kilometres from the North Pole, came to live where the hippopotamus had previously been. The extinct woolly mammoth and woolly rhinoceros were other cold-adapted mammals that flourished in southern England at this time. Although this climatic deterioration caused such great changes among the plant-eating mammals, most of the carnivores (such as the lion, the spotted hyaena and the wolf) were unaffected by it. Reindeer and woolly rhinoceros replaced red deer, narrow-nosed rhinoceros and hippopotamus in their diets (figs 10.3 and 10.4).

This spectacular faunal change does not represent the only change from temperate to glacial conditions in the British Isles during the Pleistocene. Indeed the field evidence shows a whole series of such interglacial and glacial stages accompanied by massive faunal displacements superimposed on a framework of gradual evolutionary change. Thus, although a high proportion of the mammalian remains in the earlier deposits are of extinct species, many of those from the later deposits are of forms which, even if no longer living in the British Isles, survive today in other parts of the world where their ecology can still be studied as a basis for interpreting the environmental implications of the fossil material. The fluctuations of sea-level which were associated with these climatic changes caused the British Isles to be alternatively isolated from the continent of Europe, when faunal exchanges were not possible, and connected to it. During the low sea-level of the last glacial stage most of the North Sea became dry land, providing a vast additional area of living space and allowing woolly mammoths and other animals to move as freely as the rivers would permit between England and Denmark. Ireland seems to have been isolated from Britain during most of the Pleistocene, though a land bridge allowed some British mammals to cross on at least two occasions towards the end of this epoch.

It follows that the British faunal assemblages differed from interglacial to interglacial and from cold stage to cold stage. From detailed studies of these differences the mammalian palaeontologist, with the guidance of those working in other disciplines of the Quaternary, is becoming increasingly able to identify, on faunal grounds, the climatic episode concerned.

The abundance of mammalian remains from commercial excavation sites and from coastal exposures has greatly assisted with this work.

Early studies

Interest in British fossil mammals can be followed back for more than three centuries. One of the earliest notable records is that of Plot who, in 1676, described a number of discoveries of large mammalian bones and teeth in England, including a thigh bone, generally supposed to have been that of a female giant, found during excavations at the time of pulling down of St Mary Wool Church after the Great Fire of London. Molyneux's discussion of discoveries of remains of giant deer in Ireland (written in 1677, see p. 200) was, in its thinking, many years ahead of its time. In 1729 Sir Hans Sloane (whose private collection subsequently provided the nucleus upon which that of the British Museum was built) mentioned fossil teeth of elephants from Northamptonshire in a broader treatment of this subject, which also included a lengthy account of the frozen remains already known from Siberia (chapter 9).

It was nevertheless not until the beginning of the nineteenth century that British fossil mammalian studies gained momentum. Sir Everard Home's description, published in 1817, of the bones of a rhinoceros found by Joseph Whitby at Oreston, near Plymouth, became the first scientific account of fossil remains from a British cave, and this was soon followed by William Buckland's excavation in Wirksworth and Kirkdale caves in Yorkshire and Paviland Cave on the Gower coast in South Wales. The results of this last mentioned investigation were published in *Reliquiae Diluvianae* (Relics of the Deluge) in 1823. Kirkdale Cave, Buckland believed, had been an ante-diluvian hyaena den, Paviland a human burial place. Although remains of woolly mammoth and other Pleistocene mammals were also present in Paviland Cave, Buckland refused to believe that man had coexisted with these animals and regarded the human skeleton as an intrusive burial of much later date. It was not until the present century that carbon^{14} provided an absolute date of 18 000 years, showing that it was of upper Palaeolithic age and that the animal and human remains were indeed roughly contemporaneous.

Although, at the time of publication of *Reliquiae Diluvianae*, Buckland still adhered firmly to his belief in an universal deluge, and regarded the remains of extinct mammals which he described as being of ante-diluvian age, he nevertheless later modified his views and by 1838 had become converted to the glacial theory. He did not, however, see fit to modify his views about the antiquity of man and such was his authority that this controversial topic (which could have been resolved from evidence provided by the Revd. J. MacEnery and others from Kent's Cavern, Torquay, Devon before 1830) was to remain disputed for yet another quarter of a century. It was not until 1859, when William Pengelly and Hugh Falconer published the results of their excavation in the Brixham Cavern, only six kilometres away, where they had found flint artefacts and the remains of woolly mammoth, lion and other animals beneath an unbroken flowstone floor (at about the same time that Boucher de Perthes was finding similar associations in the terrace gravels of the River Somme at Abbeville in northern France) that man's contemporaneity with the extinct fauna began to gain general acceptance.

The concept of a single ice age which had been accepted in the 1830s evolved gradually to that of four glacial–interglacial cycles fashionable in Britain during the first half of the twentieth century and, more recently, to the acceptance that such a scheme was still greatly oversimplified. By 1980 it had become apparent, from the terrestrial evidence alone, that there had been eight to ten such cycles, yet even these clearly correspond to only parts of the more complex oxygen isotope chronology. For the time being the British terrestrial sequence has been integrated with the deep sea chronology for only about the last 200 000 years, although good progress is already being made towards carrying this correlation further back into geological time.

The chronology of the British Quaternary

Today the chronology of the British Quaternary is a constantly evolving corpus of knowledge, to which new information is steadily being added. A much simplified sequence of climatic stages, based on the state of knowledge up to 1981, is shown in Table 5, p. 124.

Widely though this chronology has been accepted among Quaternary workers as a time scale to which to relate their studies, for the mammalian palaeontologist to follow this scheme is not without problems.

A special difficulty is that many of the actual type localities have produced few or no mammalian remains which can be used as a basis for broader studies. A notable exception is the type locality of the Cromerian Interglacial at West Runton in Norfolk, which has both a rich flora and an abundance of mammalian remains, representing many species. The Cromerian mammal fauna is now well known. But few mammalian remains were found at the Ipswichian type site near Ipswich, Suffolk (where even the plant remains represent only the earlier part of the interglacial), nor in the Hoxnian horizon at Hoxne in Norfolk. We can attempt to establish the mammalian faunas for these last two palaeobotanically based interglacials by looking at mammaliferous sites with floral assemblages apparently identical to those at the type localities, but we can do this with confidence only if we can be sure we know exactly how many climatic cycles occurred: and that stage, except in deep sea studies, has not yet been reached. The deposits at most of the cold stage type localities are also without mammalian remains.

continued from p. 116

depressions on the old sea bed, frost shattering of the limestone cliffs and extensive gelifluction mud flows, which cover the beach deposits shown in Figure 10.3 and dramatically alter the appearance of the valley.

The scene is set in early summer. Most of the winter snow has melted, although snow banks still remain on north facing slopes. A river, in torrent only a short time previously at the peak of the thaw, carries a now reduced water supply in a network of shallow channels flowing between sandbanks in its partly dry bed (a braided river).

The first flowers have already appeared. These include mountain avens, Dryas octopetala, *purple saxifrage,* Saxifraga oppositifolia *and catkins on a male dwarf willow,* Salix herbacea. *Lichens add colour to rocks and to a cast antler. Extensive areas of ground lack vegetation.*

Such an environment, even though it provides a good living for a few specialized animals, cannot support many individuals so that only a sparse fauna is shown here. A snowy owl is about to grab one of a group of collared lemmings, Dicrostonyx torquatus, *while two groups of reindeer,* Rangifer tarandus, *feed and walk in the valley beyond. Nearest to the observer are cow reindeer, antlers recently shed, with newly born calves. Further away are bull reindeer with antlers in velvet. The lemmings (see also p. 24 and fig. 2.12) are grouped around a summer burrow in the ground, after having wintered in a snow bank in a nest of moss and grass.*

Ideas for the scene are based on field observations in Gower, reindeer and lemmings on fossil remains in Gower caves, the migratory snowy owl on remains from a cave in nearby Devon, plants on general palaeobotanical data for southern Britain, and the periglacial features and lemming activity group on observations on Bathurst Island, NW Territories, Canada.

Both pictures painted by Peter Snowball, under the direction of the author, 1985.

Sequence		Author, date	Principal disciplines of study at type localities
Cold	Temperate		
	FLANDRIAN		
DEVENSIAN		Morgan, 1973	Physical evidence
	IPSWICHIAN	West, 1957	Plants, molluscs and insects; very few mammals
WOLSTONIAN		Shotton, 1953	Physical evidence
	HOXNIAN	West, 1956	Plants and molluscs; mammals in highest part of sequence only
ANGLIAN		Baden-Powell, 1950	Physical evidence
	CROMERIAN	West & Wilson, 1966	Plants and molluscs; abundant mammals
BEESTONIAN			Plants and molluscs
	PASTONIAN		
PRE-PASTONIAN		Funnell, Norton & West, 1979	
	BRAMERTONIAN		Pollen, molluscs, foraminifera, mammals
BAVENTIAN		Funnell & West, 1962 (shown glacial 1980)	Pollen, molluscs, foraminifera
	ANTIAN	West & Funnell 1961	
THURNIAN			
	LUDHAMIAN		
WALTONIAN		Wood, 1866	

*The principal stages names are those defined by the Geological Society of London (Mitchell *et al.*, 1973). Pre-Pastonian and Bramertonian were added by Funnell *et al.*, in 1979. A cold period of glacial intensity within the Baventian was demonstrated by West *et al.*, in 1980.

Correlation of this sequence with the oxygen isotope chronology is still in its infancy. Isotope stage 1 can reliably be equated with the Holocene, 2–4 and part 5 with the Devensian, the rest of 5 with the Ipswichian. Before this the equivalence of deposits is still uncertain. Almost certainly the Hoxnian is not stage 7, as might be expected, but may be 9 or even 11.

Table 5. *Simplified chronology for the Quaternary of the British Isles, as established from terrestrial evidence up to 1982**.

The worker in each discipline is not obliged to try to fit his findings into the palaeobotanically based sequence of interglacials shown in Table 5. Chronologies based, for example, on fossil insects or on fossil mammals, established in the context of the associated evidence from the other disciplines, would be alternative schemes of equivalent rank, important also as a basis for the elaboration of the classical scheme.

In this chapter we will not keep rigidly to the chronology shown in Table 5 but will look at the British Pleistocene with its mammals as the central theme. On the available evidence the classical scheme continues to work very well for the lower part of the Pleistocene, up to the Cromerian Interglacial, but the mammalian and other evidence suggest at least two additional but as yet unnamed warm stages after that, one between the Cromerian and the Hoxnian (as these terms are employed by palaeobotanists), and another between the Hoxnian and the Ipswichian.

The geographical distribution of British Quaternary mammalian sites

Although there is such an abundance of Quaternary mammalian sites throughout the British Isles, they can nevertheless be grouped into only a small number of categories, the simplified geographical distribution of which is shown in figure 10.5. They are the shallow water marine Crags of East Anglia; the estuarine and

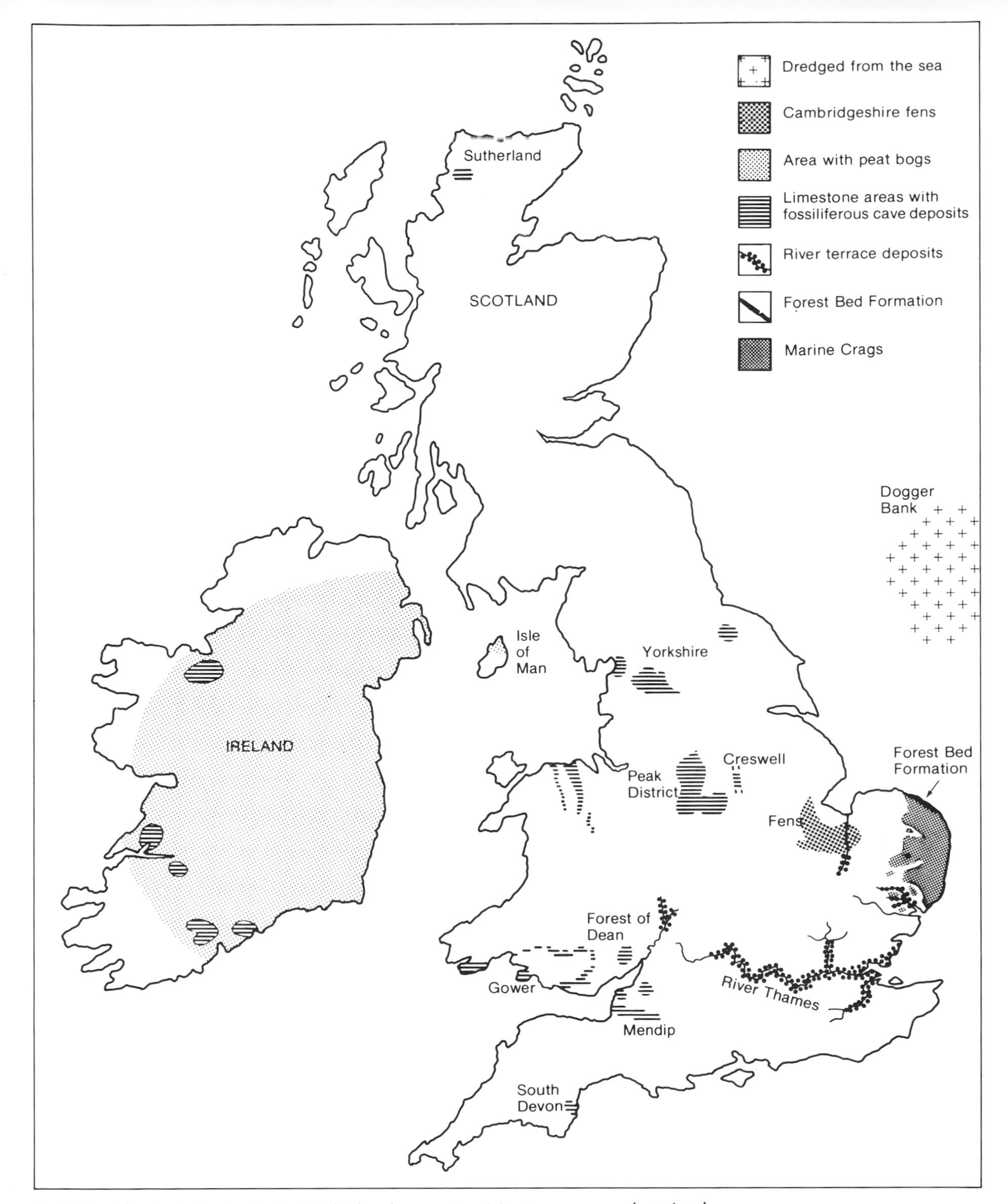

Fig. 10.5 *The principal regions in the British Isles where remains of Quaternary mammals are found.*

freshwater Forest Bed deposits of Norfolk; river terrace and associated deposits; deposits in limestone caves, including Irish caves; deposits in old sea caves and on the bottom of the North Sea; The Cambridgeshire Fens; and the Irish peat bogs, beneath which so many giant deer remains have been found.

None of the various categories of deposits just described spans the whole of the Quaternary, but neither do they all have the same stratigraphic range. They thus compliment one another. The Crags, for example, are lower Pleistocene or earlier, the cave deposits mostly middle or upper Pleistocene and the Fens Holocene (Table 5). When the information provided by all the sites is pieced together it reveals a long and detailed story of changing mammalian faunas in the British Isles.

The East Anglian Crags

The East Anglian Crags (except for the Coralline Crag, which is Pliocene) are of lower Pleistocene age; and nearly all that is known about the mammalian fauna of this period of time in Britain is based on finds from these deposits. There are two main divisions.

The Red Crag

The earliest mammal-bearing deposit of Pleistocene age in the British Isles is the Red Crag of Suffolk, which typically accumulated in the shallow waters of land-locked bays. It consists mainly of shelly sands; with a basement nodule bed containing rolled and polished mammalian fossils apparently washed out of earlier deposits by the Red Crag sea, in an already fossilized condition. The occurrence of these derived remains makes the study of the mammalian fauna very difficult and, for a long time before its true nature was recognized, the Red Crag was generally regarded as being of Pliocene age. More detailed studies of the various remains during recent years have shown what a remarkable mixture of fossils is indeed present. It includes bones of dinosaurs from the Jurassic (more than 80 million years old); bones and teeth of *Hyracotherium*, the dawn horse, and *Coryphodon* from the Eocene London Clay (*c.* 50 million years); teeth of the giant shark, *Carcharodon*, and shells which are probably Miocene (20 million or fewer years); and bones and teeth of Borson's mastodon, *Zygolophodon borsoni* and three-toed horse, *Hipparion*, from the Pliocene (less than five million years old).

Most of these derived fossils are heavily mineralized and polished and can be distinguished from the remains of animals which were living in Crag times, which are less heavily mineralized and of lighter colour. Contemporaneous Crag species include the mastodon, *Anancus arvernensis*, the southern elephant, *Archidiskodon meridionalis*; horse, giant deer, *Megaceros verticornis*, the four-tined deer, *Euctenoceros tetraceros*, fallow deer, *Dama nestii*, elk, *Libralces gallicus*, gazelle, *Gazella anglica*, wolf, fox, beaver, walrus and whale. The perfect condition of some of the faunal remains, for example of walrus and fox, leaves no doubt that they are contemporary Crag fossils. They could never have survived the rolling about by the sea that would have occurred if they had been washed out of earlier deposits.

Remains of whales, both derived and contemporary, make up a high proportion of all the mammalian fossils found in the Red Crag.

The Icenian Crag

The Icenian Crag (which includes the Norwich and later Weybourne Crags) constitutes a further series of marine deposits, which occupy a basin on the north of the Red Crag outcrop. It was apparently laid down in more open seas. Its exact relationship to the Red Crag is not fully understood, though the Icenian appears to range in age from Thurnian to Pastonian. There was an episode of glacial intensity within the Baventian. The relationship between the Norwich and Weybourne Crags is also not clear. The Weybourne Crag, typified by the mollusc *Macoma balthica*, is found at more than one stage. The youngest part of the sequence, including part of the Weybourne Crag, is of Pastonian age and is contemporaneous with the lowest part of the Forest Bed Formation. Part of the Norwich Crag of Suffolk may also be Pastonian.

In general, the fauna of the Norwich Crag is very likely that of the Red Crag. Mammals which continue include the mastodon, *Anancus arvernensis;* southern elephant, *Archidiskodon meridionalis*; horse, Falconer's deer, *Euctenoceros falconeri*; four-tined deer, *E. tetraceros*; fallow deer, *Dama nestii*; gazelle, *Gazella anglica*; elk, *Libralces gallicus*; and beaver. Rare carnivore remains add the sabre-toothed cat, *Homotherium;* Reeve's otter, *Enhydriodon reevei* and the leopard-like *Panthera pardoides*. The voles *Mimomys pliocaenicus* and *M. reidi* have been recorded from both the Norwich and Weybourne Crags. How long the mastodon survived is not altogether clear. Well-preserved remains from the Norwich Crag suggest that this animal was still living in Norwich Crag times, though some palaeontologists argue that it had already disappeared much earlier and that remains from both Crags were derived in a fossil condition from earlier deposits. The mastodon does not appear to have survived into the Weybourne Crag and was apparently extinct before the Pastonian.

The Cromer (Norfolk) Forest Bed Formation

The deposits of the Forest Bed Series, exposed mainly along the cliffs of the Norfolk coast from West Runton to Happisburgh, differ from the earlier Crag deposits, with which they partly overlap in time, in being predominantly estuarine and fresh water, with some beach deposits.

The Norfolk Forest Bed Formation covers the time-span Pastonian Interglacial, Beestonian cold stage, Cromerian Interglacial, *sensu stricto*, and the beginning of the Anglian glacial stage. Although it had long been recognized that these deposits accumulated over a very long period of time, during which there were extensive changes of climate, only recently has it become possible to distinguish the mammalian faunas of the various stages. Especially important studies have been made by Azzaroli, who examined deer remains; West & Wilson, who established the stage names listed above as a result of their palaeobotanical studies; and Stuart who studied the vertebrate remains from the Cromerian type locality at West Runton.

Although no mammalian remains are known from the intervening Beestonian cold stage, a wealth of finds from both the Pastonian (best represented at East Runton) and Cromerian of West Runton allow comparison of the faunas of these two interglacial stages. The change of fauna established from these studies is striking in its magnitude. The Pastonian fauna includes many species which we have previously encountered in the Icenian Crag, including the elk, *Libralces gallicus*; the four-tined deer *Euctenoceros tetraceros* and the vole *Mimomys pliocaenicus*. Other elements of the Pastonian fauna include the narrow-nosed ox, *Leptobos* and the early hyaena, *Hyaena brevirostris*. By Cromerian times, however, there had been extensive changes in the British mammalian fauna and, for this reason, the division between the lower and middle Pleistocene is usually drawn at the end of the Pastonian. Many of the mammals present during the Pastonian had disappeared and had been replaced by others. The four-tined deer, *Euctenoceros tetraceros* disappeared; the elk *Libralces gallicus* was replaced by *L. latifrons*; the narrow-nosed ox, *Leptobos*, by bison; the early hyaena, *H. brevirostris* by the spotted hyaena, *Crocuta crocuta*; and the vole *Mimomys pliocaenicus* by its descendant, *M. savini*. Other Cromerian species from West Runton include the Russian desman, *Desmana moschata*; the beaver and giant beaver, hamster; the pine vole, *Pitymys gregaloides*; a small wolf; Deninger's bear, *Ursus deningeri*; the Etruscan rhinoceros, *Dicerorhinus etruscus*; and the macaque monkey, *Macaca*. The southern elephant *Archidiskodon meridionalis* survived from the Pastonian. The hippopotamus, though not recorded from West Runton, made its first appearance in the British Pleistocene at other sites of Cromerian age.

The Cromerian fauna described above is fully interglacial. Towards the end of Forest Bed times there was a further deterioration of climate with ice of the Anglian glaciation spreading southwards over the Norfolk coast fossil localities. Teeth of the ground squirrel, *Spermophilus*, found a century ago in a deposit with remains of arctic plants at Mundesley, Norfolk, are interpreted as dating from the early part of this glacial advance. The deposit is immediately overlain by thick glacial deposits attributed to the Anglian ice.

Terraces and associated deposits of the River Thames

Old alluvial and marsh deposits along the flanks of rivers are commonly rich hunting grounds for remains of Quaternary mammals. Much of our existing knowledge of the changing mammalian faunas of the upper part of the Pleistocene comes from such deposits, for example along the courses of the Trent Valley of the central Midlands, the River Cam in Cambridgeshire, the Stour and Orwell in Suffolk, the Wiltshire and Warwickshire Avon and other tributaries of the River Severn. But one river stands out above all the others for the richness of its finds. This is the River Thames, which rises west of Oxford and flows through London into a wide estuary which finally opens into the North Sea. We will consider only one British mammaliferous river course here; and the Thames is selected for this purpose.

Two circumstances have combined to make this river such a rich finding place for remains of Quaternary mammals and other associated fossils. Firstly its deposits are geographically very extensive. In the London area, along the inner reaches of the contemporary tidal estuary, they have a width of about 4–10 km (fig. 10.6). Near Walton, just above the tidal part of the river, they increase to 16 km in the core of a large bend; and further narrow strips of alluvial deposits flank the upper part of the river almost to its source. Locally they are highly fossiliferous. The number of such localities is supplemented by the deposits of the Rivers Lea, Roding, Cray and Medway which share a common estuary with the Thames.

Secondly, as we have already seen, almost the whole of London, with a population of over seven million people, together with many of the surrounding conurbations, has been built upon these deposits. Interference of the underlying ground by man for economic reasons has made available to geologists deep pits and borehole records on a scale that could never be possible in less densely populated areas, where the specially initiated excavation of even a small hole in the ground may be beyond the financial resources of those interested in what lies beneath.

Most commonly the mammalian bones found in these alluvial deposits are isolated scattered specimens (fig. 10.7), although complete skeletons sometimes also occur in old marsh deposits upon their temporary surfaces. Two such skeletons have attracted special attention, a straight-tusked elephant, found at Upnor, in 1911, and the previously mentioned mammoth and straight-tusked elephant, found at Aveley in 1964.

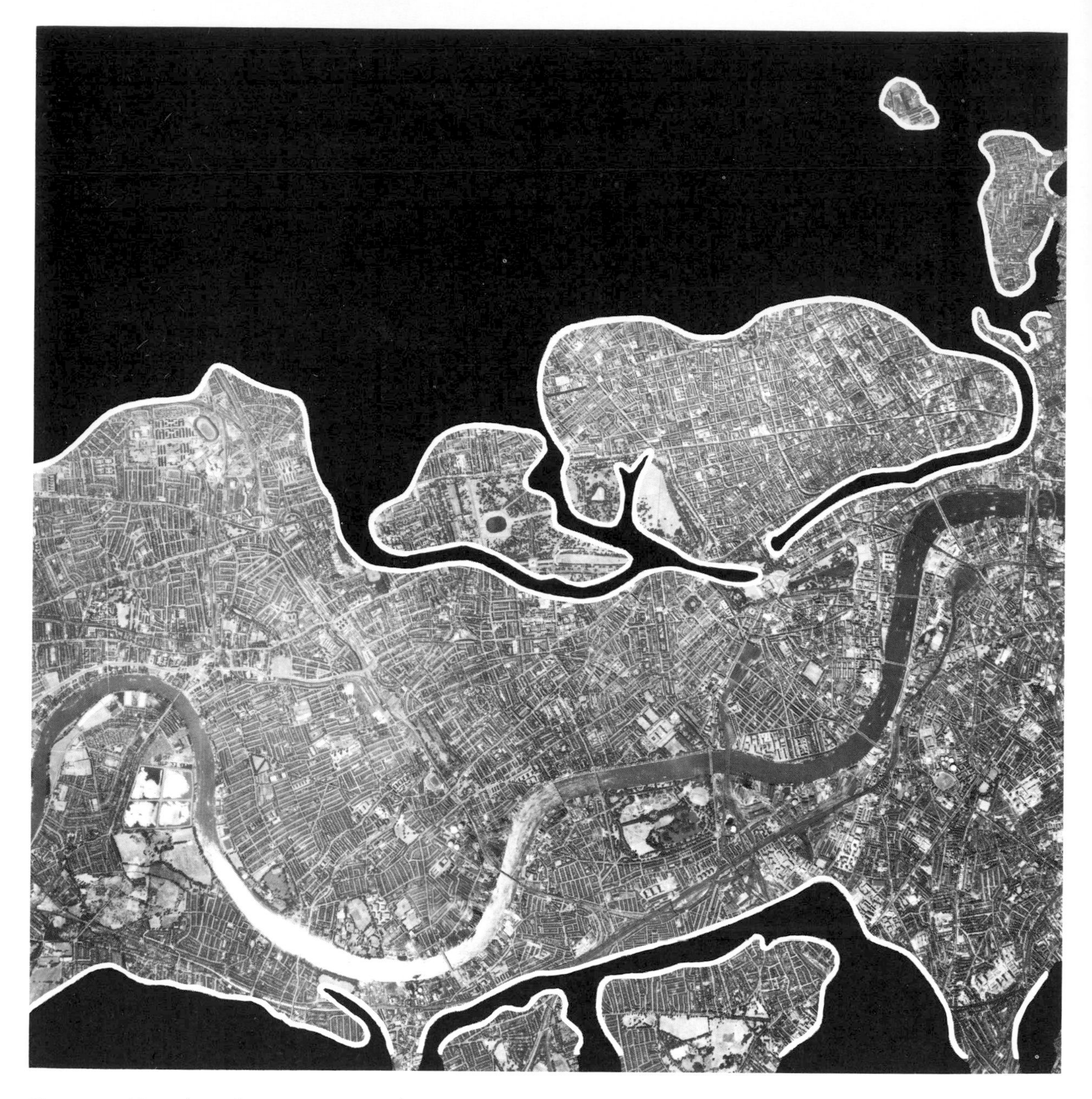

Fig. 10.6 *Vertical aerial view of part of southwest London showing the (here tidal) river Thames. Limit of alluvial deposits indicated with white lines. Area shown 8 × 8 km (Photo: Aerofilms).*

Some of the most important British Palaeolithic archaeological sites, with their abundance of stone artefacts, also come from the alluvial deposits of the River Thames. The remains of the earliest known Englishman, Swanscombe man (probably about 300–400 000 years old), was found at one of them, near Greenhithe in Kent.

Past changes in the course of the Thames

The fossiliferous deposits of the River Thames are of special interest at this point in our discussion of the British Quaternary sequence since almost all of them are of later date than the Crag and Forest Bed deposits just described and they bridge the remaining period of time up to the present day.

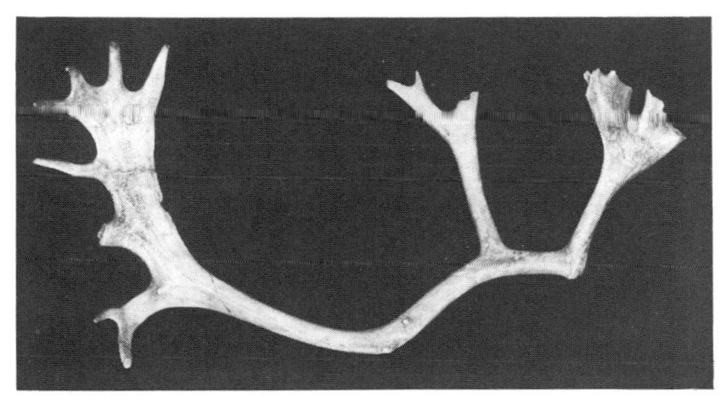

a) *Shed antler of a fine male reindeer from the Floodplain Terrace of the Thames at Twickenham. Probably about 40 000 years old. At the present day this species is of northerly distribution and is usually interpreted as an indicator of cold climatic conditions. Found before 1895.*

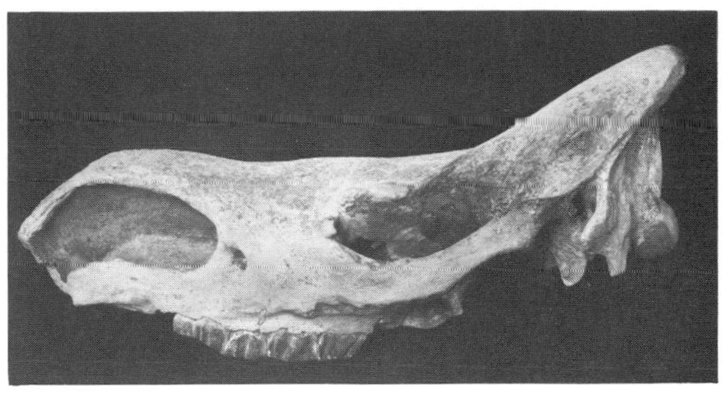

b) *Skull of a woolly rhinoceros from Whitefriars St, Strand, London. This species, commonly associated with remains of woolly mammoth and reindeer, is usually regarded as an indicator of relatively cold conditions. Found in 1903.*

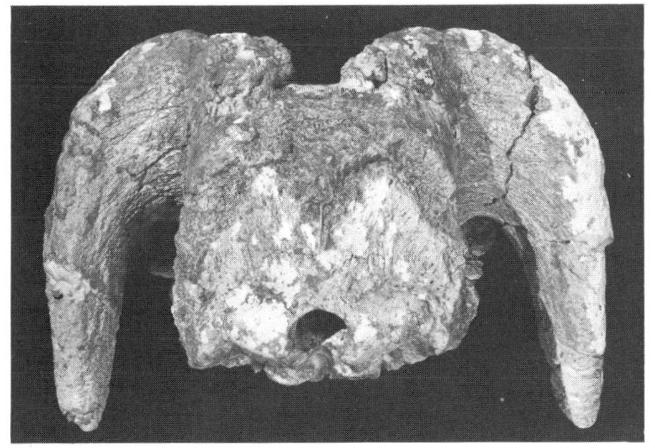

c) *Skull of a musk ox (seen from behind) from the Late Middle Terrace of the Thames at Crayford, Kent. The present-day musk ox is the most northerly occurring large land mammal in the world, able to survive under very severe conditions far north of the tree line on the arctic tundra. The Crayford specimen is probably more than 150 000 years old. Found in 1868.*

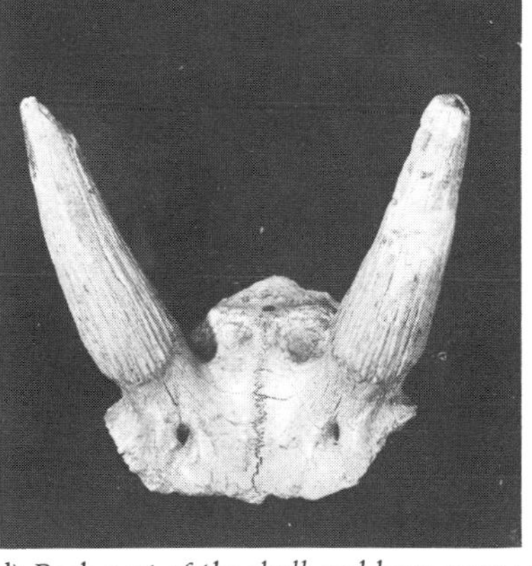

d) *Back part of the skull and horn cores of a saiga antelope from the Floodplain Terrace of the Thames at Twickenham. Probably about 12–15 000 years old. This species still flourishes in the steppes of eastern Europe southern Russia. Although it was common in Eurasia during the upper Pleistocene, (see also fig. 2.13), it is known from only three localities in the British Isles. Found in 1891.*

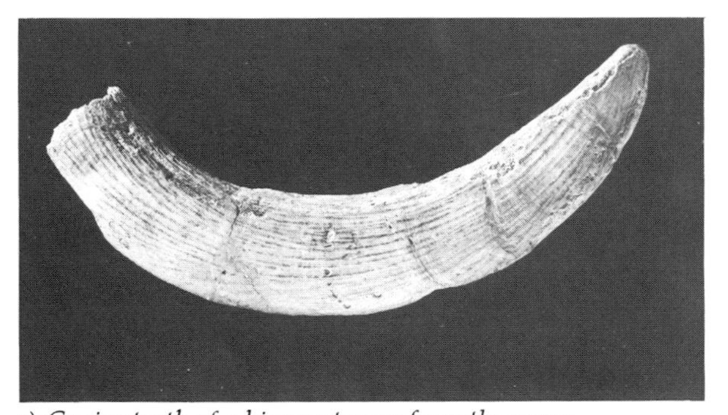

e) *Canine tooth of a hippopotamus from the upper Floodplain Terrace of the Thames beneath New Zealand House, near Trafalgar Square, London. About 120 000 years old. Associated plant and insect remains show that, at the time when this animal was living, the British climate was slightly warmer than at the present day. Found in 1959.*

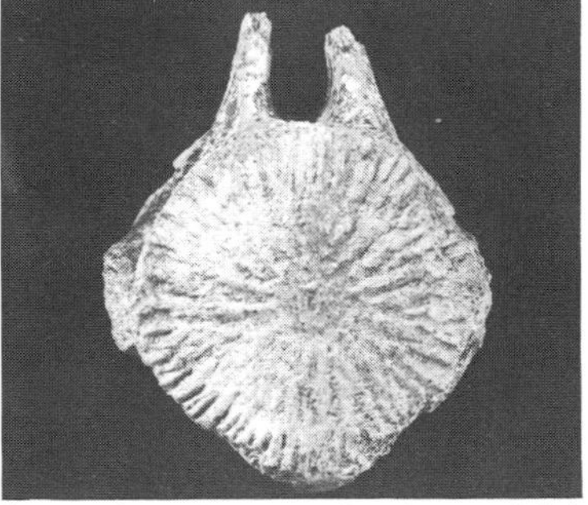

f) *Vertebra of a young bottle-nosed dolphin from old estuarine deposits of the High Terrace of the Thames at Ingress Vale, Swanscombe, Kent. Sea-level about 20 metres higher than the present day; about 300 000 years old. Found in 1901. (Photos: BM(NH)).*

Fig. 10.7 *Some mammalian remains from alluvial deposits of the estuary of the River Thames.*

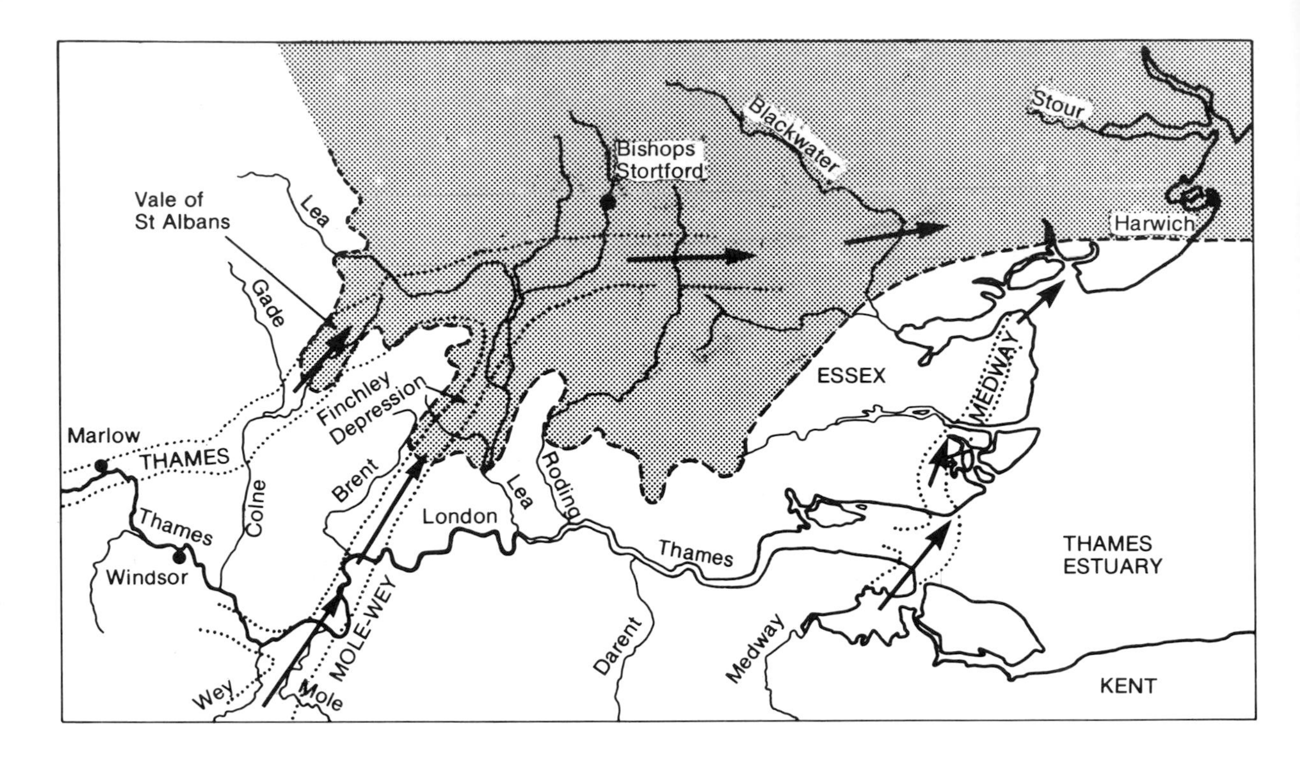

Fig. 10.8 *Diversion of the River Thames. In early Pleistocene times the lower reaches of the Thames lay further north than at the present day. The course of this river has been traced as far east as Bishops Stortford, but, beyond this, most deposits have been removed by erosion, so that it is difficult to say where it finally entered the sea; possibly somewhere near Harwich. During the Anglian glacial advance ice lobes blocked both the Vale of St Albans and the Finchley Depression, causing the eastward flowing Thames to adopt a more southerly course. At one time it was believed that, after being displaced from the Vale of St Albans, the Thames flowed instead through the Finchley Depression; though recent work by Gibbard has shown that it had been the combined Mole and Wey river which occupied this valley and that the Anglian ice also blocked it in a similar manner. When the ice retreated it left behind glacial deposits which prevented the Thames and Wey-Mole rivers from reoccupying their earlier channels. Today the drainage in these valleys flows in the opposite direction, as two small southward flowing tributaries of the Thames, the Colne and the Brent.*

Present-day rivers, shown in small lettering, former river courses in capitals. Width of area, 130 km.

Studies of the Thames show that east of Marlow this river flowed until middle Pleistocene times along a course considerably north of that which it now follows through London. It was subsequently diverted by the advancing ice of the Anglian Glaciation, which dammed its channel and caused it to find a new course further south (fig. 10.8).

As a result of this diversion, although there is a magnificent sequence of highly fossiliferous terrace deposits of middle Pleistocene and later age in the area of the present-day estuary of the Thames, earlier deposits (other than those of tributaries flowing into the proto-Thames from the south) have never been found there. The earliest fossiliferous deposits in the Thames estuary (which are at Clacton and Swanscombe) are believed to be, in part of at least, of equivalent age to the Hoxnian recognized by the palaeobotanists. A single locality with mammalian fossils of apparently Cromerian age, including the Etruscan rhinoceros, is known in the upper course of the Thames, at Sugworth near Oxford.

Processes of sedimentation in the valley of the Thames

So far we have touched only briefly upon the processes of sedimentation and terrace formation in river valleys (chapters 4 and 5). Before we can attempt to construct a chronology for the mammalian faunas of the Thames we must first look at the structure of the deposits of this river in greater detail. Sedimentary processes in rivers vary from one climatic region to another and it is not possible to consider all of them here. With the risk of oversimplification we can concern ourselves only with the most basic of these which are relevant to our study of the Thames. Firstly, there is the process whereby sediment is carried along and deposited by a river. At a time of spate all the loose material, both coarse and fine, in the bed is rolled along the bottom or picked up and carried along in suspension. The faster the water, the larger the size of the objects which can be moved. When everything has been set in motion, then the river bed itself ceases to be protected and it is at such times that downward erosion is most likely to occur.

When a flood begins to subside, then the water can no longer hold all the sediment in suspension, river bed erosion ceases, and the suspended matter gradually falls to the bottom; first the largest pebbles, then gravel, then sand, then silt and finally (but only if movement ceases entirely or if there is flocculation by salt) clay. Such a sequence of layers, which can be observed in almost any deep excavation in the Thames alluvial deposits, contains a record of the events involved and of the conditions and circumstances under which any contained fossil remains became buried. Gravel deposits of uniform particle size suggest fast flowing water near the centre of a channel and it is not surprising that bones in them are commonly isolated, abraded and rounded. Silt may represent flooding along the banks of a river; and *in situ* archaeological material or associated mammalian skeletons and more likely to be found there.

As we have already seen in chapter 5, where valley deepening occurs alluvial fragments (which are often fossiliferous) are commonly left high and dry along the sides of river valleys, where they are known as 'river terraces'. Strictly this term refers to the almost flat surface and the truncated edge only. The deposits underlying the flat are known as 'terrace deposits' and the surface of the solid rock on which they rest is the 'bench'. Older terraces are commonly situated at a higher elevation than the later ones, but this does not mean that the strata are inverted. Within each terrace the normal order of superposition applies, with the oldest layers at the bottom, youngest above (fig. 5.3, p. 46). In the most simple instances archaeological and fossil remains are likely to be older in higher than lower terraces and many chronological studies have been based on this relationship.

Often, however, the situation is not as simple as this and other factors have to be taken into consideration; most notably the position of a site in relationship to the course of the river, since processes of sedimentation are not uniform from source to mouth. In the upper part of a river erosion is commonly greater than sedimentation, which becomes more widespread in its middle part; and there are profound changes where a river reaches an estuary, where marine influences (including flocculation of suspended clay by the salt water, tidal effects and long term changes of sea-level in response to climatic change) take effect. Whereas, in the upper part of the course of a river, terraces slope gently downstream at a fairly uniform height above the present flood-plain, those in the estuary have almost horizontal aggradation surfaces, representing approximately the high sea-level of this time, which terminate inland at the head of the contemporary tidal estuary.

In an estuary the generalization that higher terraces tend to be older than the lower ones is only partly true, since all of those which are physically accessible above present-day river level represent former high sea-levels, which were the interglacial episodes. These terraces merge around the open coastline into levels of raised beaches at equivalent heights. During the repeated intervening glacial advances, when much of the sea water was locked up on the land as ice, the sea-level dropped substantially and rivers flowed along channels deep below their present-day estuaries. All deposits laid down under these conditions and all the bones in them were subsequently submerged when the sea rose again as the result of the melting of the ice and are inaccessible for examination today except by dredging. Such submerged drainage features are known by geologists as 'buried channels'.

It follows that, when trying to interpret the terrace deposits of estuaries and the upper parts of rivers, clear distinction must be made between them. A terrace situated, say, ten metres above the sloping floodplain of a river may appear to run into a terrace ten metres above the horizontal marshes along the side of an estuary (and such correlations have sometimes been assumed). Any such inference may nevertheless be erroneous since many of the terraces in the upper part of a river date from times of cold climate and are continuous with the infilling of the contemporary 'buried' channel of the estuary; whereas all the terraces in the estuary which have not been submerged tend to be interglacial and do not extend beyond the contemporary tidal head. Since the sea-level (and thus the position of the head of the estuary) has fluctuated in the past, there is a zone along the course of such rivers where both estuarine and freshwater river deposits occur interbedded together.

A further problem in the study of estuarine terrace deposits lies in their often being cut across by the channels of tributary streams; and it may be difficult to determine to which river a particular deposit belongs (fig. 10.9).

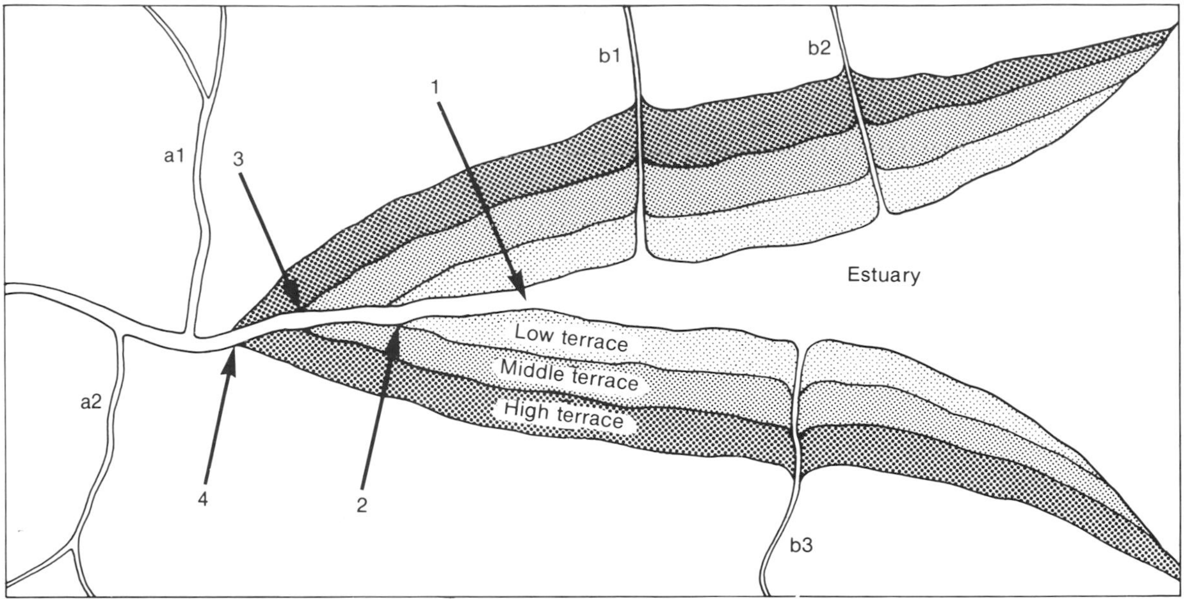

Fig. 10.9 *Sketch map showing the principal zones of sedimentation in an imaginary estuary and adjacent areas. a1, a2, tributaries of the main river; b1, b2, b3, rivers having a common estuary with the main river; 1, present-day head of tidal estuary; 2, low terrace tidal head; 3, middle terrace tidal head; 4, high terrace tidal head. Terrace deposits associated with non-tidal parts of rivers omitted.*

Even the most inland part of the course of a river is not necessarily unaffected by events at the coast. At times of low sea-level there was often great deepening of the river valleys at their seaward ends, not by down-cutting but by cutting back. If the sea-level rose again sufficiently quickly the deepened part was submerged and this process would have ceased; but, not infrequently, the cutting back had already progressed upstream beyond the limit of the new tidal estuary, with nothing to prevent it from continuing all the way to the source of the river. Thus the deepening of the upper parts of river valleys and the abandonment of alluvial deposits to form dry terraces on their flanks is commonly a secondary effect of former fluctuations of sea-level and not merely the result of erosional downcutting.

Finally attention must be paid to various deposits of terrestrial origin which are interbedded with or rest upon those laid down by the river. In the case of the Thames the most important of these are gelifluction deposits ('Coombe Rock'), which commonly contain remains of cold climate mammals and mollusca and valuably supplement the evidence from the terrace deposits. There is also some loess.

The mammal bearing deposits of the River Thames

Let us now try to establish a chronology for the mammalian faunas of this river. We will consider in turn the deposits of each of its morphological parts: the interglacial terraces of the estuary; the terraces of the upper part of the course; and the periglacial deposits which are interbedded throughout the entire system.

THE INTERGLACIAL TERRACES OF THE ESTUARY

The terrace deposits of the Thames estuary provide the main source of remains of Quaternary mammals in the London area. In principle the methods of studying these estuarine terraces are very simple. Within the tidal estuary at the present day are zones of fresh, brackish and salt water which move their position slightly downstream or upstream according to the state of the tide. The sediments laid down in the various parts of the estuary and at different levels in relation to the tidal water are readily distinguishable; and, where such deposits (for example intertidal laminated silts, with salt or brackish water invertebrate fossils, and marshland peats) are interbedded within a terrace, past fluctuations of sea-level can be reconstructed. The approximately horizontal surface of each terrace is interpreted as the highest level reached by tides during an interglacial episode. Under favourable circumstances any contained mammalian remains can be related to the sea-level prevailing at the time of their burial, which may assist with establishing their age.

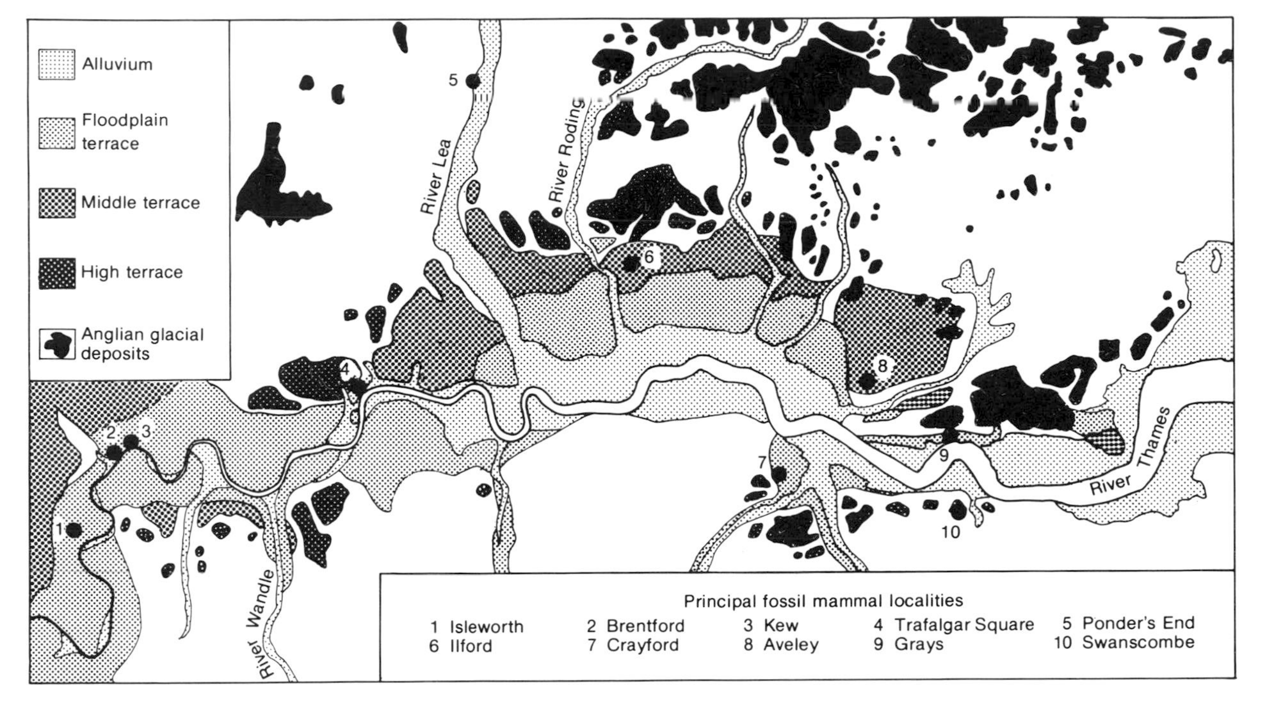

Fig. 10.10 *Simplified geological map of the terrace deposits of the River Thames in the London area, showing also some of the most important fossil mammal localities. The present-day channel is flanked on each side by alluvial deposits, of Holocene age, the surfaces of which are prevented from flooding at times of high water only by embankments. Further away from the river is a series of fragments of older terraces at various levels, representing phases of past high sea-level when the estuary was wider and extended further inland than it does today. Terrace deposits above estuary not differentiated; some high levels gravels omitted. Width of area 60 km. After British Geological Survey.*

Such studies are nevertheless unfortunately not as straightforward as this simple scheme would suggest. A special problem is the great number of sedimentary episodes (and the erosional stages between them) for which there is evidence in the Thames estuary; many more than can be accounted for in the orthodox British Quaternary chronology (Table 5). Zeuner, for example, writing in 1959, recorded twelve benches of different ages on the bedrock beneath the alluvial deposits of the London area. Often there is a complex pattern of channelling within the terrace framework, which may be difficult to interpret; and there is evidence of possibly still continuing subsidence in the seaward part of the estuary which makes necessary extreme caution when attempting to interpret altitudinal measurements on the terrace deposits of that area.

Various schemes have been established for naming the mainly interglacial terraces of the estuary of the Thames. For the purpose of the present description we will revive the terms 'High Terrace' and 'Middle Terrace' (names first used by M.A.C. Hinton at the beginning of the present century) for the higher terrace complexes, and retain the terms 'Floodplain Terrace' and 'Alluvium', of the Geological Survey maps, for the later ones. These can provide a simplified framework for our mammalian studies (fig. 10.10).

The High terrace of the Thames estuary

The High Terrace alluvial deposits reach an altitude of about 25 metres above present-day ordnance datum, believed to represent maximum high sea-level at that time, and they include a number of very important mammalian and archaeological sites.

During the excavation of a railway cutting near Hornchurch in Essex, in the 1890s, gravels of the High Terrace were exposed which could be seen to lie directly upon glacial deposits (the Chalky Boulder Clay) dating from the same ice advance that blocked the Vale of St Albans and caused the diversion of the Thames previously described. It follows that the High Terrace is later than the glacial event in question, usually referred to the Anglian. No fossiliferous deposits of pre-High Terrace age are known from the area of the present-day estuary of the Thames, suggesting that the time interval between these two series of deposits was not great.

By High Terrace times, an extensive change had occurred in the British mammalian fauna, which was now beginning to take on a much more modern appearance. Some of the Cromerian species, including the sabre-toothed cat and southern elephant had apparently disappeared, the latter species being replaced by the straight-tusked elephant, *Palaeoloxodon antiquus.* Some Cromerian rodents, including *Pitymys arvaloides* and *Microtus ratticepoides,* still survived, but *Pitymys gregaloides* had disappeared and *Mimomys* had evolved into *Arvicola.* The macaque reappears. Perhaps the most far-reaching of all the changes was the disappearance of the Etruscan rhinoceros and its replacement by Merck's rhinoceros, *Dicerorhinus kirchbergensis* and the narrow-nosed rhinoceros, *D. hemitoechus.*

How did this change of rhinoceros species occur? It could not have been by evolution in the British Isles, which was being extensively glaciated. The new forms, though probably descendants of the Etruscan rhinoceros, were undoubtedly arrivals of continental stock. Other common High Terrace mammals include lion, horse, pig, Clacton fallow deer, giant deer, red deer, roe deer, and giant ox. The hippopotamus, so common during Cromerian times, has not yet been recorded from this interglacial.

By now, man was well established as part of the British mammalian fauna. Especially important human activity sites of this age are Swanscombe and Clacton.

The High Terrace interglacial episode, as we have already seen, usually referred in part to the Hoxnian Interglacial as recognized by the palaeobotanists, was not a simple event but probably represents a considerable period of time. The Swanscombe site (which has a very rich mammalian fauna, including straight-tusked elephant, Merck's rhinoceros, narrow-nosed rhinoceros, horse, giant ox, red deer, Clacton fallow deer, giant deer, lion, cave bear, a small wolf, rabbit, vole, *Arvicola cantiana* and, from a nearby site, giant beaver and dolphin (fig. 10.7f)) is especially complex. The earliest deposits (Lower Gravel and overlying Lower Loam) contain mammalian remains and artefacts of the Clactonian industry. A horizon of magnificently preserved animal footprints (deer, giant ox, horse, possibly rhinoceros and elephant) on the surface of the Lower Loam apparently represents a time interval of some duration. The overlying Middle Gravel contains similar mammalian remains and artefacts of Acheulean type. Human skull fragments (Swanscombe man) were found in this deposit in 1938. Further deposits form the upper part of the sequence.

The Middle Terrace complex of the Thames estuary

The deposits of the Middle Terrace complex, which cover an area of tens of square kilometres in the London area, represent stages of lower sea-level than the High Terrace (probably never more than 10 metres above present level) and they apparently date from a later period of time. The rich fossil mammal localities of Ilford, Crayford and the Aveley elephant site, previously mentioned, are parts of the Middle Terrace complex.

The Ilford fauna includes straight-tusked elephant; a form of mammoth; Merck's rhinoceros, *Dicerorhinus kirchbergensis;* narrow-nosed rhinoceros, *Dicerorhinus hemitoechus;* giant ox, bison, red deer, giant deer, horse, lion and brown bear. Principal mammalian species of the Crayford deposits are the mammoth, woolly rhinoceros, musk ox (fig. 10.7c), giant ox, red deer, horse, a very large lion, a small wolf, dhole, *Cuon,* ground squirrel, *Spermophilus primigenius,* and two species of lemming.

The widespread occurrence of the warm climate mollusc, *Corbicula fluminalis* (present also in the High Terrace) throughout Middle Terrace deposits, including those at Ilford, Aveley and Crayford might suggest that the principal parts of this complex are of interglacial age; although the mammalian fauna of Crayford is predominantly composed of cold or steppe elements and it is difficult to explain the apparent association of such climatically contrasting species.

Palaeolithic man was apparently quite numerous by Middle Terrace times his artefacts (now of Levalloisian technique) occurring in special abundance in association with Pleistocene faunal remains at Crayford.

The age of the various Middle Terrace deposits has been a source of some controversy among Quaternary palaeontologists, some of whom (on palaeobotanical evidence) have argued for their inclusion within the Ipswichian Interglacial. From the evidence provided by the associated mammalian faunas, however, and also from the relationship of the various terrace deposits, it becomes increasingly difficult to place them so late in the classical chronology (Table 5). The alternative interpretation – that they represent hitherto undefined interglacial and steppe episodes between the Hoxnian and Ipswichian interglacials – would seem to fit all the available evidence very well. We will return to the problem of the mammalian fauna of the Ipswichian proper, shortly.

The Floodplain Terrace

Less extensive than the deposits of the Middle Terrace of the estuary is a series of fragments of a lower terrace (the Upper Floodplain Terrace of some writers), apparently associated with a sea-level only marginally higher than at the present day. This represents the interglacial episode already described at the beginning of this chapter (fig. 10.3.).

This terrace, which has a distinctive mammalian fauna, including hippopotamus (unknown from the High and Middle Terraces), can be recognized at many

Fig. 10.11 *Hippopotamus country in London. Air photograph of the River Thames where it passes Trafalgar Square. Indicated are the limits of the formerly marshy Holocene channel (Alluvium on the map; back to about 10 000 years old.) and the part of the Floodplain Terrace near Trafalgar Square (about 120 000 years old) where remains of hippopotamus and other interglacial mammals are commonly found during excavations at building sites. Fossiliferous localities, for example New Zealand House are marked with spots. (Photo: Aerofilms).*

separate localities extending from Brentford on the west of London, through the capital and far out along the Essex and Kent shores of the estuary. Other mammals include the straight-tusked elephant, narrow-nosed rhinoceros (*Dicerorhinus hemitoechus*), fallow and red deer, bison, giant ox, wild boar, lion, bear, spotted hyaena and beaver. The horse, so common in most Pleistocene faunas was apparently absent at least for a short period of time, and there is no evidence of the presence of man at this stage.

Most notable of all the Thames hippopotamus localities is the area around Trafalgar Square in central London (fig. 10.11) where bones and teeth discovered during building excavations (fig. 10.7e) have been found in association with plant remains (which include the southern European water chestnut *Trapa natans* and Montpellier maple, *Acer monspessulanum* previously mentioned) assigned on palaeobotanical evidence to the warmest part of the Ipswichian interglacial.

Of special interest at this site are the fossil insect remains, studied by Coope, among which are many southern species of beetles, confirming the supposed interglacial climate demonstrated by the plants and mammals. But the beetles are of special interest in another aspect. The total assemblage, which contains some very unusual species, is indistinguishable from that at the Ipswichian type site at Ipswich, and there seems little doubt, on the independent evidence of two different disciplines, that the deposits at these two sites are of the same age. There is good reason to believe that the hippopotamus fauna of Trafalgar Square would also have turned up at the Ipswich type site, had this been more productive of mammalian remains.

From the available evidence it seems probable that this hippopotamus event, which we now believe can be equated with the Ipswichian as defined by the palaeobotanists, was unique during the upper Pleistocene of Britain and that it provides a valuable stratigraphic datum in chronological studies of the Thames terraces and of the Quaternary as a whole. The same faunal assemblage is also known from many upper Pleistocene cave sites, where a uranium series date is available for it, and will be referred to again later.

The Alluvium

The Alluvium is the lowest and most recent of all the terraces of the Thames estuary and it provides a useful control with which comparison of the higher earlier terraces can be made. It is of Holocene age (i.e. not more than about 10 000 years old) and it rests in the channel cut by the Thames during the low sea-level of the Last Glaciation. The surface of the Alluvium is the present-day floodplain of the estuary and is prevented from flooding at times of exceptionally high tide only by retaining walls which have been constructed along the waters' edge to protect the buildings that have been constructed upon it.

Many excavations have been made in the Alluvium for commercial purposes, providing a detailed record of the rise of sea-level that followed the melting of the ice at the end of the Last Glaciation. This rise apparently did not take place uniformly. Excavations for Tilbury Docks a century ago revealed a series of alternating layers of estuarine silts and peat, showing that several minor rises and falls were superimposed within the principal rise of sea-level.

Abundant mammalian remains have been found in the Alluvium, although many of the Pleistocene species had by this time disappeared. Two skulls of the extinct giant ox found at East Ham in 1958 were dated from palaeobotanical evidence as 8000–5000 years old, showing that this species still survived in Mesolithic times. The beaver is another mammal, represented in the Alluvium, that has now disappeared from Britain. Man was apparently present in the area throughout most of the Holocene. A human skeleton was found in the bottom of the excavation for Tilbury Docks; and remains of domestic animals and archaeological débris are common.

THE BURIED CHANNELS

We have observed that, in the estuary of the Thames most of the terrace deposits represent high sea-levels and are therefore of interglacial age. During the cold stages the sea stood below its present level and the mammalian remains that became buried along the course of the river are now submerged and are inaccessible for collecting in the usual manner. A great number of mammoth teeth from the buried channels have nevertheless been accidently brought up from the bottom of the estuary in the course of dredging operations. Other mammalian species are less commonly represented. Not all these remains are of the same age, but they come from the now submerged deposits of successive stands of low sea-level.

THE COLD CLIMATE TERRACES OF THE UPPER COURSE OF THE THAMES AND OF THE RIVERS SHARING ITS ESTUARY

Although most of the terrace deposits of the estuary of the Thames are of interglacial date (the intervening cold stages being represented by periglacial deposits and deposits in the buried channel) the situation in the upper part of the river is very different. Above the tidal head of the present estuary the river channel of the Thames has not been drowned by Holocene rise of sea-level and there terrace deposits dating from the glacial stages (many of them highly fossiliferous) are very common. Some interglacial sites also occur in the upper part of the river.

At the present day, the Thames is tidal upstream as far as Teddington, where there is an abrupt appearance of terrace deposits with remains of reindeer (fig. 10.7a), woolly mammoth, woolly rhinoceros; even (represented by one specimen only of each) saiga antelope (fig. 10.7d) and polar bear.

Similar cold stage terraces are also widespread along the upper courses of some of the small rivers which share the estuary of the Thames. Most notable of these if the River Lea, with deposits of Last Glaciation and Holocene age. The Last Glaciation fauna of this river includes woolly mammoth, woolly rhinoceros, reindeer and collared and Norwegian lemming.

THE PERIGLACIAL DEPOSITS OF THE RIVER THAMES

Accompanying the stages of cold climate that caused the falls of sea-level described above were repeated episodes of intensive gelifluction (Coombe Rock formation) and some loess deposition. In the estuary these periglacial deposits commonly rest upon or are interbedded with interglacial terrace deposits and can sometimes be traced down into the buried channels. Although not as rich in mammalian remains as the terrace deposits they are important for the information which they provide about the cold stages of the later part of the Quaternary. Common mammalian species in the Coombe Rock of the estuary include woolly mammoth, woolly rhinoceros and horse. There is a similar relationship of periglacial and terrace deposits in the upper part of the river.

MARSWORTH

Although not strictly part of the Thames terrace narrative, this account of the deposits of that river would be incomplete without mention of the important mammal site of Marsworth, Buckinghamshire, 50 kilometres north west of London; an isolated deposit near the River Bulbourne. Commercial excavations in the early 1980s for chalk (a soft Cretaceous limestone) exposed an overburden of stratified Pleistocene deposits containing a unique sequence of mammalian faunas. The youngest of these, situated only just below ground level, contained remains of hippopotamus, narrow nosed rhinoceros (*Dicerorhinus hemitoechus*), elephant, giant deer, bison and water vole. This is a typical Ipswichian assemblage, such as we have encountered before in the Floodplain Terrace of the Thames at Brentford and Trafalgar Square, and is referred to that interglacial. Underlying this was a thick deposit of Coombe Rock, interpreted as evidence of periglacial and thus of very cold conditions; and underneath this yet another fossiliferous horizon with mammalian, insect, molluscan, and plant remains, indicating an earlier temperate episode. Mammals include a small wolf, brown bear, a lion of immense size, mammoth, horse, the extinct horse (*Equus hydruntinus*) and northern vole. Associated blocks of travertine give a uranium series date of 140–170 000 years for this earlier mammal horizon, interpreted as representing a previously unrecognized interglacial or at least interstadial episode of post-Hoxnian and pre-Ipswichian date. Its closest affinities in the Thames sequence lie in the later part of the Middle Terrace complex.

CHRONOLOGY OF THE MAMMALIAN FAUNAS OF THE RIVER THAMES

A chronology for the mammalian faunas of the River Thames, based on the evidence discussed above, is suggested in Table 6.

Deposits in caves

Even more varied than the mammalian faunas of the river terraces, which they supplement, are those from Britain's limestone caves. With processes of accumulation of the remains (already described in chapter 7) totally different from those that take place at open sites, it is not surprising that the proportions of the various mammals represented also differ. Most of our knowledge of British Quaternary carnivores comes from caves, which were used by bears and hyaenas and other mammals as places of retreat; vast accumulations of rodent bones, left by birds of prey, are often present in caves; and remains of Quaternary bats are almost unknown except from caves. The story told by the cave deposits is similar to that from the river terraces, but the emphasis is different.

In spite of the richness of the British cave faunas, the whole of the Quaternary is unfortunately not represented by them. The fossiliferous deposits are predominantly of upper Pleistocene age and they become progressively less common as they are followed back into time; a phenomenon explained by the gradual destruction of cave-containing limestone by denudation. Although not uncommon in some other regions of Europe, only a few rare instances of fossiliferous cave deposits of lower and middle Pleistocene age are known in Britain.

Most British caves (for example those of Yorkshire, Derbyshire, Somerset and Ireland) are in limestone of Carboniferous age, although there are also local occurrences in Cambro-Ordovician, Devonian, Permian, Jurassic and Cretaceous limestones. Only a few British mammaliferous cave sites can be mentioned here. The following are some of the most important.

Dove Holes, Derbyshire

The earliest fossiliferous cave deposit of Quaternary age in the British Isles, with a mammalian fauna including mastodon, sabre-toothed cat and horse. It is probably contemporary with one of the Crag deposits of Suffolk.

Westbury-sub-Mendip Fissure, Somerset

A cave, discovered in 1969 in a working quarry in the Mendip Hills, Somerset, with a rich mammalian fauna, including the Etruscan rhinoceros and Deninger's bear, suggesting an approximately Cromerian age for the deposits. Other interesting species include the sabre-toothed cat, *Homotherium*; the lynx; the European jaguar, *Felis gombaszoegensis;* an extinct dhole, *Xenocyon*, and the vole *Pliomys episcopalis*.

	Mammal localities (examples)	Climate and sea-level	Mammals
'Alluvium'	Tilbury, East Ham	Amelioration of climate; sea rises once more to present level	Beaver, wolf, giant ox and species still surviving in Britain
Buried channel deposits; terrace deposits above the estuary; and periglacial deposits	Buried channel of the Thames at Erith; Goodnestone mammoth site ('Coombe Rock'); terrace deposits above head of tidal estuary at Twickenham and in the valley of the River Lea	Extensive ice advance about 18 000 years ago not reaching Thames estuary; sea falls to very low level; Thames flows in channel deep below present estuary	Lemmings, spotted hyaena, cave lion, brown and polar bears, woolly mammoth, horse, woolly rhinoceros, reindeer, bison, musk ox, saiga antelope
'Floodplain Terrace' complex	Brentford, Trafalgar Square, Peckham and (on River Medway) Upnor	Climate warmer than at present; sea-level slightly higher	Hippopotamus, beaver, spotted hyaena, cave lion, brown bear, wolf, straight-tusked elephant, narrow-nosed rhinoceros, red and fallow deer, wild boar, bison, giant ox. Horse apparently absent
'Middle Terrace' complex; buried channel; and periglacial deposits	*Terrace deposits* 'Late Middle Terrace'; Example, Crayford 'Early Middle Terrace'; Examples Ilford, Aveley *Periglacial deposits* Examples, 'Main Coombe Rock', Ebbsfleet	Several glacial-interglacial fluctuations of climate and of sea-level up to maximum of about 6 metres above present	*Terrace deposits* Ground squirrel, cave lion, wolf, mammoth, woolly rhinoceros, horse, giant ox, red deer Cave lion, brown bear, early mammoth, straight-tusked elephant, Merck's rhinoceros, narrow-nosed rhinoceros, horse, red deer, giant deer, giant ox *Periglacial deposits* Mammoth, woolly rhinoceros, horse
'High Terrace' complex	Swanscombe (Middle Gravels) - - - - - - - - - - Swanscombe (Lower Gravel and Lower Loam) and Clacton	Sea-level rises to about 25 metres above present level during height of interglacial	Giant beaver, rabbit, dolphin, cave lion, bear, small wolf, straight-tusked elephant, Merck's rhinoceros, narrow-nosed rhinoceros, horse, red deer, fallow deer, giant deer and giant ox
Chalky Boulder Clay - - - - - - - - - -		Eastern Thames diverted southwards to approximately its present position by ice sheet advancing from the north - - - - - - - - - - East of Marlow Thames flowed along a course further north than at the present day, reaching the sea somewhere in Essex	

Table 6. *Suggested simplified chronology for the Quaternary mammalian faunas of the River Thames.*

Industries and techniques	Sequence of the Geological Society of London		Years ago
	Flandrian	HOLOCENE	10 000
	Devensian	UPPER PLEISTOCENE	120 000
	Ipswichian	UPPER PLEISTOCENE	130 000
Levallois	?	MIDDLE PLEISTOCENE	
Acheulean	?	MIDDLE PLEISTOCENE	
	Hoxnian	MIDDLE PLEISTOCENE	? 300–400 000
Clactonian	?	MIDDLE PLEISTOCENE	
		MIDDLE PLEISTOCENE	

The occurrence, however, of abundant remains of the vole, *Arvicola cantiana* (believed to be a direct descendant of *Mimomys savini*, so common in the upper Freshwater at West Runton), suggests that this locality is not exactly contemporaneous with the Cromerian type site but may in part be slightly later, possibly representing a warm episode, not recognized in the classical chronology of the Geological Society, somewhere between the Cromerian and Hoxnian interglacials, as defined by the palaeobotanists.

Joint Mitnor Cave, Buckfastleigh, Devon

A pitfall cave with a rich interglacial mammalian fauna, including hippopotamus, straight-tusked elephant, narrow-nosed rhinoceros (*Dicerorhinus hemitoechus*), wild boar, bison, giant deer, red deer, fallow deer, hare, water vole, field vole, wolf, red fox, wild cat, cave lion, spotted hyaena, badger and brown bear. This assemblage apparently represents the same interglacial episode as the Floodplain terrace of Trafalgar Square and the upper mammal horizon at Marsworth and is referred to the Last, Ipswichian, Interglacial, as defined by the palaeobotanists. It is the richest mammaliferous cave deposit of that age yet known in Britain. During this episode, the hippopotamus, after having apparently been absent from Britain since Cromerian times, spread throughout England and Wales in very great numbers as far north as Durham. The only rhinoceros represented in Joint Mitnor Cave is *D. hemitoechus*. *D. kirchbergensis*, so abundant in the High Terrace of the Thames, had apparently disappeared by this time. This interglacial was a very successful time for the spotted hyaena and, as at Trafalgar Square, there is a remarkable absence of remains of horse.

Victoria Cave, Settle, Yorkshire

Another important cave site, first excavated a century ago, with a lower cave earth with hippopotamus, straight-tusked elephant, narrow-nosed rhinoceros, hyaena and other mammals; overlain by a long series of later deposits. These show that the area was subsequently glaciated; and there was later occupation by upper Palaeolithic, Mesolithic and Romano-British man. There can be few British caves which demonstrate in such a striking manner the climatic fluctuations of the Pleistocene. The cave is situated at an altitude of 400 metres on craggy moorland, and it seems incredible, at the present day, to think of hippos walking across this wild part of the Yorkshire moors.

The establishment in 1980 of a uranium series date of 120 000 years for travertine associated with the interglacial fauna of the cave substantiated the widely prevailing view that the hippopotamus fauna of this site and presumably also of Joint Mitnor Cave and Trafalgar Square are of Last Interglacial age; and it (not unexpectedly) places this hippopotamus episode firmly in stage 5e of the deep sea oxygen isotope chronology. This is the first major step towards tying Britain's mammalian faunas into the deep sea sequence; and it is also of great importance in other lines of Quaternary studies. This Last Interglacial hippopotamus fauna is widely represented in other caves in England and Wales, though not in Scotland or Ireland. Eastern Torrs Quarry Cave, Devon (fig. 4.1, p. 36) is another example.

Stump Cross Cave, Yorkshire

Shortly after the dating of the Victoria Cave hippopotamus fauna, described above, a further important uranium series age determination was obtained from another Yorkshire cave. A mammalian fauna with wolverine (fig. 7.5, p. 77) and reindeer was shown to be slightly more than 83 000 years old. At the present day these animals are of northern distribution, and it is inferred, from the evidence of the two caves, that between about 120 000 and 83 000 years ago, there was a considerable deterioration of climate in Yorkshire. The Stump Cross cold episode nevertheless lies firmly within oxygen isotope stage 5, which continued until about 70 000 years ago. It may represent cool substage 5b).

Devensian cave sites

Dating from the most recent part of the Pleistocene, Last Glaciation cave deposits greatly outnumber all those of earlier age put together. Some important mammaliferous examples include Brixham Cave and the Torbryan Caves, Devon; Wookey Hole Hyaena Den, Uphill Cave and Bleadon Cave, Somerset; Ightham Fissures, Kent, Great Doward Cave, Herefordshire; Paviland Cave, Gower; Elderbush Cave, Staffordshire, Fox Hole Cave, Pin Hole Cave, Derbyshire; and the Inchnadamph Caves, Sutherland. The Irish cave faunas are also of Last Glaciation, or later, age. They will be referred to in the Irish section of this chapter.

As previously described, far reaching changes of the British mammalian fauna accompanied the oncoming of the Devensian, and these changes are magnificently recorded in the cave deposits. Several important warm-climate herbivores, notably the hippopotamus, the straight-tusked elephant, the narrow-nosed rhinoceros and the fallow deer had by this time disappeared. Most of the carnivores, including the lion and the spotted hyaena, continued to thrive. Species returning once more to the British Isles after being absent during the interglacial (except perhaps in Scotland) include the woolly mammoth, the woolly rhinoceros, and reindeer. The pika, *Ochotona* (a relative of the hares and rabbits) is recorded from a number of cave deposits of this age, for example the Brixham Cave. Rodent faunas from the caves vary temporally within the Devensian, the collared and Norwegian lemmings, *Dicrostonyx* and *Lemmus*, being characteristic of the coldest phases. A ground squirrel, *Spermophilus superciliosus*, found in the Ightham Fissures, is of a species different from that found in deposits of the earlier Middle Terrace of the Thames of Crayford.

Tornewton Cave, Devon

Most of the cave sites mentioned above have only relatively short sequences of deposits; mainly because, once sedimentation commenced, they were soon filled to the roof. Long stratified sequences of deposits in caves are therefore rather unusual. Two exceptions are of special chronological importance and will be considered in greater detail here. They are Tornewton Cave and Kent's Cavern, both near Torquay in Devon. In its upper levels Tornewton Cave was found to have a typical Devensian fauna with woolly mammoth, woolly rhinoceros, reindeer, lemmings and other cold species. A few flint flakes of upper Palaeolithic technique show that man was present during at least part of this time. These upper levels were underlain by a deposit containing great quantities of remains of spotted hyaenas, both adult and juvenile, together with many droppings. These animals had apparently used the cave as a den. Associated remains of hippopotamus, narrow-nosed rhinoceros and fallow deer point to this horizon dating from the same interglacial as that represented in Victoria and Joint Mitnor Caves and at Trafalgar Square. Beneath this deposit was yet another fossiliferous layer with many remains of brown bears, which animals had apparently occupied the cave before the hyaenas. The associated fauna from this level includes two notable cold climate species, the reindeer and wolverine; some steppe species, including common hamster, *Cricetus cricetus* and the steppe lemming, *Lagurus lagurus* (the only record to date of this last mentioned genus in Britain) and also various carnivores and horse. From its stratigraphical position, immediately underlying the interglacial horizon, there can be no doubt that this fauna, though superficially like that of the Devensian, is of pre-Ipswichian age, probably dating in part from oxygen isotope stage 6 of the deep sea record.

Kent's Cavern, Devon

Excavations in this cave, which go back to the 1820s, have established long series of stratified deposits, which vary in detail from one chamber to another. The principal deposits are as follows:

5. Black Mould
4. Granular Stalagmite
3. Cave Earth
2. Crystalline Stalagmite
1. Breccia

The precise chronological range of these deposits is uncertain, though there is evidence that the Breccia, which has a lower Palaeolithic industry, may be relatively early. The occurrence of a cave bear, the sabre-toothed cat and the rodent *Pitymys gregaloides* suggest a horizon somewhere in the cave that may be of about the same age as the Westbury Fissure. The Cave Earth has a typical Devensian fauna including woolly mammoth, woolly rhinoceros, horse, red deer, reindeer, giant deer, spotted hyaena and cave lion, associated with upper Palaeolithic artefacts. The Granular Stalagmite and Black Mould have Mesolithic and later industries and are post-Pleistocene. It follows that the Crystalline Stalagmite probably represents a long time interval when flowstone formation was the only form of sedimentation in the cave, possibly because the cave was sealed so that no animal remains or sediment could reach it from outside. The Ipswichian hippopotamus fauna, which would be expected to appear between the Cave Earth and Breccia is entirely lacking from the cave. The lower Palaeolithic artefacts from the Breccia are probably the earliest human artefacts yet found in the British Isles.

Pontnewydd Cave, North Wales

The most recent chronologically important excavation to have been undertaken in a British Cave is that of the National Museum of Wales in Pontnewydd Cave, where a series of deposits has been dated by the uranium series method to the period of about 100 000–220 000 years ago (i.e. approximately stages 5–7 of the oxygen isotope record). Fossils include human teeth (the earliest hominid specimens known from the British Isles, except for Swanscombe man, previously mentioned) and the remains of various mammals including the rhinoceros *Dicerorhinus kirchbergensis* and a leopard-sized cat (animals that apparently did not persist into the upper Pleistocene) and also other species such as the reindeer, which still survive in northern regions today. In terms of Pleistocene chronology this site, like Victoria Cave, is unusual in that it has been possible to relate its deposits to the deep sea oxygen sequence, by means of absolute dating, without any palaeobotanical studies being involved. For the time being the relationship of Pontnewydd Cave to the classical glacial–interglacial sequence has not been directly demonstrated.

Mammalian fossils associated with the sea

As in other parts of the world, the British Isles experienced repeated fluctuations of sea-level during the Quaternary. Studies of mammalian faunas from deposits that can be related to these fluctuations provide a further important source of chronological information.

Since the northern part of the British Isles was repeatedly heavily glaciated, whereas the southern part of England lay outside the glaciated area, the effects of eustatic and isostatic changes vary regionally. Scotland is an area of recent isostatic uplift; the south of England and the North Sea are areas of eustatic submergence, with only the higher interglacial shorelines above present-day sea-level (chapter 5).

There are three main groups of sites with mammalian remains which we must consider here: the coastal caves of Gower in South Wales; the Scottish raised beach localities and the submerged floor of the North Sea.

The coastal caves of Gower

The Gower coast has long been famous for the many mammaliferous caves which are situated there; Bacon Hole, Minchin Hole (fig. 10.12) and Ravenscliff Cave being the most important. As early as the 1850s Hugh Falconer remarked on the relationship between the mammal-bearing and raised beach deposits of the various caves.

Recent excavations have now provided a very detailed record of the stratigraphy of these sites. A fossil beach with many marine shells (the 'Patella Beach', best developed in Minchin Hole), representing a sea-level marginally higher than at the present day, is immediately overlain by deposits with an interglacial mammalian fauna with narrow-nosed rhinoceros, fallow deer, lion and other species. Hippopotamus is an additional member of the fauna, from Ravenscliff Cave only. The sea-level in question is interpreted as an interglacial eustatic high, with the mammalian fauna only marginally younger, within the same warm episode. This assemblage is closely reminiscent of that from Joint Mitnor and Victoria Caves and Trafalgar Square; and the sea-level marginally higher than at the present day is also in agreement with the evidence from this last mentioned site in the Thames estuary. The principal interglacial horizons of these caves can probably also be referred to oxygen isotope stage 5e, a conclusion provisionally confirmed by uranium series dating of flowstone. Other bone deposits in the caves provide information about the mammalian fauna of a cold stage preceding the interglacial horizon and of the Devensian cold stage which followed it.

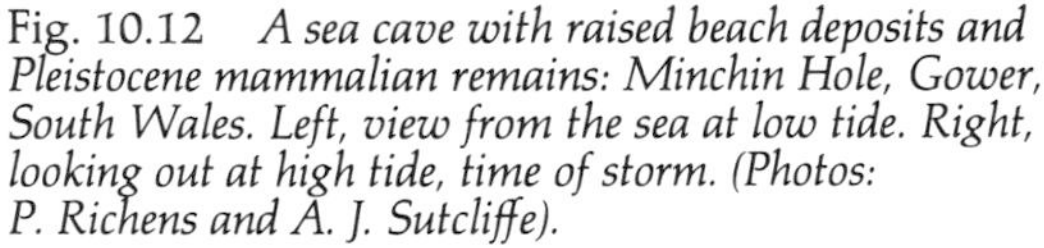

Fig. 10.12 *A sea cave with raised beach deposits and Pleistocene mammalian remains: Minchin Hole, Gower, South Wales. Left, view from the sea at low tide. Right, looking out at high tide, time of storm. (Photos: P. Richens and A. J. Sutcliffe).*

The Scottish isostatically raised shore lines

With the melting of the ice at the end of the Devensian glaciation Scotland was extensively affected by isostatic uplift (fig. 10.13). Elevation was greatest at Rannoch Moor, Western Highlands, and the old shore lines decrease in altitude away from this centre. One of them (of Lateglacial age, about 13 500 years old) falls in southeast Scotland as much as 1 metre every 2–3 kilometres from an altitude of over 40 metres to less than 20 metres east of Edinburgh (fig. 10.13). Four beaches of Holocene age are also present at lower altitudes. Further south, in England, where they have been submerged by eustatic rise of the sea, all these shore lines disappear below present-day sea level. Remains of seals and whales have occasionally been found associated with these beach deposits, especially along the shores of the Firth of Forth. Most of these are of Holocene age, though remains of sperm whale and seal found at an altitude of 30 metres at Camelon, near Grangemouth, may be Lateglacial.

The submerged floor of the North Sea

The lowering of sea-level that accompanied the glacial episodes of the Quaternary allowed plants and animals to spread out onto areas of land that are currently sea floor. The remains of these are sometimes recovered as a result of fishing and diving operations. The low sea-level that accompanied the Devensian and the early part of the Holocene was the most recent of these events and nearly all such finds from around the British Isles date from this period of time, when the southern part of the North Sea was dry; and a broad tract of land, with a rich flora and mammalian fauna, joined eastern England to the continent of Europe, as far as Denmark.

Mammals whose remains have been trawled from the North Sea include woolly mammoth (fig. 10.14), woolly rhinoceros, horse, red deer, giant ox, bison, reindeer and giant deer. Bear, spotted hyaena, wolf and beaver are represented by single specimens. The fishing area now known as the Dogger Bank (believed

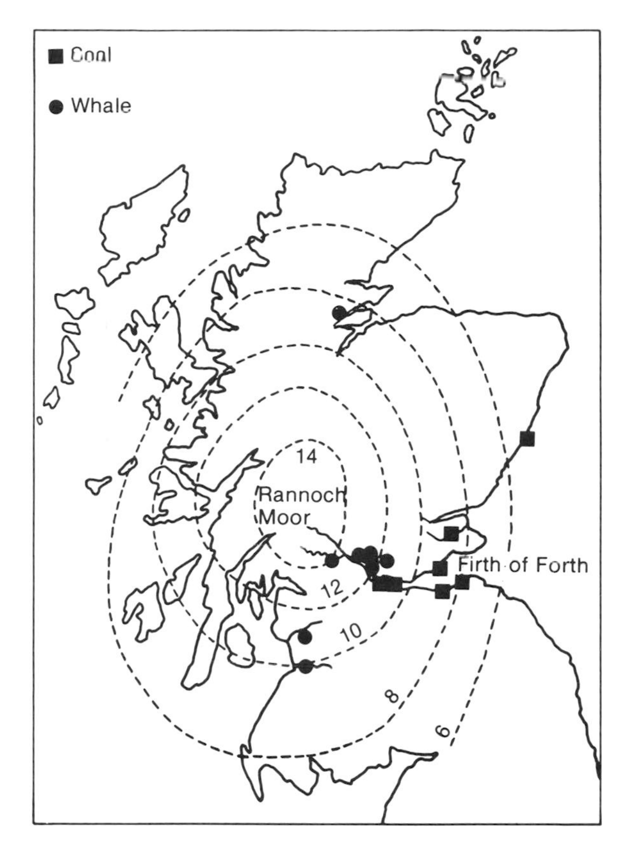

Fig. 10.13 *Map of Scotland, showing the altitude (2-metre isobase interval) above present sea-level of a Holocene shoreline, formed about 6500 years ago. The shoreline has been up-domed due to glacial isostatic uplift centred over Rannoch Moor, where the last Scottish ice sheet was thickest. (After Sissons, 1983). Also shown are localities where fossil remains of whales and seals have been found associated with raised shoreline and estuarine deposits of various ages. (After Delair, 1969). Most are of Holocene age, though those from Camelon are apparently late Pleistocene.*

Fig. 10.14 *Lower jaw of a woolly mammoth, on which an oyster shell has grown, trawled during the nineteenth century from the bottom of the North Sea. (Photo: BM(NH)).*

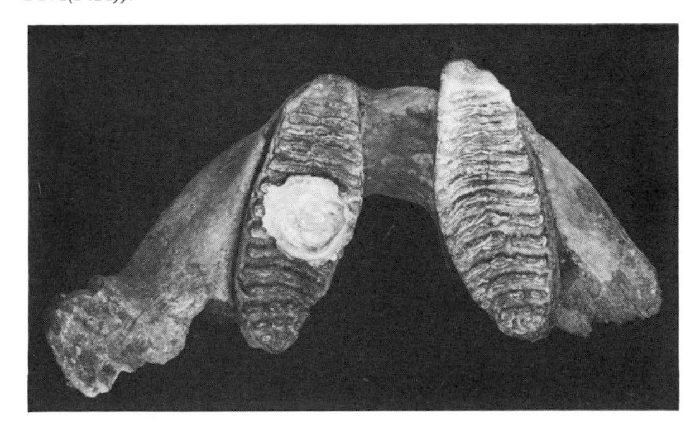

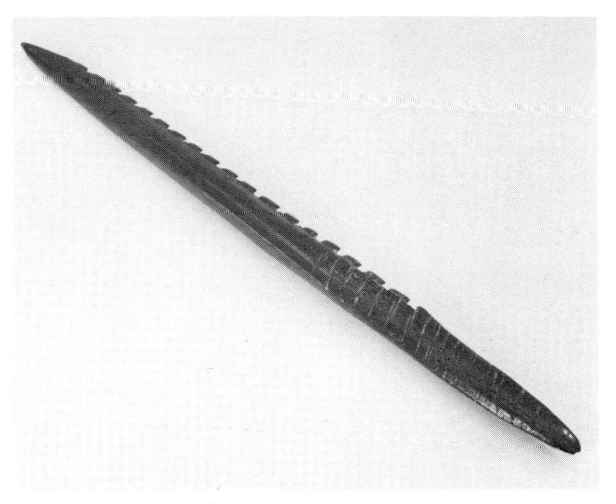

Fig. 10.15 *An antler harpoon found inside a block of peat recovered by the sailing trawler* Colinda *from a depth of 40 metres between the Leman and Ower Banks of the Dogger Bank, in the North Sea, 40 kilometres from the Norfolk Coast, in 1931, shows that man was present there in Mesolithic times, about 9000 years ago. Similar harpoons have been found at sites of equivalent age in England and in Denmark. (Photo: Norfolk Museums Service (Norwich Castle Museum)).*

to be a Devensian moraine), part of which still has a depth of less than 20 metres below sea-level, has been an especially fruitful source of such finds.

The recovery by trawlers of blocks of Holocene peat, one of them containing an approximately 9000-year-old Mesolithic antler harpoon head (fig. 10.15) shows that there was a time lag after the melting of the Devensian ice, when the exposed land was colonized by man, before the sea returned once more to its present level. By this time tundra had been replaced by oak forest. Soon afterwards the Dogger Bank became an island and finally disappeared beneath the sea.

Deposits of Holocene age

The Cambridgeshire Fens and Star Carr, Yorkshire

Since the Holocene is the most recent part of the Quaternary, information about its mammalian fauna is especially abundant. By this time many of the large mammals characteristic of the Devensian had disappeared from the British Isles. Notable disappearances were the woolly mammoth, the woolly rhinoceros, the bison, the cave lion and the spotted hyaena. Both lemmings apparently disappeared towards the end of the Devensian. The reindeer and lynx may have survived into the Holocene, but probably not for very long.

Most extensive of Britain's Holocene deposits are the Cambridgeshire Fens, low lying alluvial and peaty areas near the North Sea bay known as the Wash. Widespread drainage operations for agriculture commonly lead to the discovery of mammalian remains, for example of giant ox, brown bear and beaver.

Excavations at the Mesolithic occupation site of Star Carr, Yorkshire, show that 9500 years ago Yorkshire supported the European elk, *Alces alces* (which probably disappeared soon afterwards) red deer, roe deer, giant ox, (which survived in England until the Neolithic, on the continent of Europe until the seventeenth century), the wild boar, the wolf, marten and beaver. The beaver is believed to have survived in England until the twelfth century and there are extinction dates for the wolf of about AD 1500 in England, 1740 in Scotland and 1770 in Ireland. The brown bear, *Ursus arctos,* may have survived until the tenth century. There is a carbon date of 2700 years for a skeleton from a cave in northwestern Scotland.

Today the extensively cultivated British Isles support only a limited mammalian fauna. The wild cat survives is the mountainous areas of Scotland, the polecat in Wales and the pine marten in Scotland, Wales and Ireland. The hedgehog, red fox, stoat, weasel, badger, red deer and various bats are widespread. The otter and red squirrel, though still surviving, have declined greatly in numbers during the present century and so have at least two species of bats.

There are also 14 introduced species of mammals, some of which have caused considerable commercial damage. These include the rabbit (probably imported by man for meat in the twelfth century); the black and brown rat; the house mouse; the American grey squirrel, the mink and recently the South American coypu, which has become well established in East Anglia. Three species of deer (sika, muntjac and Chinese water) have become wild after escape from captive herds in parks in England; and most remarkable of all, there exist several groups of red-necked wallabies (a marsupial from Tasmania and eastern Australia) which are holding their own in a wild condition. The best-documented of these, in Derbyshire, became wild about 1940 and has been spreading slowly, with several setbacks during bad winters, ever since that time.

Ireland

Lastly, in this chapter on the British Isles, we must move across the water to Ireland, which has its own remarkable story of Quaternary mammals to tell us. Being isolated from the British mainland throughout most of the Pleistocene, this island was apparently generally inaccessible to mammals and it was only towards the end of this epoch that a few species somehow managed to cross. Thus, not only is the great sequence of glacial–interglacial faunas, so well represented on the mainland, lacking from the Irish evidence, but even those mammals which did cross were few in species. The fauna nevertheless takes on unique interest because of its island situation; and it is with this aspect that we will mainly concern ourselves here. It is also of interest for the period of time that it represents – the later part of the Last Glaciation. We will take the Irish evidence as our principal example of this part of the Pleistocene

The Irish giant deer

Before we consider the Irish fauna as a whole, mention must be made of the giant deer or so-called Irish elk, *Megaceros giganteus* (fig. 10.16). Because its well-preserved remains are more numerous in Ireland than those of any other mammal, and because of the gigantic size of its antlers, this particular deer has become famous throughout the world as an Irish species. Indeed so Irish has it become that (even though the usually fragmentary remains of this genus have been found in deposits of much earlier age in many other parts of the northern hemisphere, as far east as Japan and in France, where it was portrayed by the Palaeolithic cave artists; and even though it is known that the Irish variety did not arrive in Ireland until almost the end of Pleistocene), it would be unthinkable in popular imagination to accept that it arose anywhere other than in Ireland!

Remains of giant deer greatly outnumber all other mammalian remains found in Ireland, but a few other mammals also managed to spread there from the British mainland. We will return to a discussion of the other species shortly. Nearly all the Irish mammalian remains come either from beneath peat bogs or from caves.

Peat bogs

Peat bogs are a common feature throughout Ireland (fig. 10.17). One of the most famous is Ballybetagh Bog near Dublin, which is situated at the bottom of a steep sided valley and may have been a crossing place for migrating animals. In the years 1876 and 1877 alone, three complete skeletons and twenty-six skulls of giant deer are reported to have been excavated there.

Fig. 10.17 *A drained bog near Balbriggan, Co. Dublin. Drainage ditches, such as those shown here, are commonly the finding place of giant deer remains. Professor G. F. Mitchell, leading Irish Quaternary geologist, is seen examining the deposits. (Photo: A. J. Sutcliffe).*

Fig. 10.16 *Mounted skeletons of a male and female giant deer from Ireland, in the Fossil Mammal Gallery of the British Museum (Natural History). The giant deer was an animal of great size, the antlers on the male frequently attaining a span of over three metres. The female, as in all deer except the reindeer, had no antlers.*

	Mainland stage names	Sea-level	Mainland mammalian fauna available for colonizing Ireland	Irish stage names	Environment in Ireland	Irish mammalian fauna	Years ago Present day
HOLOCENE	FLANDRIAN	Low, rising to present level		LITTLETONIAN	Holocene amelioration of climate. Growth of peat bogs	Extinction of wolf in Ireland about AD 1770. Mesolithic Man (Toome Bay and early Larnian cultures) arrives about 8000 years ago. Red deer, wild boar, horse and other mammals arrive in Ireland	10 000
UPPER PLEISTOCENE	DEVENSIAN glacial stage	At least 100 m lower than today	Including woolly mammoth, woolly rhinoceros, giant deer, reindeer, bison, brown bear, spotted hyaena, lion, wolf, lemmings and other species	MIDLANDIAN glacial stage	Deterioration of climate (Nahanagan stadial). High valleys of Ireland ice-filled		*c.* 10 500
					Ammelioration of climate (Woodgrange interstadial)	Probable second wave of mammalian immigration. Giant deer very abundant; disappears at the end of this episode. Reindeer also present in Ireland	14 000
					Ice disappears from Ireland Ice spreads over northern Ireland; ice caps in the Wicklow Mountains and SW Ireland. Greatest extent 18 000 years ago		25 000
					Amelioration of climate	First known arrival of mammals in Ireland; mammoth, brown bear, spotted hyaena, wolf, giant deer, reindeer and lemmings (Castlepook Cave *c.* 33 000 years ago)	45 000
					Climate mainly cool. No mammalian remains known from Ireland		120 000
	IPSWICHIAN interglacial	Marginally higher than today	Including hippopotamus and other warm-climate species	LAST INTERGLACIAL	Ireland isolated from the British mainland. Rich interglacial mammalian fauna apparently failed to cross, in spite of conditions in Ireland being suitable for them		130 000
	'WOLSTONIAN' glacial stage	Very low		MUNSTERIAN glacial stage	Ireland almost completely glaciated. Sea falls to a low level but no mammals appear to have been able to cross from England		

Table 7. *History of mammalian fauna of Ireland, as far is known from the fossil evidence, compared with sequence (very much simplified) of the British mainland.*

Similar finds have come from many other Irish bogs. From the earliest times those finding giant deer remains in such deposits have pointed out that they come not from the bog peat, which lies near the surface of the ground, but from an underlying marl. The sequence has been observed at very many sites, which have also been studied by palaeobotanists. Always the age of the skeletal remains beneath the bogs is found to be same – from a minor warm episode known as the Lateglacial interstadial, which occurred between about 11 800 and 11 000 years ago. Another mammal whose remains are not uncommon beneath the bogs is the reindeer. There is a record of brown bear from Derryheel Bog, Ottaly.

Cave deposits

A greater variety of Quaternary mammals is represented by the finds from Ireland's limestone caves. For a long time it was uncertain whether the more diverse mammalian fauna represented in the cave deposits was of the same age as the finds from beneath the bogs, representing animals that reached Ireland during the same wave of immigration, or whether more than one period of immigration was confused. Mitchell, writing in 1969, pointed out that the Irish mammalian fauna included species that are characteristic of more than one environment, making it unlikely that all the remains were contemporaneous.

It was not until 1974 that the first absolute age became available for any of the cave remains. A head of a femur of a mammoth from Castlepook Cave, County Cork, provided a carbon^{14} date of 33 050 $\pm$ 1200 years old, thus confirming the supposition that the Irish mammalian fauna dates from at least two periods of time. Associated remains from the same cave represent brown bear, spotted hyaena, wolf, red and arctic foxes, stoat, giant deer, reindeer, collared and Norwegian lemmings, and possibly the wood mouse.

Irish mammalian chronology

With dates now available for at least two mammalian episodes in Ireland, it is possible to look at the significance of these faunas in the broader context of the climatic and mammalian chronology of the British Isles as a whole. This relationship, where the stage names that have been established for the Irish sequence are shown alongside the mainland stage names, is summarized in Table 7.

Seen within a framework of changing climate and sea-level the story of the Irish mammalian fauna outlined above (with its paucity of species and with evidence of immigration and extinction on more than one occasion) provides a very feasible extension of the mainland chronology. It is nevertheless astonishing that no mammalian remains earlier than about 40 000 years old are yet known from Ireland. Although this could be explained by lack of fossiliferous deposits, inability of the mammalian fauna to cross the barrier of the Irish sea is another possibility. If Ireland was indeed previously uninhabited by mammals, then we are faced with the incredible prospect of a vast tract of land with ample rainfall and a luxuriant vegetation flourishing during the Last Interglacial (the Ipswichian of the British mainland, 120 000 years ago) yet without any mammalian fauna to consume it; whilst a short distance away great herds of hippopotamus and other mammals competed for food. Perhaps future discoveries will fill this apparent gap in the geological record.

In Ireland, as on the British mainland, the Last Interglacial was followed by a general cooling of the climate, punctuated by a series of genial and less genial phases. The earliest known mammals (as represented by the Castlepook Cave fauna, listed above, apparently managed to cross from the British mainland during a minor temperate stage, about 45–30 000 years ago. Castlepook Cave is of special interest in being the only locality in Ireland where remains of spotted hyaena have been found.

About 25 000 years ago ice again spread extensively over the British Isles, including much of Ireland, and was at its greatest extent about 18 000 years ago. A broad belt of southern Ireland remained free of ice, though climatic conditions must have been severe there. It is not known what effect this glacial episode had on the Irish mammalian fauna, but some of the species listed above are unknown from later deposits and may have disappeared, as the result of it.

By about 11 800 years ago, before which time the last of this ice had disappeared from Ireland and conditions had become more genial, though not fully interglacial, it appears (from a study of fossil plant remains) that Ireland was an oceanic sector of the sub-arctic Irish region of northwest Europe, with copses of brown birch, *Betula pubescens* and stretches of open country covered by heaths or by grasses and herbs. Other plants included the juniper, golden willow and crowberry. It was at this time that many of marl deposits beneath the peat bogs, which have been so productive of remains of giant deer, and less commonly of reindeer, accumulated in open water pools lying in depressions left by the melting ice. It is not clear whether these animals were descendants of ancestors which had come to Ireland 20 000 years previously and had survived the intervening glaciation or whether they represented a new wave of immigration. Remains of giant deer found in deposits of similar age

	Years ago	Mammalian episodes	Other important mammalian localities	Typical mammalian species
HOLOCENE	10 000		Fens, Star Carr, Thatcham	European elk, giant ox, red deer, wild boar, wolf beaver
UPPER PLEISTOCENE		Lea Valley mammoth fauna	Brixham Cave, Fisherton Brickearth, Ightham Fissures, Paviland Cave, Pin Hole Cave and Wookey Hole Hyaena Den	Lemmings, spotted hyaena, cave lion, brown bear, woolly mammoth, horse, woolly rhinoceros, red deer, reindeer, bison, musk ox
	c.43 0000	Isleworth reindeer/bison fauna		
	c. 85 000	Stump Cross wolverine fauna		
	120 000			
	130 000	Victoria Cave hippopotamus fauna	Trafalgar Square, Joint Mitnor Cave, Ravenscliff Cave	Hippopotamus, straight-tusked elephant, spotted hyaena, narrow-nosed rhinoceros, fallow deer
MIDDLE PLEISTOCENE	170 000	Marsworth Lower Channel fauna		Vole (*Arvicola*), lemmings, ground squirrel, cave lion, brown bear, mammoth, straight-tusked elephant, horse, Merck's, narrow-nosed and woolly rhinoceros, red deer, giant ox, bison
		Ilford and Aveley (Middle Terrace of the Thames) mammoth-horse faunas	Crayford, Hutton Cave, Pontnewydd Cave, Brundon, Stutton	
			Hoxne	Lemming, horse, red deer
	? 3–400 000	Swanscombe (High Terrace of the Thames) fauna	Clacton, Grays	Straight-tusked elephant, Merck's and narrow-nosed rhinoceros, horse, red and fallow deer
		Westbury-sub-Mendip, stages 2–3, fauna	Ostend	Desman, vole (*Arvicola*), lemming, Deninger's bear, wolverine, sabre-toothed cat, jaguar, horse, Etruscan rhinoceros, red deer, bison
	? 6–7000 000	West Runton Freshwater Bed fauna	Sugworth	Desman, vole (*Mimomys*), giant beaver, Deninger's bear, Etruscan rhinoceros
LOWER PLEISTOCENE	c. 1 000 000	East Runton Fauna		Four-tined deer, *Euctenoceros*

Cave sequences shown by vertical arrows in column 4: Bacon Hole; Tornewton Cave ('Reindeer Stratum', 'Hyaena Stratum', 'Glutton Stratum'); Kent's Cavern ('Granular Stalagmite', 'Cave Earth', 'Crystalline Stalagmite', 'Breccia').

Table 8. *Simplified chronological table of the Quaternary mammalian faunas of the British Isles. Space prevents inclusion of comprehensive faunal lists in column 5 (for which refer to the text).*

Some significant faunal changes	Climate	Scotland	Ireland	Recognized pollen stages (England)	Deep sea oxygen isotope stages (schematic)
	Interglacial	Creag nam Uamh bear, 2700 years		(Flandrian)	1
Extinction of mammoth Extinction of woolly rhinoceros and spotted hyaena	Major glacial episode	Corstorphine lemming	Giant deer fauna 11–10 000 years		2
Decrease in abundance of reindeer and bison; increase of woolly mammoth and woolly rhinoceros	Interstadial		Castlepook Cave fauna *c.* 33 000 years		3
					4
Appearance of wolverine	very cold				5 a b c d
Disappearance of hippo fauna					
Arrival of hippopotamus fauna	Interglacial			Ipswichian	e
	Glacial episode				6
Appearance of steppe species	Imperfectly understood glacial-inter-glacial sequence with steppe episode in later part				
Presence of lemming					
	Interglacial			Hoxnian	?11
Etruscan rhinoceros replaced by Merck's and narrow-nosed rhinoceros	? Including part/all of Anglian glacial				
	Including an interglacial episode				
Vole, *Mimomys*, evolves into *Arvicola*	?				
	Interglacial			Cromerian	
Major faunal break. Disappearance of four-tined deer *Euctenoceros* and other species	Including a cold episode			Includes Beestonian	
	Interglacial			Pastonian	

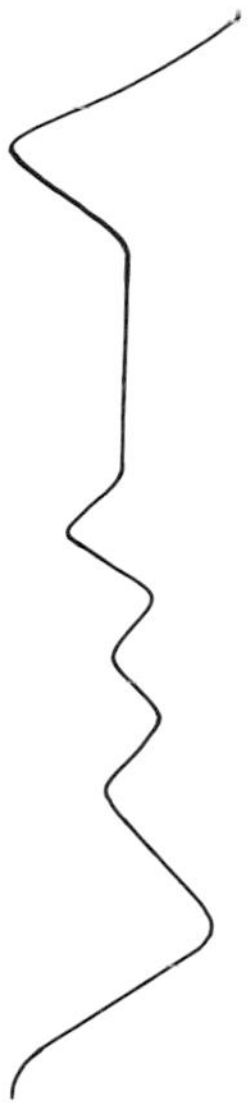

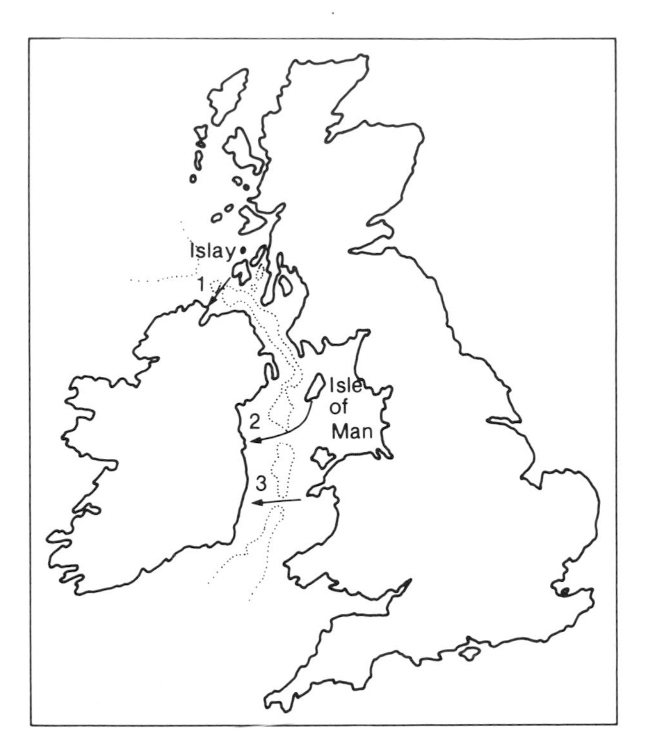

Fig. 10.18 *Possible mammalian migration routes from the British mainland to Ireland. The present-day 50-fathom (91-metre) submarine contour is shown as a dotted line. A fall of sea-level of 55 metres would link Islay, Scotland to Northern Ireland; 84 metres would link North Wales to Ireland at Arklow; and 90 metres would link England and the Isle of Man to Ireland near Dublin. During the last glacial advance the sea fell by over 100 metres so that land bridges at all of these places are feasible. Although the route via Islay has least depth of water over it at the present day, this area has been subjected to extensive Holocene isostatic uplift of the land (Fig. 10.13), following the melting of the Pleistocene ice sheets, and may not have been exposed during Devensian times. The occurrence of giant deer remains of Lateglacial interstadial age (11–12 000 years old), on the Isle of Man, may represent an intermediate point along one actual migration route.*

on the Isle of Man, between England and Ireland, suggest that, the sea still being low, an immigration route via the Isle of Man may have been open at this time (fig. 10.18). Whatever the origin of these animals, on the rich soils left by the retreating glaciers and in the absence of competition from other herbivores, they flourished in vast numbers.

About 11 000 years ago there was renewed cooling of the climate. An ice cap over 100 kilometres long formed on the mountains in western Scotland (the Loch Lomond advance) and the high valleys and corries of northern Ireland were partly filled with ice. Gelifluction occurred on a small scale and the plant cover (which now included the dwarf willow, *Salix herbacea* – a typical snow patch plant – mountain avens, *Dryas octopetala*; alpine meadow rue, *Thalictrum alpinum*; and mountain sorrel, *Oxyria digyna*) became discontinuous. Mitchell, who has made an exhaustive search for evidence that the giant deer in Ireland survived into this cold episode, had no success, suggesting that the animal probably disappeared from Ireland as a result of it. If the giant deer was unable to survive the cold conditions which began 11 000 years ago it is even less likely to have survived the much more severe conditions of 18 000 years ago, when ice covered much of Ireland. Thus, the giant deer of the peat bogs were almost certainly new immigrants from the British mainland.

The end of the Loch Lomond advance about 10 250 years ago, brings us to the Holocene (the Littletonian of Ireland) and to an amelioration of climate extending to the present day. Peat began to form on top of the lacustrine marls containing the giant deer remains. Most of the Pleistocene species of mammals had disappeared, though the wolf continued to flourish. New arrivals included man and various domestic and accidently introduced mammals. There is no evidence that Palaeolithic man ever reached Ireland. The earliest evidence of man's presence is of Mesolithic date (Toome Bay and early Larnian cultures), about 8000 years ago and more than 2000 years later than the last known occurrences of giant deer there. Man was apparently not responsible for the extinction of this animal in Ireland, although he certainly exterminated the wolf about 1770.

Chronology of the British Quaternary

Table 8 shows a simplified listing of the mammalian faunas of the British Isles, based on the evidence discussed above.

Chapter 11 ***The East African Rift Valley***

One of the greatest wonders of the world is the African Rift Valley, a complex graben (downfaulted valley) structure which extends from Mozambique to Syria, a distance of more than 5000 kilometres (or an eighth of the way round the world). In its northern part, along the Red Sea and the Gulf of Aden, Africa and Saudi Arabia are drifting apart as part of the process of continental drift.

The East Africa rift system has two main branches; the Western Rift, to the west of Lake Victoria; and the Eastern Rift, which reaches its most spectacular development (up to 60–70 kilometres across) in Tanzania and Kenya, where it is known as the Gregory Rift (fig. 11.1).

As the result of geological studies, the history of the structural development of the rift is now very well understood. Land movements apparently began during the late Cretaceous, more than 60 million years ago, with rifting and associated volcanism developing during the Oligocene (more than 30 million years ago) and continuing to the present day. During this time the basal rocks became locally flexed, with two large domes (the Kenya Dome and the Ethiopian Dome) becoming areas of high altitude, while elsewhere (most notably in the Turkana and Afar regions) there was corresponding subsidence.

Only from the air is it possible to appreciate fully the present-day magnitude of the rift, which is bounded on either side by innumerable strikingly fresh parallel step faults, in some areas as many as two to three faults every kilometre (fig. 11.2). Associated with this faulting, which created planes of weakness up which lava could readily force its way to the surface, there were long periods of volcanism, a process which still continues in some areas (fig. 11.3).

In consequence there exists today, along much of the Rift Valley, a contrasting relief of mountains (most of them volcanic) overlooking chains of lakes in the graben hollows. Five of these mountains are of gigantic size. Kilimanjaro (5900 metres) is a volcano of Plio-Quaternary age. Mt Kenya (5200 metres), the denuded plug of which towers as a pinnacle from its centre, is Plio-lower Pleistocene. Meru (4500 metres) last erupted in 1877 and has calderas, lava flows and ash cones of remarkable freshness. Mt Elgon (4300 metres) is a Mio-Pliocene volcano. Ruwenzori, though also of great height (5100 metres) is not volcanic but is an upfaulted block of Precambrian basement rock.

In addition to the large volcanoes, there are many hundreds of other volcanic structures along the floor of the Rift Valley and on its flanks, one of the most remarkable being the 19-kilometre-wide Ngorongoro Caldera in Tanzania, which was active in late Pliocene times (2·9–2·0 million years ago).

In the lowest parts of the Rift Valley are innumerable lakes, some of them (especially those in the Western Rift) being of great size. Many of the lakes in the Eastern Rift (where evaporation sometimes exceeds inflow from rivers) are highly saline; one of the most extreme examples being Lake Magadi in Kenya, which has no outflow and has dried out except in a few places, leaving in the bottom a 60-metre-deep deposit of solid soda, upon which a car may safely be driven.

During the past there have been many other lakes in the rift, the configuration of which has changed repeatedly as the result of changes of precipitation, faults and volcanism. In consequence there exist today, at many places, thick and widespread accumulations of old lake deposits at various heights above and sometimes far from existing lakes.

For palaeontologists there can be few places where circumstances have been so favourable for their studies. Lake basins and the river channels associated with them act as traps for animal and plant remains which are buried there in abundance; and the fossil remains of organisms are constantly being brought to light by 'badlands' erosion which now affects so much of the area. Where the remains had been quickly entombed by ash from volcanoes, their state of preservation is sometimes astonishing; especially striking being the insects and plant fragments in the Miocene deposits of northwest Kenya. The richness of the fauna which still survives allows a ready comparison with the fossil and recent evidence. Bones of mammals, birds and fish, shells of molluscs, and plant remains are still being buried by the same processes that occurred throughout the Pleistocene, allowing straightforward taphonomic comparisons. Furthermore, many of the fossiliferous deposits are interbedded between lava flows or contain minerals from contemporary ash falls that can be accurately dated by the potassium–argon method. Already many hundreds of dates have been established, extending from the Miocene to the present day, to which studies of the changing mammalian faunas of East Africa (including a long range of hominid species) can be related.

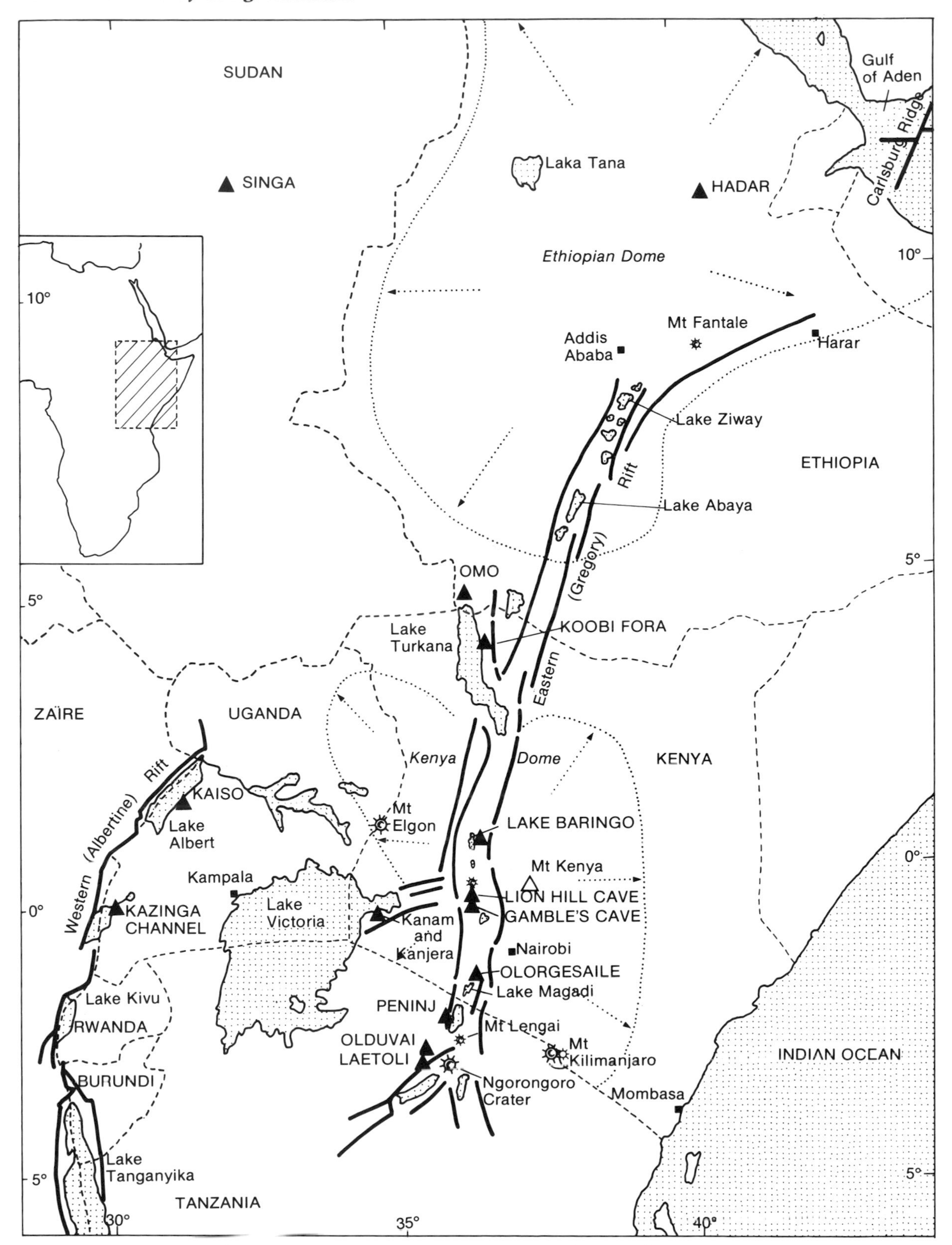

Fig. 11.1 *Map showing the Eastern and Western Rifts of East Africa, and some of their more important Plio-Pleistocene fossil mammal localities (solid triangles).*

Fig. 11.2 *Step fault scarps in the Gregory Rift, near Nairobi, Kenya. (Photo: A. J. Sutcliffe).*

Fig. 11.3 *Schematic cross-section of the East African Rift Valley showing the relationship between fault blocks, volcanic and lake deposits. (Drawn by U.M.K. Sutcliffe).*

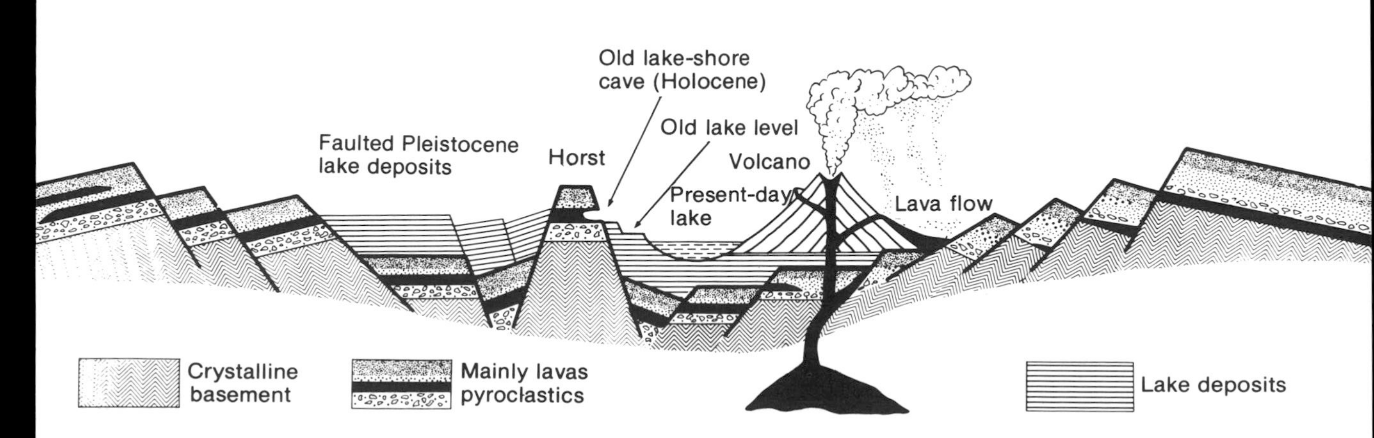

As in other parts of the world, the Quaternary was a time of major fluctuations of climate in East Africa, for which there is both palaeontological and physical evidence. Moraines on Kilimanjaro, Mt Kenya, Ruwenzori (all of which still carry glaciers at their summits) and Mt Elgon show that, in the not very distant past, the snowline on these mountains was lower and the glaciated areas were more extensive. The floral and faunal zones on the sides of the mountains (in descending order moorland, bamboo forest, deciduous forest, savannah) were correspondingly lowered.

At the lower altitudes there is also evidence of substantial past fluctuations of rainfall, reflected by changes in the levels of the East African lakes. Great caution must nevertheless be exercised in the study of these old lake levels since changes resulting from tectonic movement are commonly superimposed upon those of climatic origin.

Although, in this book, we are primarily concerned only with the Quaternary we must bear in mind that the beginning of this period is no more than an arbitrary time boundary defined by man. Some deposits which twenty years ago, were regarded as of lower Pleistocene age (such as the lowest beds in Olduvai Gorge) are now referred to the upper Pliocene. In this chapter we will depart from our usual practice of restricting our studies to the Quaternary, so as to consider also some of the evidence from the immediately preceding Pliocene beds, since without so doing we would be looking only at the latest part of the story.

The first fossil mammalian finds from East Africa; and East Africa's place in the general context of the African continent

When compared with studies in some other parts of the African continent, interest in East African fossil mammals began very late. Apparently the first time that such remains were noticed there was in 1911 when Professor Kattwinkel, a doctor from Munich with an interest in sleeping sickness, who was travelling in Tanzania (then German East Africa), noticed fragments of fossil bone being weathered out of the sides of the massive erosion gully which was subsequently to become famous as Olduvai Gorge. In 1913, Professor Hans Reck led a palaeontological expedition to Olduvai with personal support from the Kaiser himself and, numbering the beds there in ascending order 1–5, made the first serious collection of fossils. He was to return in 1931 with Louis Leakey, son of a British missionary working in Kenya, who had already embarked on an ambitious programme of archaeological work in the Rift Valley nearer to Nairobi, and who had obtained his Ph.D. in Cambridge in 1930. Leakey's subsequent palaeontological and archaeological achievements, and that of other members of his family, were to become a legend within his own lifetime. But, before we follow their work more closely, let us first consider briefly the state of palaeontological knowledge in other parts of Africa in the 1930s to 1950s, which provided the setting for it.

Most previous studies had been concentrated in two areas; along the North African coast and in South Africa. The findings from North Africa (an area separated from East Africa today and intermittently throughout the Quaternary by the inhospitable Sahara Desert) differed in many ways from what was later to be found in this new area. In particular, the mammalian fauna included many species with European affinities which never reached East Africa.

South Africa was already world famous as the finding place of australopithecine (man-ape) remains. The first specimen (fig. 11.4), a skull of a juvenile individual, had been discovered in a limestone quarry at Taung, in the Cape Province, in 1924. With the finding of additional australopithecine remains at other sites, it gradually became apparent that two different forms were represented among the South African material. The first of these, which included the Taung child, already named *Australopithecus africanus* by R. A. Dart, was a relatively slender (gracile) form. Subsequently this species was also recorded from the limestone caves of Makapansgat and Sterkfontein. An almost complete skull of an adult female from the last mentioned site, found by Robert Broom in 1947 (fig 11.4), was popularly named 'Mrs Ples', short for Broom's genus *Plesianthropus,* though it is now generally considered unnecessary to separate this from *A. africanus.*

The second australopithecine, *A. robustus* (recognized by some authorities as a separate genus, *Paranthropus*), was far more heavily built. The first specimens, from the Cave of Kromdraii, were described by Broom in 1938. Later additional remains were found, only a short distance from Sterkfontein and Kromdraii, in the Cave of Swartkrans.

Associated with the australopithecine fossils from various sites were also great quantities of remains of other mammals. Those from Sterkfontein, for example, included buffalo, gnu, reedbuck, springbok, warthog, the extinct giraffid *Libytherium,* a chalicothere, leopard, sabre-toothed cat, hyaena, jackal, mongoose, baboons, shrews and mice. True horses have been found at Swartkrans and Kromdraii, but not

Fig. 11.4 *Pioneers of South African australopithecine studies; left, Robert Broom with 'Mrs Ples' in the Transvaal Museum; right, Raymond A. Dart with the skull of the Taung child, type specimen of* Australopithecus africanus. *(Photos: Transvaal Museum and P. Nagel via J. McGuire).*

at Sterkfontein, although the three-toed horse, *Hipparion,* is known from that site.

From the time that the two australopithecines were first distinguished in the South African cave sites their relationship was to become a subject of discussion and controversy. Was one form earlier in time than the other and possibly ancestral to it, or did they live contemporaneously? Unfortunately, no means of obtaining absolute dates for remains from limestone caves existed at that time; and the associated mammalian fauna, which now provides a good stratigraphic guide to the sequence of events in South Africa, had been only briefly studied. It was necessary to await later work in East Africa and the more recent studies of Brain, Cooke, Dart, Robinson, Vrba, McGuire and others in South Africa before the haze could begin to clear. We will return to South Africa again at the end of this chapter.

The work of Louis and Mary Leakey

When Louis Leakey began his archaeological studies in Kenya in the 1920s, he little anticipated the far-reaching significance which his work would later assume. There was very little upon which he could build, for he was among the first to excavate there; and he could hardly guess that within his own lifetime potassium–argon dating would provide absolute dates for fossiliferous sites in the Rift Valley going back for millions of years; and which, in turn, would provide a chronological framework to which palaeontological findings in other parts of Africa could begin to be related.

Although archaeologically very important, Leakey's description of his earliest field work (1926–9) contains only a brief mention of mammalian finds – in an appendix by Dr A. T. Hopwood of the British Museum (Natural History) – and these were mostly of late Pleistocene and Holocene age, with few or no extinct forms among them. Hopwood nevertheless observed that two other known mammalian assemblages from East Africa must be of earlier date. These were Reck's collection from Olduvai; and another collection which had been obtained by E. J. Wayland, Director of the Uganda Geological Survey, at Kaiso in Uganda. Hopwood had already described this last mentioned fauna in 1926; he had shown it to include *Hipparion,* the three-toed horse, and an extinct chalicothere, and he had supposed it to be of Pliocene age – an interpretation confirmed by more recent studies. A sequence of mammalian faunas was beginning to be established.

An important aspect of Louis's work at this stage was his great interest in the evidence for past fluctuations of climate which seemed to be decipherable from the geological evidence. The four-stage glacial–interglacial sequence, generally accepted as a chronological framework on which to hang Quaternary events in Europe, could not be applied to the fossil-bearing deposits of East Africa as (even though there was evidence of formerly more extensive glaciation on the mountains) all the fossiliferous sites lay in hollows in the Rift Valley and could not be directly related to any glacial sequence. A possible alternative was to try to establish a pluvial–interpluvial chronology based on the evidence of changing lake levels, which were interpreted as manifestations of changing rainfall. The resulting stage names are shown in Table 2, p. 61. Commenting on this sequence in 1931 C. E. P. Brooks even went as far as to suggest a pluvial–glacial correlation, with Leakey's Gamblian Pluvial corresponding to the Würm and Riss glaciations of Europe; his Kamasian Pluvial to the Mindel and Günz. It was to be many years before the validity of this interpretation could be put to the test. although, when this became possible, it was evident that this scheme did not provide the hoped for chronology. The lake deposits of the Rift Valley are extensively faulted, confusing the evidence for the earlier pluvials, with only the latest high lake levels being associated with undeformed shore lines acceptable as evidence of pluvial conditions. One such high lake level stage which can still be attributed to a period of increased precipitation is the Gamblian Pluvial, although, with a date of about 9000 years now established for its maximum, it is seen to fall within the Holocene and is not the time equivalent of any part of the Last Glaciation. As we have already seen, it is now generally accepted that the high rainfall stages of the low latitudes correspond in most parts of the world to interglacial (not glacial) stages; and in East Africa a time scale, with absolute dates based on potassium–argon dating, has now superceded the earlier climatically based chronology.

From the time when Louis commenced his archaeological work one of his greatest wishes had been to visit Reck's fossil locality at Olduvai Gorge, the mammalian remains from which were apparently much earlier than those that he had been finding in Kenya. Reck had not found any stone tools there, although he had found a crouched human skeleton in one of the lower strata of the gorge (his 'bed 2') which he believed to be contemporary with the deposit containing it. Louis was convinced that another search would reveal hand axes, although it was difficult to find funds for an expedition and it was not until the end of 1931 that he was at last able to reach Olduvai, travelling for eight days across country with constant risk from wild animals and problems in obtaining water. Other members of the party included Hans Reck, who had been specially invited to come from Germany, Vivian Fuchs (later of Antarctic fame), A. T. Hopwood from the British Museum (Natural History), Sir Edmond Teale and Donald MacInnes.

Olduvai gorge is a 'badlands' erosional channel, which has been cut by a seasonal river across the lacustrine and volcanogenic sediments of this otherwise relatively flat part of the Serengeti Plains (fig. 11.5). Throughout most of the year the river bed is entirely dry, though after heavy rain a torrent flows through it, eroding the sides of the gorge and exposing fossils preserved at all levels in the adjacent sediments. Upon arrival at the gorge, the 1931 expedition at once got off to a good start. Within 24 hours Louis had found the first hand axe there, and further work during the remainder of the expedition produced many more artefacts and an abundance of mammalian remains.

Louis's next visit to Olduvai was in 1935, accompanied by 22-year-old Mary Nicol, who was to become his wife the following year. But Mary was not to remain a subsidiary partner in Louis's field work for long. She was already a developing archaeologist of some calibre, who had toured in England and France and South Africa; and in subsequent years her ability to locate rare specimens in the field (her most notable discoveries being the skulls of the Miocene ape *Proconsul* on Rusinga Island in 1948 and of the australopithecine *Zinjanthropus* at Olduvai in 1959) outstripped that of Louis. At Olduvai Mary was to become more and more involved in all aspects of the investigation. During the later years of Louis's life and after his death in 1972 she continued to direct the field work and to organize the findings for publication on her own.

After this first visit in 1931, the Leakey's involvement with the site was for a long time necessarily spasmodic. There were other fossil sites that needed investigation, including the notable discovery at Olorgesailie, near Nairobi, in 1942. There, many thousands of hand axes lying weathered out of the underlying Pleistocene lake deposits, were found to be associated with fossil mammalian remains. The Second World War, and shortage of funds for both field work and publication, meant that it was necessary to wait until 1951 for the first major description of the gorge. This included both a summary of the conclusions of Reck and a description of the fauna by Hopwood, who accepted the entire mammalian assemblage as a single faunal unit of later than lower Pleistocene age. Leakey distinguished the stone industries as a single evolutionary sequence beginning with a primitive chopper tool assemblage (the 'Oldowan' culture) in Bed I through ten subsequent stages to the Acheulean of Bed IV.

In 1959, there came a great turning point in Louis and Mary Leakey's fortunes at Olduvai. Since 1931 work there had been an uphill battle against every conceivable obstacle. In the early days there had been the problem of actually getting to the gorge and of finding water there, although these hindrances had diminished with the improvement of tracks to the site. The greatest difficulty was still that of finance, which

Fig. 11.5 *Olduvai Gorge, Tanzania, showing the dry river bed and the finely stratified fossiliferous lacustrine sediments exposed along its side. (Photo: Bob Campbell).*

limited both the size and length of expeditions and the publication of findings. There did not exist, at that time, the great teams of specialists who were later to take up the study of the different aspects of East African palaeontology and archaeology; and the Leakeys had to try to cover nearly every field of study on their own, with only scanty specialist help to assist them.

On 17 July, 1959, there occurred an event that was to change everything. Louis was not feeling well and stayed in camp while Mary was exploring the gorge, where she encountered a much broken australopithecine skull weathering out of the surface of Bed I (fig. 11.6). On hearing of the discovery Louis forgot his illness and was soon at the finding place. After careful recovery of all the fragments (a process

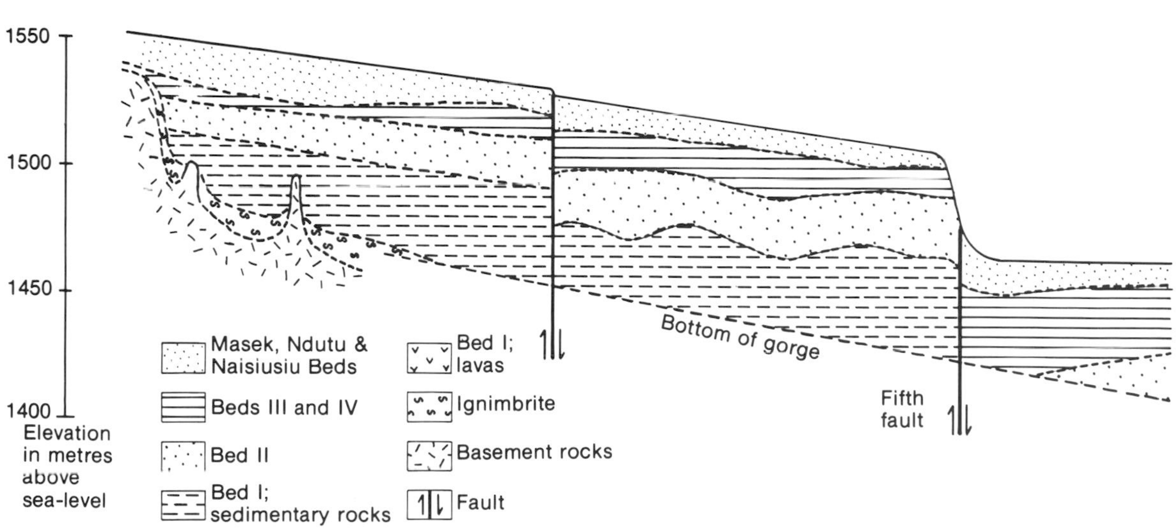

1550
1500
1450
1400
Elevation in metres above sea-level
Bottom of gorge
Fifth fault
Masek, Ndutu & Naisiusiu Beds
Beds III and IV
Bed II
Bed I; sedimentary rocks
Bed I; lavas
Ignimbrite
Basement rocks
Fault

◀ Fig. 11.6 *Excavations at Olduvai Gorge, 1960. The concrete block marks the place where the skull of 'Nutcracker man'* Australopithecus (Zinjanthropus) boisei, *was found in 1959. (Photo: A. J. Sutcliffe).*

that was carefully filmed) and subsequent laborious reconstruction it was found that, although the skull lacked the lower jaw, it was in other respects almost complete, more so than any of the South African australopithecine skulls except for 'Mrs Ples'. The new find was a robust australopithecine with immense cheek teeth, which soon earned for it the name 'Nutcracker man' (fig. 11.7). Its latin name was to be *Zinjanthropus boisei*, though it is now more generally accepted as *Australopithecus boisei*.

The implications of the new find were astonishing. Throughout the layer in which the skull was lying were also many stone tools of the Oldowan culture, subsequently shown to be part of an *in situ* activity level of whatever animal had made them. Was *Zinjanthropus* the manufacturer of the tools, or had they been made by a more advanced, but as yet unknown, homimid who might even have preyed upon *Zinjanthropus?*

At last, work at Olduvai began to attract widespread interest in the USA and, after a lecture tour at the end of 1959, Louis heard that the National Geographic Society had awarded a substantial grant for further work there. Louis and Mary could now employ a large labour force and undertake work on a more realistic scale. During 1960 alone, they organized 92 000 man hours of work, more than twice the time that had been spent during the previous thirty years, helped by their eldest son Jonathan, who was by then 19 years old and had just left school.

Fig. 11.7 *Louis Leakey with the skull of 'Nutcracker man'. (Photo: A. J. Sutcliffe).*

The *Zinjanthropus* skull was only the first of many surprises. Absolute dating, based on the potassium–argon method, gave the lower part of the Bed I deposits (Leakey continued to use Reck's numbering, now replaced by Roman numerals) an age of 1·75 million years, almost twice as long as the South African evidence had so far indicated. The unexpected discovery of the remains of *Homo*, which were definitely *in situ*, at a level slightly below the *Zinjanthropus* horizon, meant that australopithecine and *Homo* had coexisted. The first find, later named as a new species, *Homo habilis*, was found by Jonathan Leakey in the lowest part of Bed I on 2 November, 1960. Only a month later Louis found a human cranium in Bed II. This was subsequently determined

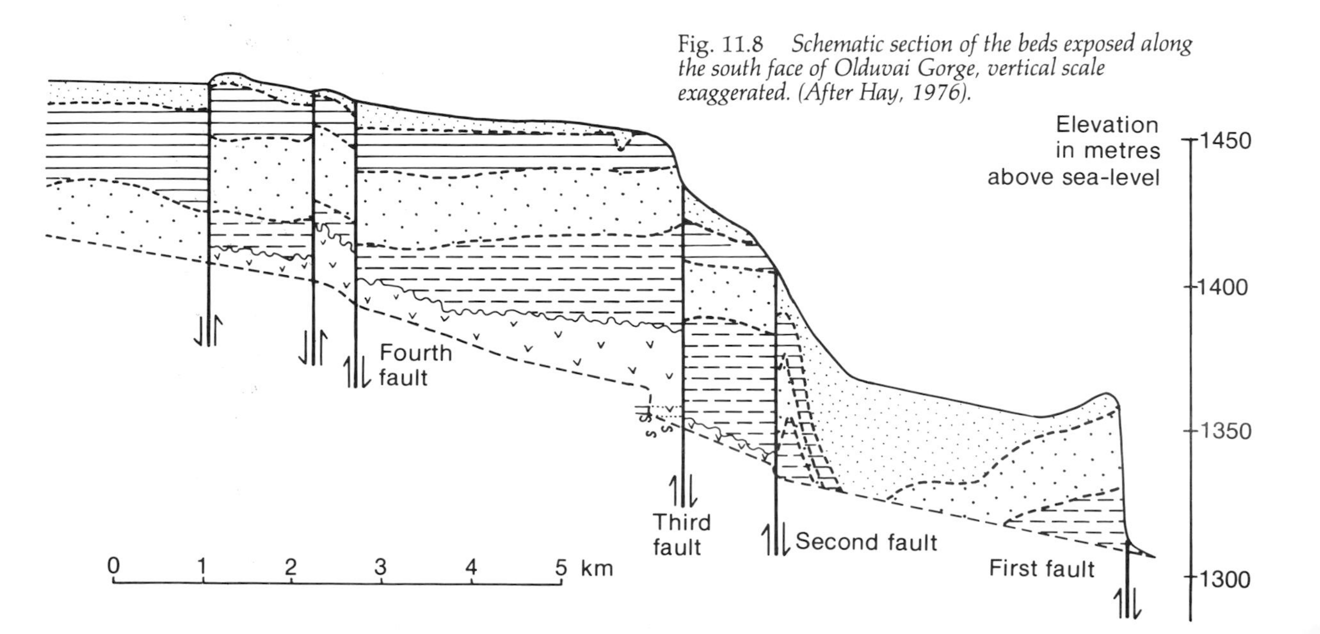

Fig. 11.8 *Schematic section of the beds exposed along the south face of Olduvai Gorge, vertical scale exaggerated. (After Hay, 1976).*

as belonging to *Homo erectus*, a species already known from Java and China, where it had previously been regarded as a separate genus, *Pithecanthropus*. No longer was it possible to consider the australopithecines, or at least the later ones, as possible ancestors of *Homo*. It would be necessary to look back to much earlier times for a common ancestor.

The large-scale excavation which the Leakeys had begun in 1960 could at last be extended to a systematic examination of the entire stratigraphic sequence at Olduvai and of the fossil remains preserved in the different layers. The first bed to be examined was Bed I, in which *Zinjanthropus* had been found the previous year, then Bed II. By 1968, it was possible to start on Beds III and IV; and work on these and overlying layers continued under Mary's sole direction until 1973, the year after Louis Leakey's death. By this time specialist help was less difficult to obtain than it had been 20 years previously. One of the specialists who was brought in by the Leakeys, whose contribution was to be of special importance, was Richard L. Hay of the University of California. His study of the geology of the gorge was carried out concurrently with the excavations and formed the basis upon which the Leakeys established their archaeological succession. Like the Leakeys, Hay continued to use Reck's system of numbering, though now with certain modifications at the top of the sequence. Bed IV he subdivided into Bed IV and the Masek Beds; Bed V was subdivided and renamed the Ndutu and Naisiusiu Beds (fig. 11.8).

Olduvai Gorge – outline of the findings

With the specialist phase of work at Olduvai now having continued for over 20 years, many completed reports have already appeared and it gradually becomes possible to piece together the findings as a unified and very detailed story. Hay's description of the deposits of the gorge is one of the most advanced reports of its sort ever published. It provides not only a series of absolute dates for the various beds but also a continuous picture of the changing palaeogeography of the area; and thus of the local environment of the humans and animals living there. Principal of the many sediment types are lavas and tuffs, erupted from nearby volcanos, lake and lake margin deposits, river and alluvial fan deposits and mudflows. For the palaeontologist and archaeologist, the lake margin deposits are of special interest since, in such situations, occupation horizons with their associated artefacts and mammalian remains were often quickly swamped by lake waters and buried without disturbance. There are other deposits in the gorge that contain fossil remains of fish, suggesting deeper water; others which are entirely unfossiliferous. While this long series of sediments was accumulating, rift valley faulting was active and a subsidence known as the Olbalbal Depression appeared at the eastern end of the present-day gorge. Five faults associated with this movement, which cut the Pleistocene sediments of the gorge, are today exposed in the cliffs along its sides. This down-faulting led to a major change in the drainage pattern of the area and the beginning of gorge formation by streams following the gradient into the Olbalbal Depression. The latest sediments exposed in the present-day gorge are seen to rest in an earlier infilled channel, showing that more than one cycle of gorge erosion has occurred during Olduvai's history.

From his study of the lateral extent of the deposits of various beds, Hay was able to prepare a series of maps, showing the geographical changes (and hence the changes of environment for man and animals) that had occurred in the region around the gorge from late Pliocene times until the present day. It is now seen that the deposits exposed in Olduvai Gorge represent a period of unusually continuous terrestrial sedimentation, with accurate dating at most levels, over a period of more than two million years, back to the end of the Pliocene. A considerable part of this sequence contains mammalian remains and archaeological activity levels; so that there exists at Olduvai an opportunity for following the changes of industries and mammalian faunas throughout most of the Pleistocene in a manner previously unprecedented at any single site anywhere else in the world. At the same time, it becomes necessary to abandon any immediate further attempt to tie the relatively simple Olduvai sequence into any pattern of glaciations and interglacials at the higher latitudes. As we have previously seen the deep sea evidence suggests that, during the period of time concerned, there were apparently as many as seventeen major cold–warm cycles, yet no such detailed pattern of changing precipitation can be recognized at Olduvai. Recent oxygen isotope studies of carbonates from Olduvai, which suggest a minor decrease of rainfall at the end of Bed I times, about 1·7 million years ago, and another very dramatic decrease during Masek times 0·5–0·6 million years ago, nevertheless provide a useful first step towards such a comparison.

The sequence of archaeological industries and the stratigraphic range of the hominid species are now fairly well known. A robust *Australopithecus* first appears in Bed I about 1·8 million years ago and continues to near the top of Bed II about 1·2 million years ago. *Homo habilis* is first recorded at about the same time but is unknown in the upper part of Bed II. *H. erectus* appears after a time gap near the top of Bed II, about 1·2 million years ago, and survives at least until

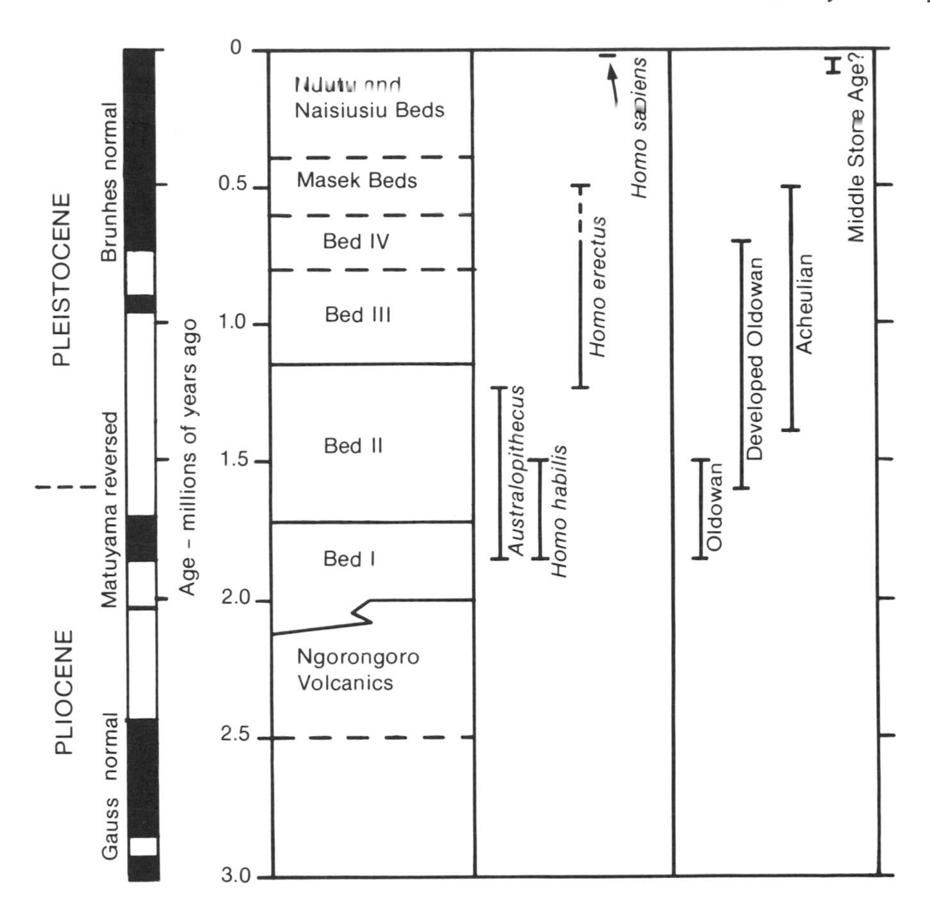

Fig. 11.9 *Time range of the archaeological and hominid remains from Olduvai Gorge. The end of the Olduvai palaeomagnetic epoch falls in the lower part of Bed II. (Modified after Hay, 1976).*

Bed IV, about 700 000 years ago. The time gap between *H. habilis* and *H. erectus* throughout the upper part of Bed II is tantalizing, since the relationship of these two forms of man cannot yet be determined from the Olduvai evidence. Did the former species die out, move elsewhere, or evolve into *H. erectus?* Olduvai man, discovered by Reck, and now regarded as *H. sapiens,* has a carbon date of about 17 000 years and must be contemporary with the Naisiusiu Beds.

The development of the various stone tool industries can be followed in parallel to the changes of hominid species and an attempt can be made to relate the two (fig. 11.9). Whilst it is feasible that *Australopithecus* could have been responsible for the stone tools and other living floor débris of Beds I–II, *H. habilis* was also present in these levels and seems more likely to have made them. The number of animal bones at the 'living sites' suggests to some workers that *H. habilis* not only hunted, but did so successfully. However, others have raised doubts about the reality of these living sites at Olduvai Gorge. *H. erectus* becomes a likely manufacturer of the Acheulian. The occurrence of a third industry at Olduvai, the Developed Oldowan, from Beds II–IV, which apparently existed contemporaneously with, but separate from, the other industries, presents a problem requiring further study.

Lastly, we must consider the rich mammalian fauna of Olduvai Gorge. Mammalian remains have been found at most levels, being especially abundant in Beds I and II. They accumulated under diverse circumstances and range from complete associated skeletons of animals, which died on the lake shore and were subsequently covered by silt laid down by rising waters (fig. 4.2, p. 36) to fragmentary bone remains in the hominid activity horizons.

Although the Olduvai deposits accumulated over such a long period, the changes in mammalian faunas are not so striking as those which occurred in Europe, where cold and warm climate species alternated repeatedly as the result of geographical movements related to climatic change. With recent detailed attention from specialists working on the different groups, an increasingly detailed picture of successive mammalian faunas of Olduvai is nevertheless steadily being assembled. One of the most striking aspects of these studies is the great number of mammalian species represented, already more than a hundred. Among the more interesting are the extinct *Deinotherium* (a relative of the elephant family with down-turned tusks in its lower jaw); the elephant, *Elephas recki;* the three-toed horse, *Hipparion;* a large true horse, *Equus oldowayensis;* two species of rhinoceros, probably ancestral to the living white and black rhinos; the extinct chalicothere, *Ancylotherium;* diverse antelopes and members of the buffalo family; hippopotamus; diverse pigs; an extinct giraffe and the antlered giraffids *Libytherium* and *Sivatherium;* an extinct baboon, *Theropithecus;* bush baby, *Galago;* many carnivores including the extinct sabre-toothed cat, a large feline, an ancestral spotted hyaena, civet, jackal, mongoose, otter, shrews, elephant shrews and a hedgehog; diverse rodents including the porcupine, *Hystrix,* the jumping hare, *Pedetes;* various rats and mice; and (represented by a few specimens only) at least two species of bat.

Some of these mammals are of unusual interest. The hippopotamus of the lowest part of Bed I is relatively unspecialized, but, in the higher strata, a distinct species has been discovered, *Hippopotamus gorgops.* Eyes were placed especially high on the head, a character which apparently developed in geographical isolation from hippos in other areas and which provided it with a special advantage when all but its eyes and nostrils were submerged in water. Scarce remains in Bed II show that a pigmy hippopotamus was also present at Olduvai for at least a short time.

Giantism and dwarfism are found in some of the species present. Civet remains in Beds I and II are of an animal nearly twice as large as its present-day relative; there is similar giantism in the baboon *Theropithecus,* and in some of the pigs. The large bovid *Pelorovis* had a horn span of two metres. Dwarf mammals include a small giraffe, *Giraffa stillei,* from Beds I–IV. Some of the mammals (for example *Hipparion* and *Sivatherium*) survived at Olduvai long after they had become extinct in Europe and Asia.

Several of the mammalian groups (for example the shrews and rodents) are ecologically significant. The replacement of *Sylvisorex olduvaiensis,* the dominant shrew in the lower part of Bed I, by *Suncus* in its upper part indicates increased aridity, a conclusion supported by the appearance of gerbils and naked mole rats which favour a dry savannah habitat.

There is even better evidence of ecological conditions in the Olduvai area from the bird remains. Over fifty species of fossil birds have been recorded. Especially interesting are the aquatic forms, which include gulls, cormorants, pelicans, ducks, storks and herons. Flamingos indicate that some of the water was brackish or saline.

Remains of crocodiles, snakes, fish and even slugs provide additional ecological information. The general picture which emerges, is of wet conditions in early Bed I times, drier conditions at the top of the bed and a return to wet conditions once more at the bottom of Bed II. Similar palaeoecological studies have not yet been completed for the upper strata at Olduvai.

When interpreting such data, special care must be given to distinguishing evidence that might be conclusive of climatic change from that which might be explained by other causes. Fossils in the hominid activity zones may have been brought in by man from scavenging or hunting areas some distance from the gorge. Faulting and the consequent drainage of the lake basin could cause locally increased aridity and changed fauna and flora, but this change would not be climatically caused. Although these alternative explanations should be borne in mind, at Olduvai the main faulting did not occur until the middle of Bed II and it seems reasonable to suppose that the wet–less wet–wet sequence inferred for Bed I was climatically rather than tectonically caused.

The post-Olduvai era of studies

Although the excavation at Olduvai had provided an unprecedented sequence of deposits spanning over two million years, together with a rich mammalian fauna, remains of both *Australopithecus* and *Homo* and much archaeological evidence, it still left many problems, especially anthropological problems, unresolved. What was the relationship between *H. habilis* and *H. erectus* and did *Australopithecus* and *Homo* have a common ancestor somewhere in pre-Pleistocene times? Let us make a brief excursion into the Pliocene in order to see what happened there.

It has long been recognized that the fossiliferous deposits of the Rift Valley represented the accumulation of a long period of time. The mammal-bearing deposits of Rusinga Island are lower Miocene, about 18 million years old, those of Fort Ternan middle Miocene, about 14 million years old. As late as 1967 no mammaliferous deposits in the Rift Valley were accepted as being of Pliocene age, although this was partly because those known (including some which had previously been accepted as Pliocene) were regarded as being lower Pleistocene.

With excavations, accompanied by new radiometric dating, by various parties since 1967 at Laetoli in Tanzania, Lake Turkana in Kenya and at Omo and Afar in Ethiopia, much new information has now been obtained to fill this 'Pliocene gap'.

After completing work at Olduvai, Mary Leakey transferred her attentions in 1975 to Laetoli, only 30 kilometres away to the south, which she had first visited with Louis in 1935, and where East African australopithecine remains had first been recognized by Kohl Larsen, in 1939. Beds of the Laetolil sequence underlay Bed I at Olduvai and this relationship was subsequently confirmed with a potassium–argon date of 3·59–3·77 million years for the tuffaceous sediments exposed at Laetoli. Mammalian remains were present in abundance, also more hominid specimens, at first thought to be *Homo* but subsequently interpreted as a very early australopithecine.

Among the mammals are *Deinotherium*, elephant, *Hipparion*, rhinoceros, chalicothere, antelope, buffalo, pigs, giraffe, *Sivatherium*, baboon, hyaena and sabre-toothed cat. But the mammalian bones were not the most exciting discoveries at Laetoli. As excavation proceeded it was realized that depressions on the surface of several beds were the most beautifully preserved footprints of animals (fig. 11.10).

As these were carefully exposed it became apparent that nearly the whole of the mammalian fauna, already represented by skeletal material, had left its record on the once muddy ground, including the hominid, which is clearly shown to have been bipedal. There is even the track of a centipede and there are the well-preserved eggs of birds. It is inferred that the tracks were preserved as the result of quick burial by ash from an eruption of the nearby volcano Sadiman.

Meanwhile, even before Mary had finished work at Olduvai, other palaeontologists and archaeologists were beginning to appear on the scene and to carry the investigations to new sites further afield. Now a new generation of Leakeys, notably Louis and Mary's son Richard (now Director of the National Museum in Nairobi) and his wife Meave took up the work (fig. 11.11).

In 1967 there commenced a large-scale international investigation of the highly fossiliferous deposits along the Omo River in southern Ethiopia. Over 50 tonnes of fossils were collected from a sequence of deposits over 1000 metres thick and covering the time-span from 4 million years ago, almost to the present day—completely bridging the time gap between Olduvai and Laetoli. The mammalian faunas show a striking change during this time, though this appears to be predominantly a gentle transformation resulting from progressive evolution, with no real faunal breaks. Interesting mammals in the lowest levels are *Anancus* (a mastodon), *Stegodon* and *Hexaprotodon*, a hippopotamus with six (three pairs of) incisors in the upper and lower jaws. The hippopotamus in the later part of the sequence is *H. gorgops*, already noticed at Olduvai, which is of different ancestral stock and which has two pairs of incisors in the upper jaw and only one in the lower. Several of the groups of mammals, most notably the horses and rhinoceroses, show a trend from the earlier to later strata towards more high-crowned teeth, interpreted, with other lines of evidence, as an indication of a change of climate from

Fig. 11.10 *Fossil animal footprints at Laetoli, Tanzania; an elephant track and, left, footprints of antelopes. (Photo: M. D. Leakey).*

Fig. 11.11 *One of Richard Leakey's earliest excavations was in the region of Lake Baringo, northern Keyna where, in 1966, he excavated an entire elephant skeleton in a single block of sediment and transported it back to Nairobi for display in the Museum. (Photo: Bob Campbell).*

wet to less wet and of the ecological setting from bush and scrub and of tall grass to more open savannah and shorter grass. Important hominid finds include an australopithecine of gracile appearance in the lower part of the sequence and a robust australopithecine appearing about two million years ago. *Homo* is claimed to appear at about the same time.

Soon after the commencement of the excavations at Omo, which is situated at the north end of Lake Turkana, yet another area came to the fore for study. Richard Leakey, who had repeatedly flown over Lake Turkana on his way to Omo from Nairobi noticed that there were vast areas of potentially fossiliferous 'badlands' on the east side of the lake which had apparently never been properly investigated for fossils. A helicopter landing confirmed that fossils were indeed abundant. In 1968 Richard withdrew from the Omo excavations and started work in this new area near a place called Koobi Fora. Among Richard's party was the American Anna K. Behrensmeyer, who conducted the most detailed taphonomic studies of a fossil bone assemblage made up to that time. As in the case of the earlier sites there was a rich mammalian fauna, with remains of both *Australopithecus* and *Homo* and also many artefacts. The hominid remains were among the best and most long ranging yet discovered. A skull of *Homo*, 'skull 1470', found in 1972 was at first believed to be 2·6 million years old, and thus more than half a million years older than Louis Leakey's earliest *Homo (habilis)* from Olduvai; though a subsequent revision of the potassium–argon age on which this find had been dated and application of other methods suggests that it is probably only about 1·8–1·9 million years old. i.e. approximately the same age and probably conspecific with the Olduvai finds. As at Olduvai, *H. erectus* follows *H. habilis;* and *Australopithecus* is long-ranging, surviving until at most 1·5 million years ago, when it coexisted with *H. erectus.* In 1984, Richard's party found an almost complete skeleton of *H. erectus c.* 1·6 million years old, in West Turkana.

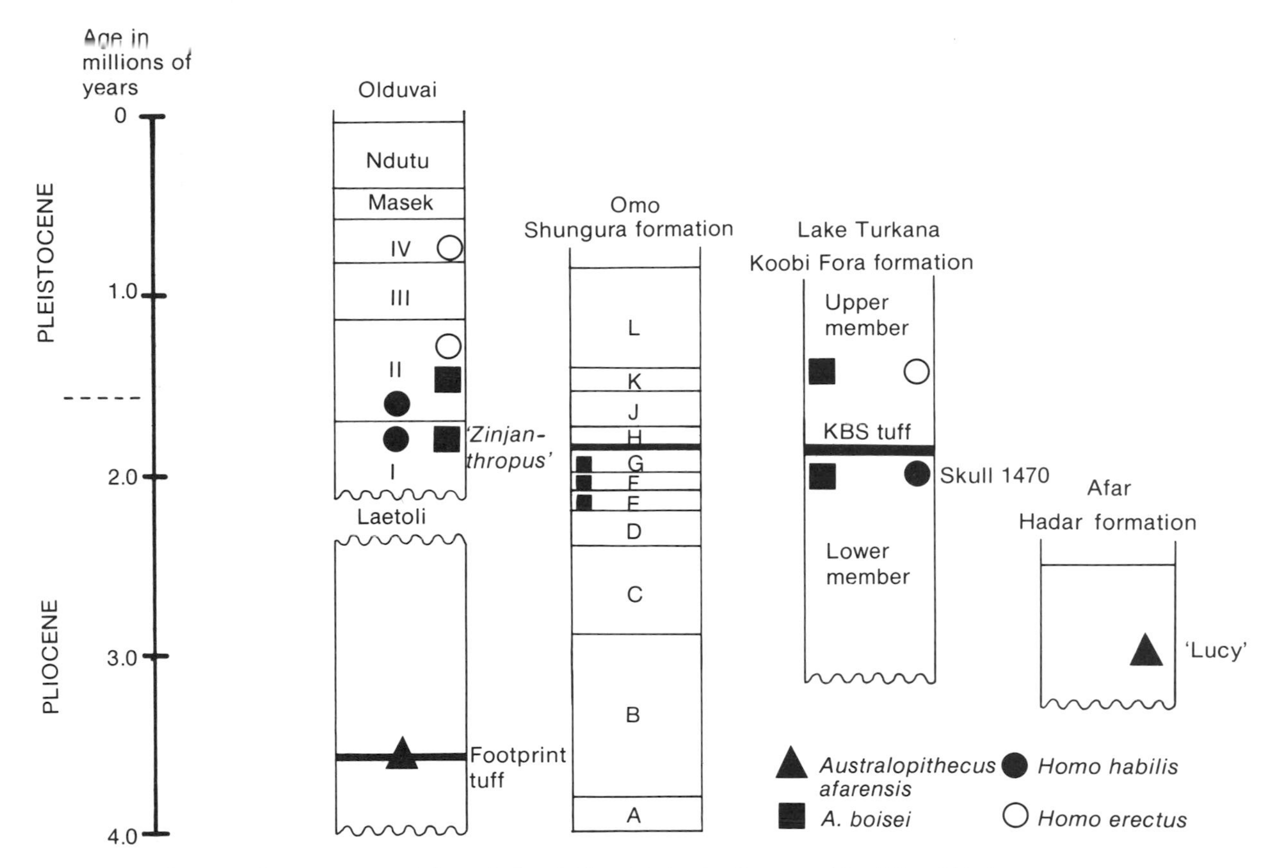

Fig. 11.12 *Stratigraphic comparison of the main hominid containing fossil mammal localities of East Africa. (After Drake* et al., *1980 and other sources).*

We need consider only one more African mammalian locality, Hadar in the Afar Depression of Ethiopia, where excavations were commenced by a Franco-American Party in 1973. In addition to many mammalian remains, one find – a substantial part of a skeleton of a small female australopithecine, subsequently named 'Lucy' – created a special sensation. Previously no associated skeleton of any hominid earlier than about 100 000 years old had been found anywhere, yet here was one about 3 million years old, much older than any other previously discovered associated hominid skeleton, though not quite as old as the dissociated remains from Laetoli. Lucy was referred to a new species, *Australopithecus afarensis* and subsequent excavations revealed parts of other associated skeletons, possibly of a family overwhelmed by the waters of a flash flood.

Putting the story of the African hominids together

We have seen how, during a period of twenty years, the story of hominids in Africa has developed from an unclear picture of undated remains that could not be fitted into any absolute chronology (but which was generally supposed to have occurred within the last million years or so) to a much more detailed record where the South African finds have been supplemented by an abundance of material from East Africa, for which a really good chronological framework, based on potassium–argon dating and other methods, is now available. In consequence, the hominid evidence has been taken back more than three million years.

The fossil record is nevertheless still very incomplete and many more problems about man's ancestry remain to be resolved.

One of the most fascinating aspects of work during this period has been the correlation of the sequences of deposits and faunal remains from the various East African sites, to provide a single unified story of events with which the South African evidence can now begin to be compared. In 1980, according to the latest dating evidence, the main East African sites are believed to be related as shown in figure 11.12.

It now becomes possible to place all the important

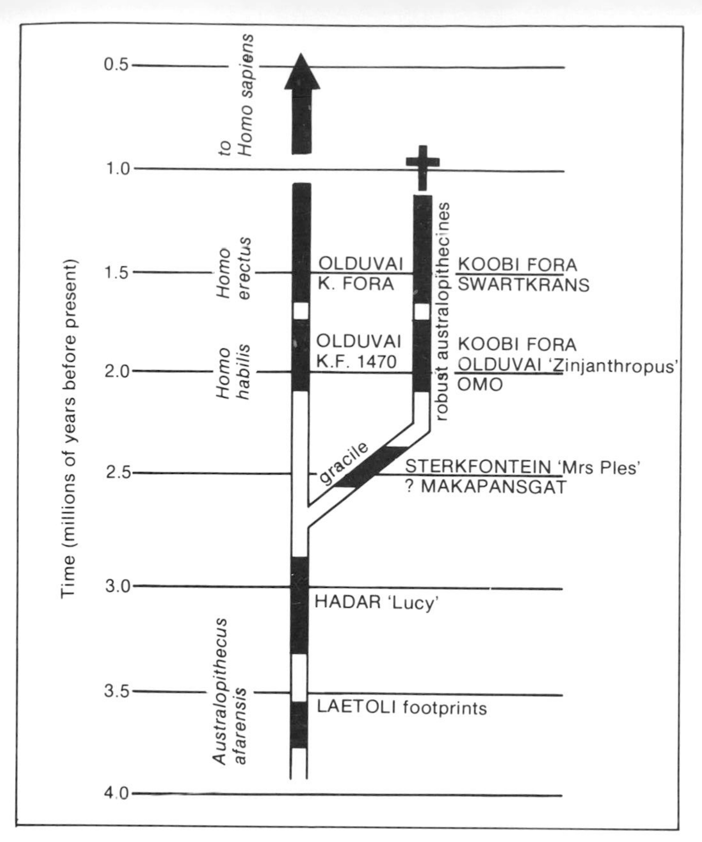

Fig. 11.13 *Possible relationship of the African Pliocene and Pleistocene hominids. The dark portions indicate periods from which hominid fossils are well known. (After Johanson & White, 1979).*

East African hominid remains in a good chronological framework. The earliest are those from Laetoli (about 3·5 million years) and Hadar (about 3 million years), which are currently usually regarded as conspecific (*Australopithecus afarensis*), although some authorities have suggested that the amount of variation among the remains is greater than would be expected and that more than one species may in fact be present. *Australopithecus boisei* is recorded at about 1·9 million years at Olduvai and Koobi Fora. *Homo habilis* (which may include skull 1470 from Koobi Fora) appears at about the same time with *H. erectus* at about 1·5 million years (Koobi Fora). *Australopithecus* survived until about the time of arrival of *H. erectus,* with whom it co-existed for a time.

The problem of working out an evolutionary lineage or tree for East African and South African hominids is nevertheless still immense, because there remain so many gaps in the fossil record. A recent discussion suggests a probable relationship as shown in figure 11.13. According to this interpretation the Laetoli and Hadar australopithecine is close to the ancestral line which, between 2·5 and 3·0 million years ago, divided into two distinct lineages which are represented by fossil remains from both South and East Africa. One of these gave rise to *Homo habilis, H. erectus* and finally *H. sapiens.* The other gave rise to the gracile, and subsequently to the robust australopithecines, which finally became extinct about 1·2 million years ago.

Although the South African cave breccias have still not been radiometrically dated, the gracile australopithecine from Sterkfontein and Makapansgat is provisionally placed somewhat later than those from Hadar, but earlier than *Australopithecus (Zinjanthropus) boisei* from Bed I at Olduvai, perhaps at 2·5 million years old. The robust australopithecine from Swartkrans is placed latest of all, at perhaps 1·6 million years old. The South African australopithecines thus turn out to be substantially older than seemed credible before the first radiometric dates were obtained for the deposits of the Rift Valley.

With the hominid evolution now proved to extend back for more than three and a half million years in Africa the often asked question re-arises: Was Africa the birth place of man? Many anthropologists argue that this was so. The remains of *Homo erectus* (previously *Pithecanthropus*) of Java could be earlier than some from Africa, but are unlikely to be more than about 2 million years old; and no certain australopithecine remains are yet known from that part of the world. For the time being Africa, with a lead of at least one and a half million years, seems most likely to have been the place of man's origin. Only when the fossil record of Asia is more complete can this finally be proved or disproved.

Chapter 12 The New World

In both North and South America the Quaternary was a time of astonishing diversity of mammalian species. Before we examine the assemblages from some selected localities it is first necessary to consider the cause of this diversity, which can be understood only within a framework of earlier events. We must look briefly back to pre-Quaternary times.

Palaeogeography of the New World

Geographical studies have shown that, until lower Eocene times, about 50 million years ago, North America was connected to Europe by a land bridge, probably through Greenland (fig. 12.1). Similar mammalian faunas, including *Hyracotherium* (formerly known as *Eohippus*), the 'dawn horse' (which later evolved in North America to *Equus*, the horse) and the primitive hoofed mammal, *Coryphodon*, existed on both continents. By middle Eocene times, as the result of the two continents moving away from one another by continental drift, this land bridge had disappeared; and subsequent evolutionary changes caused the European and North American faunas to become progressively less and less alike. By the end of the Eocene, about 40 million years ago, the first Asian mammals (for example ancestral hares) were beginning to arrive in North America, probably from the direction of the present-day Bering Straits. There is evidence of successive waves of Asian mammals crossing to the New World until the end of the Pleistocene, with peaks during the Miocene and Plio-Pleistocene. The Proboscidea (elephants and mastodons) are examples of an Old World group whose representatives spread to the New World on several occasions. Both cusp-toothed and ridge-toothed mastodons apparently first reached North America in the Middle Miocene. 'Shovel-tusker' mastodons followed a little later. True elephants (mammoths) did not arrive until the Pleistocene.

Meanwhile, in South America, the mammalian fauna had been developing along entirely different lines. From late Cretaceous until Pliocene times this continent was isolated from North America by a seaway where the Panamanian land bridge is now situated. Although, at first, there was some similarity between the faunas of the two continents (the opossum, a marsupial, for example, was present on both) faunal exchange was very severely restricted. For nearly all of this time the mammals of South America evolved almost independently and developed into a remarkable indigenous fauna including the edentates (sloths, glyptodonts and armadillos) and litopterns (including the Miocene *Thoatherium*, which had side toes reduced even more than in a modern horse). A few families of mammals that were probably of North American origin did manage to reach South America while it was still separated from North America, possibly by island hopping (some rodents and monkeys arrived during the Oligocene, a procyonid carnivore during the Miocene), although the failure of

Fig. 12.1 *Simplified map, illustrating the history of land bridges and seaways between South America, North America, Asia and Europe since the Cretaceous. 1. Land connection between Europe and North America until the earliest Tertiary, about 56 million years ago. 2. Intermittent land bridge from Asia to North America. Successive waves of migration from about 40 million years ago until the early Holocene, less than 10 000 years ago. 3. Land connection between North and South America until the late Cretaceous, about 60 million years ago; these continents then being separated by a seaway until the late Pliocene, about two million years ago, when the 'Great Interchange' commenced.*

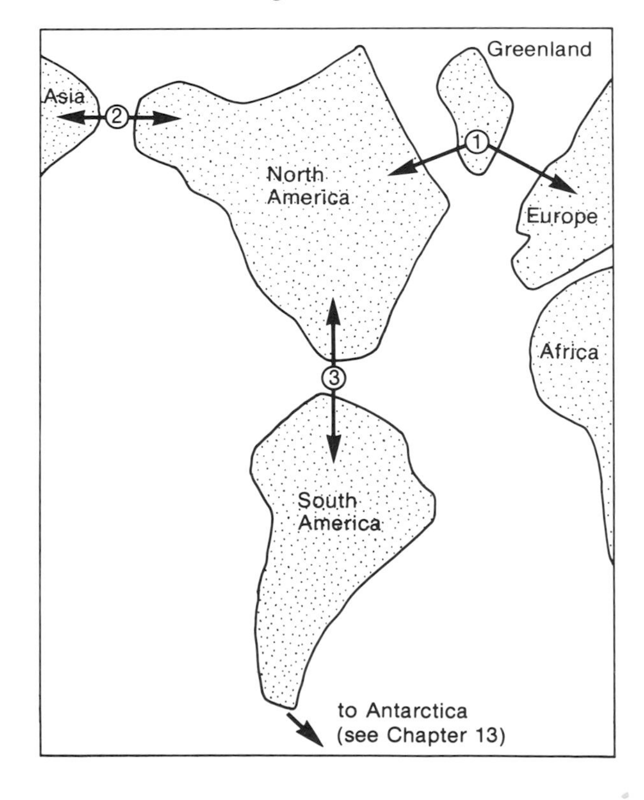

all the other North American species to spread at the same time suggests that crossing was difficult and that no continuous land bridge existed.

At the end of the Pliocene, about two million years ago, a new land connection developed across the Panamanian Isthmus, allowing mammals from each continent to cross to the other, an event which was to become known as the 'Great Interchange'. Its effect on the ecological balance of the faunas of the two continents was immense and has long been a favourite topic of study among palaeontologists. Among the mammals which spread from North to South America were mastodon, horse, tapir, deer, llama, peccary and various placental carnivores including puma, sabre-toothed cat and bear. Some of these were themselves only recent arrivals in North America from Eurasia. Before the end of the Pleistocene, puma and horse were in Patagonia. Among the South American species that went north were ground sloths, a glyptodont, armadillos, opossum and some rodents.

The possibility of a land bridge, allowing movement of animals between South America and Antartica at an early date, long before the Quaternary, will be discussed in chapter 13.

In their movements across the Panamanian and Beringian land bridges some mammals succeeded in penetrating further from home than others. By the end of the Pleistocene the ground sloth, *Megalonyx*, had reached Alaska. The glyptodonts reached only the southern part of the North American continent; and the litopterns, although still surviving in South America, failed to establish even a bridgehead there. In North America the opossum had become extinct at the end of Oligocene and was replaced by new stock from South America during the Great Interchange. This animal is still spreading north at the present day. Some mammals survive today only in their new homes, for example the porcupine, *Erethizon*, (of South American ancestry) in North America and the spectacled bear (of North American ancestry) in South America. Some mammals, probably in consequence of eustatic fluctuations of sea-level, found themselves shut off on islands. Examples include ground sloths on some of the Caribbean islands and elephants (which became much reduced in size) on islands near the coast of California.

By Pleistocene times the mammalian faunas of North and South America had become a mixture of quite extraordinary variety. We may divide the various forms into a number of groups, according to their geographical ancestry. Firstly there are the Pleistocene descendants of very early North and South American ancestors. The classic example is the living horse *Equus*, which evolved on North American soil over a period of 50 million years from an ancestor close to *Hyracotherium*, mentioned above. Likewise, the South American sloths, armadillos and litopterns are descendants of very early South American stock. Secondly there are the Pleistocene descendants of species which, though themselves immigrants to the American continents, arrived there so long ago that there was still time for them to evolve into later forms. The mastodon, *Mammut*, for example, which survived on the North American continent until at least the end of the Pleistocene, was a descendant of Miocene immigrants from Asia. Lastly, there were the mammals that did not arrive in North or South America until the Pleistocene (or very late Pliocene): for example the true mammoth, brown bear and bison from Asia to North America; and (as previously mentioned) the mastodon, horse and sabre-toothed cat from North to South America; and the ground sloths, glyptodonts and opossum from South to North America.

So far we have considered mammalian movements on the American continents only within an assumed framework of land masses similar to those that we know today, which were either connected to or isolated from one another by seaways. But, especially in the Beringian area, the story was never as simple as this. During Pleistocene times the Bering Straits did indeed open and close repeatedly as a result of eustatic changes of sea-level related to the advance and retreat of the world's ice sheets. During the interglacials sea-level was high, seals and whales could move freely between the Pacific and Arctic oceans, and movement of terrestrial mammals between Siberia and Alaska was much restricted. At times of glacial advance sea-level fell drastically, a wide land bridge appeared, and interchange became possible once more. But mammals arriving in Alaska did not find the whole of both American continents open for their colonization. Great ice sheets stood in their way. Similar conditions probably prevailed during each of the glacial maxima.

Since the ice advance which reached its climax 18 000 years ago (Wisconsinan Glaciation of North American nomenclature, stage 2 of the deep sea record) obscured much of the evidence left by earlier glaciations, the details of this episode have been reconstructed with the greatest precision and will be the focus of our attention here. At this time, as we have seen (in chapter 2), two huge ice sheets formed in Canada – the Laurentide in the Hudson Bay region and the Cordilleran to the west. Until recently it was believed that the corridor between these ice sheets closed for several thousand years as the two masses abutted together, forming a continuous barrier (which was also a barrier to mammalian migration) from the Pacific to the Atlantic. More recent studies now suggest that, although closure may have been complete during some of the earlier glacial advances, during

Wisconsinan times the corridor may have in fact have been blocked for only a millenium or less; or not at all. Ice dammed lakes and adverse ecological conditions along the length of the corridor would nevertheless have discouraged or attenuated faunal exchange; and the general concept of a barrier to migration still remains valid. Even at the times of glacial maxima central Alaska and part of the Yukon remained unglaciated and were faunally a continuation of eastern Siberia, which also remained mostly unglaciated. The vast area of unified land (which included also a 1500-km-wide belt of present-day sea floor) that appeared, has been named Beringia by palaeogeographers (fig. 2.13, p. 25). It had a rich mammalian fauna, with both Asian and American elements, and was probably a centre of mammalian diversification in its own right.

Among the many mammals that may have found themselves temporarily shut off from the rest of the American continents in Beringia during Wisconsinan times were the American mastodon, woolly mammoth, camel, bison, musk ox, horse and probably the ground sloth, *Megalonyx*. The yak and saiga antelope, from Asia, just managed to establish a bridgehead on the American side of the Bering Straits at about this time.

The Wisconsinan was also the time when man first arrived on the American continents. Although it is generally agreed among archaeologists that he came from Asia, the precise chronology of his arrival is still only imperfectly known. An admittedly controversial carbon date of 27 000 years for an artificially worked bone scraper found in the Yukon suggests that man had already arrived there long before the Last Glacial maximum 18 000 years ago. But that locality is situated on the Siberian side of the ice barrier. Did man manage to cross what is now Canada and begin to populate the area to the south before two great ice sheets formed; did he manage to make his way along the corridor; could he have by-passed the ice by travelling along the coast or by boat; or did he have to await its partial retreat before he could pass? Several carbon dates from both North and South America (for example 19–17 000 years for an occupation level in Meadowcroft Rockshelter, Pennsylvania, and 13 000 years for the mastodon butchering site of Taima-Taima in Venezuela, see fig. 4.3, p. 37 and chapter 14), suggest that man was present south of the ice before the ultimate amelioration of climate.

The Pleistocene was a time of the extinction of many of the large mammals on both American continents. By the early Holocene, the mastodon, sloths, glyptodonts, sabre-toothed cats and even the horse were extinct in both Americas. In the meanwhile the horse had nevertheless managed to cross the Bering Straits into the Old World, where it still survives at the present day. It was not until the arrival of Europeans during historic times that this animal was reintroduced. In some places herds, descended from individuals which have since turned feral, live once more in the wild state.

The study of Pleistocene mammals in the Americas goes back for more than two centuries. Since mastodon bones are very large and since the American mastodon survived until a very late stage in the Pleistocene (perhaps only 10 000 years ago) this animal attracted very early attention. The mastodon excavated by the family Charles, Titian, and Rembrandt Peale, in 1801, acquired a special place in American history through its immortalization in Rembrandt Peale's enchanting painting showing the excavation in progress (fig. 12.2). Thomas Jefferson, President of the United States from 1797–1814, lost no opportunity in obtaining fossil bones and teeth. In 1807 he made arrangements for excavations to be carried out at Big Bone Lick, a well-known mastodon locality in Kentucky and, for a time, a room in the White House was specially set aside for his collection.

Having considered the geographical background of the American mammalian faunas let us now examine some selected fossil mammal localities in more detail. Two examples are chosen here, both of them dating from quite late during the Wisconsinan. They are the asphalt pits of Rancho La Brea, California, and the Cave of Ultima Esperanza, Patagonia.

The Asphalt pits of Rancho La Brea

Recently described as a 'gigantic fossil time capsule', the La Brea asphalt deposits, on the outskirts of Los Angeles, have proved to be one of the richest sources of Pleistocene mammalian remains in the world. It has been estimated that, over the years, more than a million bones have been excavated there. Not only have the magnificiently preserved skeletal remains of a great variety of mammals and birds (some of them very spectacular species) been found in astonishing local concentrations, but the evidence provided by the associated insect and plant remains and by geological studies has provided palaeontologists with an unrivalled opportunity for palaeoecological and taphonomic studies.

History of the investigations

Documented knowledge of the La Brea asphalt pits extends back as far as 1769, when Father Crespi described coming upon several 'springs of pitch', the hardened residue of which made excellent fires. José Longinos Martinez, travelling through the region

Fig. 12.2 *Excavating the Peale mastodon; found near Newburgh, New York, in 1799 and excavated in 1801. The painting, by Rembrandt Peale, gives a delightful insight into the techniques employed for excavating a swamp during the early nineteenth century. Water is being removed by a chain of buckets lifted by a man-powered treadmill; whilst on the left labourers can be seen throwing mud out of the hole up a series of platforms. Of the three men holding the scroll that on the left is Dr Charles Peale and the other two are believed to be Titian and Rembrandt Peale. (Photo of painting belonging to Mrs Bertha White, by courtesy of the American Museum of Natural History).*

in 1792, described how animals were still becoming accidentally entrapped in seepages of asphalt in a manner similar to that which we now know probably also occurred in Pleistocene.

> *'In the vicinity of San Gabriel are other pitch springs, and near the Pueblo de Los Angeles more than twenty springs of liquid petroleum, pitch, etc. To the west of the said town, in the middle of a great plain of more than fifteen leagues in circumference, there is a large lake of pitch, with many pools in which the bubbles or blisters are constantly forming and exploding. They are shaped like conical bells and make a little report when they burst at the apex. I tried to examine the holes left by the bubbles, but they explode in such rapid succession that they give one no opportunity to do so.*
>
> *The variety of these great masses of tar, the movement that one sees in all of them at once, the pitchy smell, the sight of that great lake of strange matter, and all these phenomena together present an astonishing and frightful picture, reminding one of those painted of the infernal cavern. If one stands upon the more solid masses of pitch one seems to be rising insensibly from the ground. The plain where one stands sinks in the form of a cone with the apex downward. In hot weather animals have been*

seen to sink in it, unable to free themselves because their feet were stuck, and the lake swallowed them. After many years their bones come up through the holes, as if petrified. I brought away several specimens of them.

For a great distance around these volcanoes there is no water, and when the heat of the sun forces birds to seek it they alight upon the lake, mistaking it for water. All the birds that do so are caught by the feet and wings until they die of hunger and thirst. Rabbits, squirrels, and other animals are deceived in the same way, and for this reason the gentiles (the unconverted Indians) keep a careful watch at such places in order to hunt without effort.'

The subsequent development of the asphalt deposits during the nineteenth century as a source of material for making roads resulted in the discovery of ever increasing quantities of bones and an explosion of scientific interest in the locality. In 1906 John Merriam, of the University of California, conducted excavations from which he obtained the fossil remains of elephant, horse, bison, camel, ground sloth, dire wolf and sabre-toothed cat. He noted that those of carnivores were proportionately more abundant than might be expected and concluded, like Longinos before him, that the asphalt pools had acted as traps for unwary animals secondarily attracted (figs 4.7, p. 40, and 12.3).

Fig. 12.3 *Probably the earliest artist's reconstruction of a Pleistocene scene at Rancho La Brea. A sabre-toothed cat,* Smilodon, *disputes with two dire wolves,* Canis dirus, *the carcass of an elephant mired in the asphalt. A similar fate is already befalling the dire wolves. Painted by R. B. Horsfall in 1911, two years before the commencement of the main excavations of the Los Angeles County Museum. (From Scott,* A history of land mammals in the western hemisphere*).*

F. CHAUVENET
NUITS
FRANCE

Fig. 12.4 *Rancho La Brea, about 1915. In the foreground is an overgrown and flooded former excavation for asphalt, known as the Lagoon. Beyond are some of the derricks of the Salt Lake Oil Field and, in the distance, are the Santa Monica Mountains, where Hollywood is situated. (Photo: Natural History Museum of Los Angeles County).*

Fig. 12.5 *Excavation of part of a skeleton of a ground sloth and other bones from one of the pits at Rancho La Brea, about 1914. (Photo: Natural History Museum of Los Angeles County).*

Further, larger scale and more careful, excavations followed, the greatest being that conducted from 1913–15 by the Natural History Museum of Los Angeles County (figs 12.4 to 12.6). The investigation of a number of different asphalt 'pits' resulted in the recovery of the remains of over 4000 individual mammals of 40 different species and also those of over 100 species of birds. Scientific studies of the remains commenced almost at once (Stock, writing in 1956, was already able to list over 200 publications) and they still continue at the present day.

Although the Los Angeles Museum excavation produced such an abundance of mammalian and bird remains, relatively little attention was directed at this stage to the smaller fossil material, such as plant and insect remains, which was associated with them. Some insect remains had been collected incidentally with the skeletal remains, but their importance was not realized until the 1940s, when Dr W. D. Pierce began working on specimens that he had retrieved from the brain cavities of some of the skulls.

In 1969, for the purpose of expanding the information obtained from the early investigations, the Los Angeles Museum (which in 1945 had drilled 87 test cores at Rancho La Brea) commenced a further and very successful excavation at one of the sites ('pit 91')

Fig. 12.6 *Mr Lytle sorting bones and skulls of sabre-toothed cats, found during the Rancho La Brea excavations of 1913–15, in the basement of the Los Angeles Museum: an early photograph. (Photo: Natural History Museum of Los Angeles County).*

Fig. 12.7 *Excavation in Pit 91 at Rancho La Brea, about 1970. Note the special care which is being taken. Only small implements are being used for the most delicate stages of excavation; and all the sediment is being packed into cans for sieving elsewhere. (Photo: Natural History Museum of Los Angeles County).*

which was known to have been only partly investigated (fig. 12.7). One of the problems of recovering the smallest fossils came from the asphalt itself, which could not be broken down with water. Breakdown was achieved, however, by boiling the sediment in the solvent chlorozene, whcih freed the small bones and teeth, insect remains, fragments of wood, seeds, molluscs and other fossils ready for sieving, drying and sorting by hand in the laboratory. Smaller samples of sediment were also examined in greater detail for pollen and other remains that would pass through the sieves.

Meanwhile, during the two centuries that had elapsed since the locality first began to attract attention, there had been great man-made changes in the area. As late as 1915, although there existed a forest of derricks associated with the extraction of oil with which the asphalt was associated (fig. 12.4), it was still mostly open country. In that year 9 hectares of land around the asphalt pits were presented to the County of Los Angeles by its owner, Mr G. A. Hancock, with a request that their scientific features should be preserved and exhibited. The area was declared a public park, named after its owner. Development was by

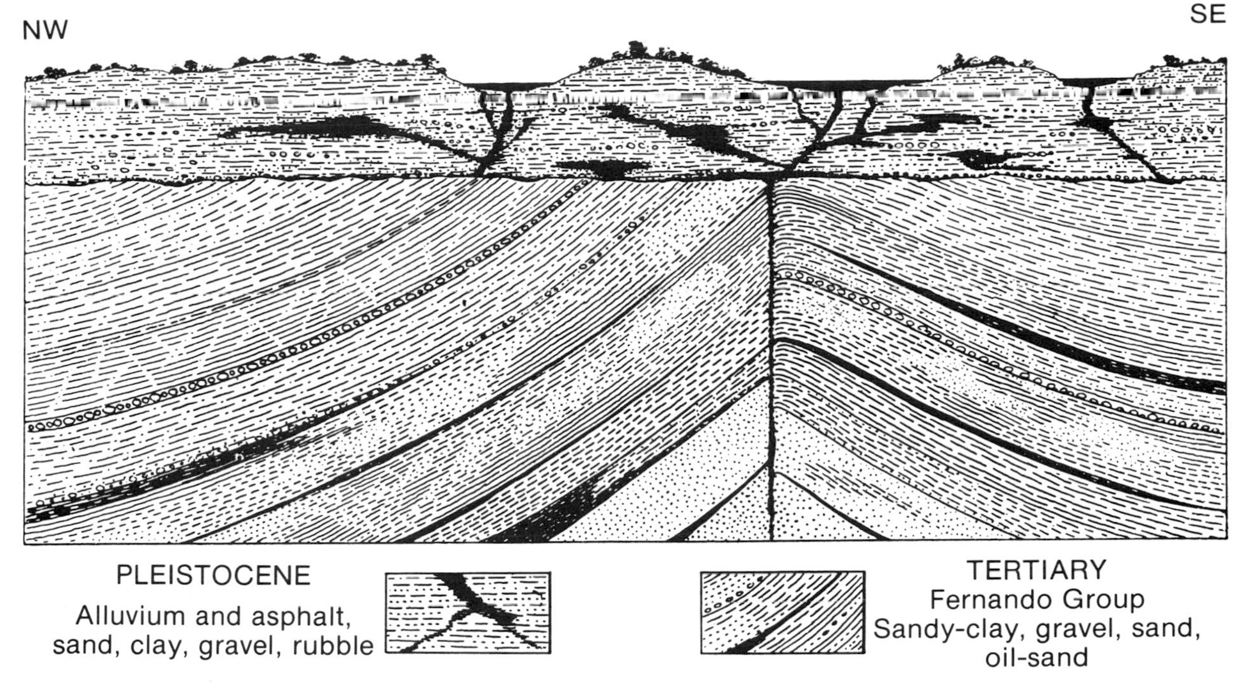

Fig. 12.8 *Generalized cross-section showing the geological structure at Rancho La Brea. (From Stock, 1965).*

stages, culminating on 13 April, 1977 (by which time the park was entirely surrounded by the outskirts of Los Angeles) when a new purpose-built museum, the George C. Page Museum, was opened there to explain to the public the story of the asphalt deposits and to display some of the finest skeletons.

Let us consider in greater detail how the evidence from the various excavations has been interpreted and the conclusions that have followed. How did the asphalt deposits accumulate and what sort of mammals are represented in them? How did the remains find their way into the asphalt in such quantity and how old are they? Was man present in the region at that time? What do the plant and insect remains tell us about the environment of the mammals?

The origin of the asphalt deposits

The asphalt deposits of Rancho La Brea, in which so many fossil mammalian remains have been found, owe their occurrence directly to the existence of the nearby Salt Lake Oil Field and to chance circumstances of geological structure that made it possible for some of the oil to find its way to the surface at this locality. These seepages provided the first indication to oil speculators at the end of the nineteenth century of the wealth lying beneath their feet.

There are two main series of sedimentary strata in the region: an earlier series of folded and faulted Tertiary deposits (the Miocene Puente shales and sandstone of the Pliocene Fernando group), unconformably overlain by the Pleistocene marine San Pedro formation followed by the predominantly terrestrial Palos Verdes Sand (fig. 12.8). The oil originates in the Tertiary part of the sequence, in which are numerous impermeable layers, and the concentration of oil and gas escaping on the surface at Rancho La Brea is interpreted as evidence of considerable fracturing of the underlying rocks at the locality.

On coming in contact with the air oil gradually becomes more viscous or even solidifies when the temperature is low, as the result of oxidation and the loss of its volatile constituents (it is then known as asphalt), a process which was already taking place in Pleistocene times and still continues on a small scale today. At places where asphalt seeps onto the surface, insects and small mammals still sometimes become mired; and thus become potential fossils, in the same manner that must have occurred frequently during the Pleistocene.

The earliest excavations of the Los Angeles

Museum were often crater shaped and it was commonly inferred that the resulting pits represented asphalt pools of inverted conical shape, into which the unwary animals had become mired and submerged. It was also assumed that because of convection currents in the asphalt there was no meaningful stratification of the deposits in the pits; and that the absence of any associated skeletons was the result of dispersal of the individual bones by this process. More recent studies, especially by Woodard and Marcus, now suggest that, although some of the largest bone accumulations might still be explained in this way (for example 'pit 9', which contained 15 of the 21 individual mammoths known from La Brea), the popular idea of pools of liquid asphalt must be substantially modified. They found that, in fact, the same sequence of stratified deposits continued into the walls of many of the excavations, the sides of which had been of an entirely artificial nature. They concluded that the fossiliferous strata at Rancho La Brea had accumulated principally as a gradual build up of fluviatile and other sediments on a series of old land surfaces, during a period of time when asphalt seepage had been taking place discontinuously. The rolled condition of some of the bones suggested local stream action; and Pierce's studies of the insect remains showed (since different species of insect denote different stages of decomposition of a carcass) that some of the bones had remained unburied on the surface of the ground for as long as seven months.

Stream action by itself, however, although it may have re-worked some of the bone deposits, was insufficient to account for the great accumulations of remains at Rancho La Brea. When obscured by leaves and other débris then asphalt seepages would have been a hazard to mammals which might accidentally become mired by the feet, although the larger and more powerful ones would usually have been able to escape. After death, trapped animals would usually fall on one side so that the limbs and the underside of the trunk became buried in asphalt, while the rest of the carcass remained exposed. Predatory mammals and birds would remove exposed parts of the carcasses (as a result of which bones of the lower limbs are more abundantly preserved than other bones), sometimes becoming trapped themselves.

Some of the asphaltic bone deposits had apparently been secondarily impregnated with seeping asphalt only after deposition, and it was evident that the distribution of fossil bones was to some extent controlled by the disposition of the asphalt, since this acted as a preserving medium, whereas bones in non-asphaltic parts of the deposits were often in a poor state of preservation and might even have disappeared.

The mammals and other vertebrates of Rancho La Brea

The entire list of mammalian species represented in the asphalt deposits is truly astonishing. Extinct forms include the American mastodon, Jefferson's mammoth, the Shasta ground sloth, Jefferson's ground sloth; Harlan's ground sloth, western horse, tapir, two species of bison, extinct peccary, two species of camel, small tar-pit antelope, sabre-toothed cat, American lion, dire wolf and giant short-faced bear. Species that still survive in North America include mule deer, pronghorn antelope, puma, bobcat, coyote, wolf, grey fox, black and grizzly bears, badger, two genera of skunks, long-tailed weasel, ornate and desert shrews, black-tailed jackrabbit, two cottontails, ground squirrel, pack rat, gopher and various small rodents. The improved recovery techniques of the latest phase of excavation have led to the addition of mole and bat to this list.

Following the completion of the 1913–15 excavations it was possible to make a detailed study of the relative proportions of the various mammalian species. This work was undertaken by C. Stock, who displayed his results as a series of pie diagrams (fig. 12.9).

The most astonishing aspect of this census, previously commented upon by Merriam, is the disproportionate over-abundance of carnivores which make up nearly 90 per cent of all the individuals represented from the various excavations. In a normal mammalian community, carnivores comprise only a small proportion of the fauna (otherwise they would soon starve), so that some selective process must have influenced their entombment at Rancho La Brea.

Most abundantly represented of the carnivores is the dire wolf, closely followed in numbers by the sabre-toothed cat, *Smilodon*. Remains of coyote are also common, those of the American lion a poor fourth. The occurrence of such immense quantities of remains of these animals has led to considerable discussion of their habits. Did they habitually scavenge to the extent suggested by the fossil evidence and can the extent to which they hunted or scavenged be established from their anatomical structure? Merriam & Stock, writing in 1932, concluded that *Smilodon* usually killed its prey by stabbing with its knife-like upper teeth. Its front legs, however, were especially sturdy, indicating great striking and grasping strength and a lesser degree of fleet-footedness than in a present-day lion or tiger. The dire wolf was an animal with a large head with powerful teeth capable of breaking bones. Its legs were of slightly different proportions from those of living timber wolves, suggesting that it too was less fleet of foot.

In the light of the most recent studies of the

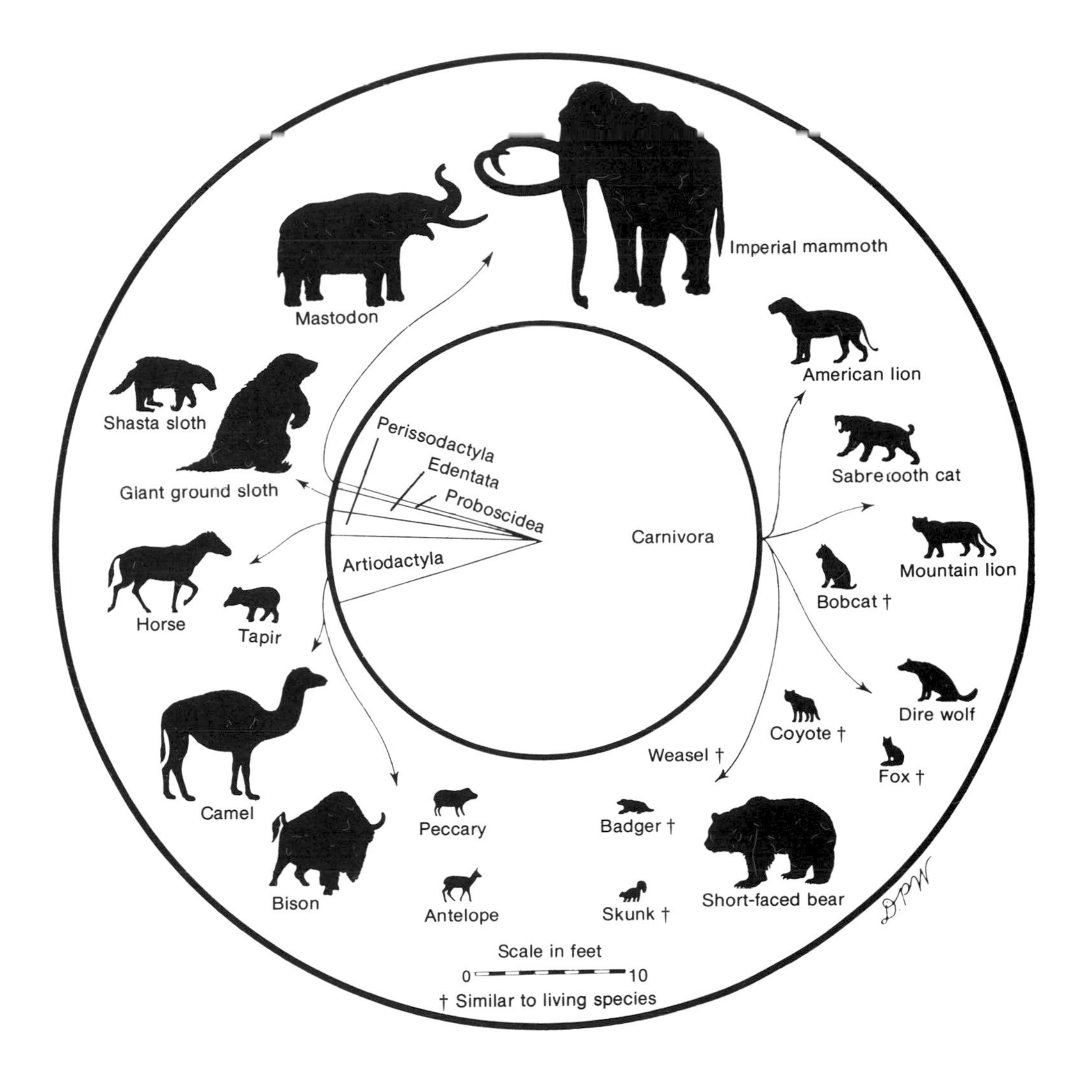

Fig. 12.9 *Diagram illustrating the relative member of individuals of the various mammalian species (rodents and other small forms omitted) from the Rancho La Brea asphalt deposits. (From Stock, 1956).*

Fig. 12.10, pp. 178–179 *Artist's reconstruction of the landscape around the* Mylodon *Cave at Puerto Consuelo, Ultima Esperanza, Patagonia, South America, about 12 000 years ago, based on geographical data, on mammalian bones and dried skin and on plant remains (including ground sloth dung contents) excavated from the floor of the cave.*

A jaguar scatters a small herd of the extinct South American horse, Onohippidium *(height at shoulders about 1·25 metres), while a family of ground sloths,* Mylodon *(height, all four feet on the ground, about 1 metre) is involved in various activities outside the cave.*

The body of water which can be seen in the mid distance is an arm of the sea, known as Eberhardt Fjord, beyond which are mountains (foothills of the Andes), carrying small glaciers, which terminate in the sea. Only a few thousand years before Andean ice had extended over and beyond the cave.

The scene shown is one of recent de-glaciation, with a cool moist climate (800–1000 mm annual precipitation), the flora dominated by grasses and sedges. Also prominent are 1–2-metre-high trees of the evergreen southern beech, Nothofagus betuloides *and the crowberry,* Empetrum rubrum. *The cave has a more open aspect in this scene than at the present day, its entrance now being largely obscured by trees of the deciduous southern beech,* N. pumilio *and* N. antarctica.

Painted by Peter Snowball, under the direction of the author, 1985, from data provided by D. M. Moore and E. C. Saxon.

behaviour of present-day carnivores (especially Kruuk's study of hyaenas in Africa) a division into hunters and scavengers has largely disappeared. Most living hunting carnivores will also scavenge when they can and it would be surprising if those at Rancho La Brea had not taken advantage of the opportunity existing there. The asphalt seepages were nevertheless exceptional local phenomena, which could not provide an unlimited source of food, and it is not necessary to postulate that the most abundantly represented carnivores feeding there were primarily scavengers. Miller, who made a study of the age distribution of the various individuals of *Smilodon* excavated at Rancho La Brea based on the examination of over 2000 skulls and skull fragments, writing in 1968, recorded that animals of all ages were represented and that no particular age group had been more liable to entrapment than others.

No less remarkable than the mammalian remains from Rancho La Brea are those of the birds. Because of delicate construction the bones of birds readily disintegrate, so that their survival in any Pleistocene deposit is a rare occurrence. The La Brea collection, however, (studied over many years by Dr Hildegarde Howard) is remarkable not only for the perfection of the remains but also for their quantity and for the abundance of species represented (nearly 140); which provide a detailed picture of the avian fauna which coexisted with the mammals around the asphalt seepages. Apparently birds were especially vulnerable since only a small amount of asphalt on their wings would prevent them from flying.

Most spectacular is the extinct condor-like vulture, *Teratornis merriami*, which had a wing span of nearly four metres and ranks among the largest known birds of flight. Among the other birds of prey are five extinct species of vulture and four extinct eagles.

For those concerned with determining how the mammalian remains of Rancho La Brea accumulated the bird fauna is of considerable interest, since there is a disproportionate abundance of day-flying birds of prey, suggesting that they, like the mammalian carnivores, were attracted by the sight of carcasses. Remains of night-flying birds of prey such as owls were less abundant than would be expected, suggesting that the asphalt may have partially hardened at night as the result of cooling, thus becoming less of a hazard of entrapment. The occurrence of remains of ducks and geese suggests that there were small ponds or water courses in the La Brea area and this is further confirmed by the remains of pond turtle, and of frogs and toads.

The plant remains

The plant remains from the latest excavations in pit 91, among which are numerous species not previously recorded from Rancho La Brea, indicate a number of distinct plant communities: a pine forest association including Montery Cypress, which would have flourished under cool humid coastline conditions with fog and at least 50 cm rainfall; a coastal redwood element with *Sequoia*; an inland dry foothill plant assemblage; a shallow fresh water pond assemblage; a stream bank assemblage; and a drier grassland situation. This mixture of plant communities from different habitats is interpreted as the result of transport of plant débris by streams which flowed through a series of different environments before reaching La Brea; not as evidence of any change of climate while the deposits of this particular site were accumulating.

The age of the La Brea mammal remains

Most of the mammalian species represented at Rancho La Brea were widespread in North America during the later part of the Pleistocene, just before the wave of extinction that caused the disappearance of so many of them. They include some immigrants of not very long standing on that continent, most notably the bison (from Asia) and the three ground sloths (originally from South America). On faunal grounds a late Wisconsinan age would be expected.

For a long time it was not possible to establish an absolute date for the La Brea remains, because impregnation by asphalt introduced dead carbon and made radiocarbon dating impossible. It was generally assumed, however, that their burial may have continued over a considerable period of time. Since 1967 it has been possible to obtain reliable carbon dates by a new procedure which allows the removal of petroleum and humic acid contaminants. Dates obtained by this new method for bones of *Smilodon* range from over 33 000 years to to less than 12 000 years old. These dates and similar dates for other mammalian species and plants confirm the supposed Wisconsinan age of the deposits and the theory of spasmodic outpourings of asphalt over a long period of time.

Pleistocene man at Rancho La Brea

Man's contemporaneity with the extinct fauna in the La Brea region was for a long time a matter of uncertainty.

Parts of a skeleton of a 25-year-old woman found in one of the asphalt pits in 1914, the only human remains found at La Brea, were regarded by Stock as being perhaps a few thousand years old, but not tens of thousands. A radiocarbon date obtained in 1970 showed them to be 9000 ± 80 years old, less than

300 years younger than the latest remains of *Smilodon.* But these remains are not necessarily from the first man or from the last *Smilodon* to live in the area. Did the two coexist in California, even if only briefly?

During recent years, a re-examination of the old collections of the Los Angeles Museum by Dr G. J. Miller has brought to light a series of limb bones of *Smilodon* and of the American lion and a foot bone of a bison (total five specimens) showing diagonally cut paired grooves, penetrating through the outer part of the bone into the marrow cavity, which are very difficult to explain other than as the work of man. The purpose of these artefacts is not clear, but man, it seems, had already reached California before the extinction of these animals there.

The Great Cave of Ultima Esperanza

Until the 1830s, only a few extinct mammals were known from South America. These included the mastodon, *Glyptodon* (a relative of the armadillo with a bony carapace rather like the shell of a tortoise) and the huge ground sloth *Megatherium.* For the first extensive collection of such remains it was necessary to wait until the 1830s. In 1831 HMS *Beagle* set out from Plymouth under the command of Captain Fitzroy on a voyage for the purpose of surveying the southern parts of America and of subsequently circumnavigating the world. This continued until October 1836. On board was the young naturalist Charles Darwin, then only 22. Throughout the voyage Darwin collected extensively, and, among the many natural history treasures that he brought back was a large collection of South American Pleistocene mammals. This contained remains of most of the major groups of mammals now recognized from South America and several were new to science. Of special interest were *Toxodon platensis,* a herbivore the size of a rhinoceros and the camel-like *Macrauchenia patachonica.* There were also fossil remains of ground sloths, glyptodonts and, surprisingly, of horses. When the Spanish had arrived in South America during the sixteenth century no horses were present, yet here was definite proof that these animals had existed there in the remote past.

It is ironical that, working round the south coast and up the west coast of Patagonia, the *Beagle* passed within a short distance of a cave which was later to become the best-known fossil mammalian site in South America, without Darwin ever knowing of its existence. Although long known to Indians, and a place of periodic prehistoric occupation, this was not discovered by Europeans until about 1895, when it was visited by Herman Eberhardt, who had recently established a farm nearby. The cave was of spectacular dimensions; with a vast opening 120 metres wide and 30 metres high, leading into a chamber 200 metres long, the irregular floor of which was covered with sediment which varied locally in dampness from moist to very dry and dusty. It was subsequently to become known by a variety of names; Eberhardt Cave, after its discoverer; *Mylodon* Cave, after its most famous fossil; and Last Hope Cave (Cueva Ultima Esperanza), after its locality (fig. 12.10 and fig. 2.7, p. 20).

In 1896, Eberhardt discovered mammalian bones and an extraordinary piece of dried skin with short yellow-brown hair in the cave. These he subsequently showed to Otto Nordenskiold, the Arctic explorer. The following year a further fragment was discovered by Dr F. P. Moreno, hanging on a tree. Although at first unable to identify the animal concerned, he noted that there were many small nodules of bone in the skin (fig. 12.12), rather like those from the extinct ground sloth *Mylodon*. Perhaps *Mylodon* still survived in this region, undetected?

The interest which a published description of this skin aroused led to two excavations being conducted in the cave in 1899, the first by Erland Nordenskjiold (nephew of Otto, name with additional j), the second by Dr Rudolph Hauthal, geologist of the La Plata Museum. Although Nordenskjiold had been the first to excavate in the cave, he did not publish his findings until 1900, whereas Hauthal's report appeared in 1899.

The two excavators dug at slightly different places and, there being local differences in the stratigraphy of the deposits, the interpretations they placed on their findings were conflicting. Both encountered thick layers of ground sloth dung, which suggested to them that these animals had actually lived in the cave. They also found the remains of other mammalian species, including the extinct horse (*Onohippidium*), *Macrauchenia*, guanaco, jaguar, dog and man. Hauthal found two man-made bone points below a layer of sloth dung. He concluded that man and *Mylodon* had coexisted in the cave and, believing that he could identify chopped hay, he suggested that *Mylodon* had been kept and fed in a corral in the cave by its human inhabitants. Nordenskjiold, on the other hand, finding a deposit with butchered bones of guanaco lying *above* a layer of sloth dung, refuted both Hauthal's claim that the artefacts were beneath it and also that the cave had been used as a corral. Both assumed, however, that man and *Mylodon* had coexisted in the neighbourhood.

The interpretations which Hauthal and Nordenskjiold had placed on their findings were to remain a subject for discussion and speculation for a further three quarters of a century before any further scientific excavation was to be undertaken in the cave. Meanwhile, not only had there been many improvements of scientific technique, but the timing of man's spread into South America and the reasons for the extinction of so many large mammals at the end of the Pleistocene had become focal points for attention among archaeologists and palaeontologists. In both respects the cave of Ultima Esperanza is of unique importance. Its extreme southern situation places it almost at the geographical limit of man's spread into the southern hemisphere. Thus a date for man's first arrival there would have a far-reaching impact on the interpretation of human occupation sites throughout both American continents. Furthermore, it appeared that *Mylodon* had survived in the area until after man's arrival, making the site an important one for extinction studies.

As late as 1970 the cave still presented innumerable unanswered questions. What was the climatic history of the region? By what process had the skin been preserved? What was the sequence of deposits in the cave, about which Hauthal and Nordenskjiold had disagreed? How old were the various deposits; and what was the relationship between man and *Mylodon* in the cave?

Past climatic history of the area round the cave
Today, the snow-capped Andes can be seen from the entrance of the cave at a distance of about 20 kilometres. The cave lies within the moraine belt of the Last Glaciation and only a relatively small climatic deterioration would bring glaciers over it once more.

Fig. 12.11 *Looking out of the Cave of Ultima Esperanza; Stone Hill on the left, old excavation pits in the foreground. Figures standing in front of the Stone Hill provide scale. (Photo: M. Bruggmann, 1976).*

The dates of the most recent glacial advances in the Andes have recently been studied in some detail, for example by Mercer, who found that, a little further north, the Chilean glaciers reached their maximum extent about 19 400 years ago, shrank to half their volume by 16 000 years and then re-advanced to a smaller maximum after 14 800 years ago. Climatic evidence suggests that any deposits in the cave which are not of glacial or pre-glacial origin are likely to be later than the last retreat of the ice, i.e. less than about 14 000 years old.

Preservation of the *Mylodon* skin

The preservation not only pieces of skin of *Mylodon*, with the hair still in place, but also of claws of this animal, of the hoofs of *Onohippidium* and of great quantities of *Mylodon* dung, requires explanation. Most caves are damp and bacterial decomposition soon leads to the destruction of the soft tissues of any animal remains which find their way underground. Even if a cave is too dry for this to occur (and some instances of the preservation of soft tissue under arid conditions are indeed known – for example the skin, hair and dung of ground sloths in Rampart Cave, Arizona) some beetles can destroy tissues with very low water content. Why had this not happened in Rampart Cave and at Ultima Esperanza? There are two other factors that may have assisted preservation, locally, in the latter cave. Nordenskjiold recorded that, at one place, there was a one-metre-thick bed of well preserved dung separated from an overlying bed of gravel by a layer of magnesium sulphate ('Epsom Salts'; another form of magnesium sulphate is kaiserite, which is strongly hygroscopic and might be expected to draw water out of the surrounding sediments), which he suggested had protected it from decay. Investigation of the centres of lumps of dung for bacteria showed them to be completely sterile, though some grew mouldy when exposed to moisture. The dung layer, he believed, had once been more extensive, but had decayed in the parts of the cave which were damp.

In 1970, Mr Reginald Harris, a specialist in freeze-drying at the British Museum (Natural History) made a study of one of the pieces of *Mylodon* skin in the collection of that institution and, finding even the nuclei of the cells preserved, concluded that the skin had been naturally freeze-dried at a time of very cold climate. Such an explanation is entirely feasible, since conditions inside the cave during the period of glacial retreat must have been extremely cold (fig. 12.13).

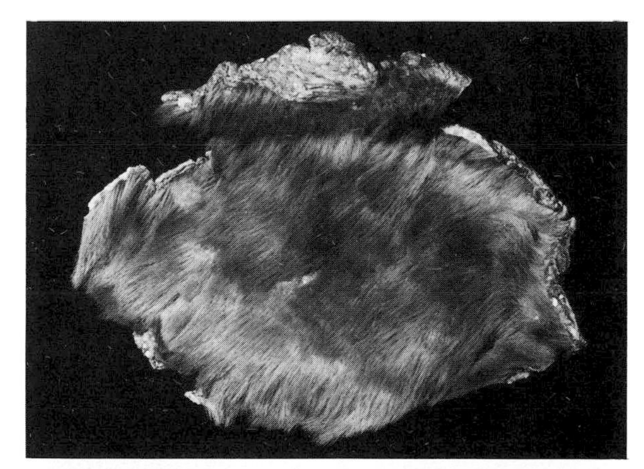

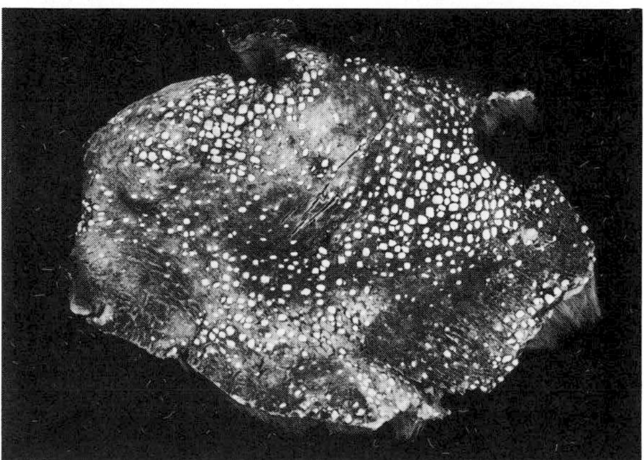

Fig. 12.12 *The fragment of* Mylodon *skin from the Cave of Ultima Esperanza collected by Dr Moreno in 1897 and subsequently sent to Dr Smith Woodward in London; top outside, above inside. Longest dimension 500 cm. Note the bony ossicles within the skin. It was figured in the* Proceedings of the Zoological Society of London *in 1899 and is still preserved in the British Museum (Natural History). (Photo: BM(NH)).*

Fig. 12.13 *Microscopic section of part of one of the fragments of skin of* Mylodon *from the cave of Ultima Esperanza. The uppermost layer is the epidermis, with a longitudinally cut hair (right) projecting through it. Below are the dividing cells of the dermis. Such magnificent nuclear preservation is likely to be the result of freeze drying. Section prepared by Mr R. Harris and Mr D. W. Cooper. (Photo: BM(NH)).*

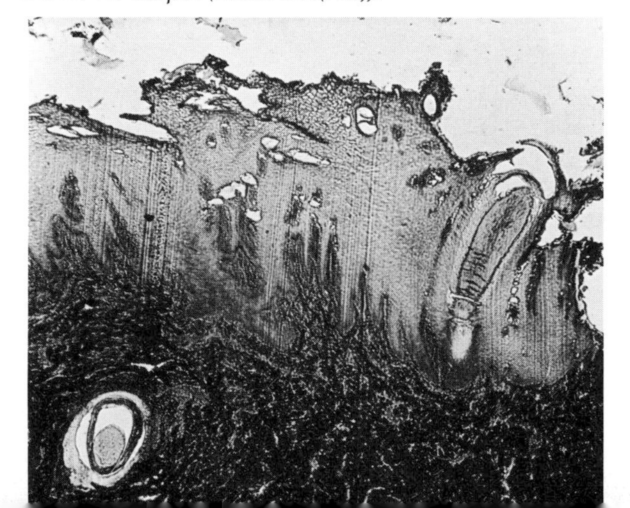

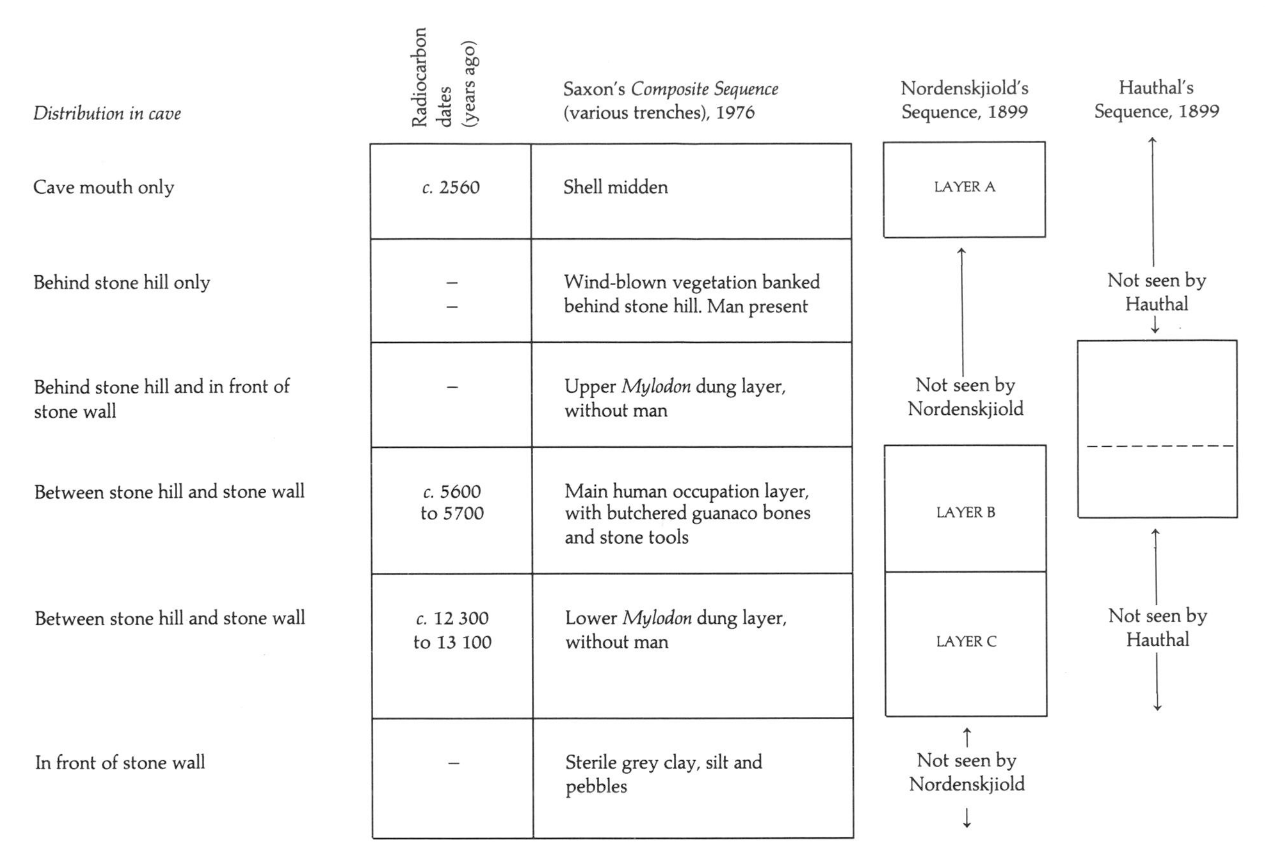

Distribution in cave	Radiocarbon dates (years ago)	Saxon's *Composite Sequence* (various trenches), 1976	Nordenskjiold's Sequence, 1899	Hauthal's Sequence, 1899
Cave mouth only	*c.* 2560	Shell midden	LAYER A	↑
Behind stone hill only	– –	Wind-blown vegetation banked behind stone hill. Man present	↑	Not seen by Hauthal ↓
Behind stone hill and in front of stone wall	–	Upper *Mylodon* dung layer, without man	Not seen by Nordenskjiold	
Between stone hill and stone wall	*c.* 5600 to 5700	Main human occupation layer, with butchered guanaco bones and stone tools	LAYER B	
Between stone hill and stone wall	*c.* 12 300 to 13 100	Lower *Mylodon* dung layer, without man	LAYER C	↑ Not seen by Hauthal ↓
In front of stone wall	–	Sterile grey clay, silt and pebbles	↑ Not seen by Nordenskjiold ↓	

Table 9. *Comparison of the sequences of strata found in the Cave of Ultima Esperanza by Nordenskjiold, 1899, Hauthal, 1899 and Saxon, 1976.*

Perhaps all these processes worked together – freeze-drying; the hygroscopic effect of the salt; perhaps the salts discouraged carnivorous beetles from laying their eggs on the sloth skin; perhaps it was too cold for the beetles. The result was the survival of some of the most remarkable fossil mammalian remains ever found in a deposit of Pleistocene age.

The sequence of deposits in the cave and their age

It was not until 1976 that the problem of Hauthal's and Nordenskjiold's conflicting excavation conclusions were resolved. During that southern summer Dr Earl Saxon, a biological archaeologist sponsored by the University of Durham in England, searched the floor of the cave for unexcavated parts of the deposit. His team excavated a series of trenches there, employing the latest scientific methods.

Saxon encountered a sequence of stratified deposits, more complex than that recorded by either Hauthal or Nordenskjiold. Whereas these excavators had each encountered only one layer of dung, Saxon found two. Here, apparently was the answer to the problem. Both these early excavators had investigated only part of the total sequence of deposits – in unrelated parts of the cave! Hauthal had found the upper layer of *Mylodon* dung, with human occupation below it. Nordenskjiold had found the lower layer with the same occupation level above it.

On the basis of his excavations and from the radiocarbon dates of bones and dung from various levels of the cave, Saxon established a more detailed sequence of events than either Hauthal or Nordenskjiold had been able to do (Table 9).

The earliest deposit on the floor of the cave was interpreted as glacial till. This was overlain by the lower layer of sloth dung, with radiocarbon dates of approximately 12 500 years for a sample of the dung itself and 13 100 years for a contained *Mylodon* bone. The glaciers had apparently already retreated from the cave by this time. Next followed a deposit with remains of guanaco, stone tools and an age of 5300 years, but devoid of remains of *Mylodon*. Then there was the second layer of dung without evidence of the presence of man, unfortunately so far without a radiocarbon date; then the vegetation sequence; and finally the shell midden, with an age of 2500 years.

A study of plant remains from the 1976 excavation by Professor D. M. Moore shows that, at the time of accumulation of the lowest fossiliferous deposits in the cave, there existed locally a flora associated today with cool wet sedge grasslands, as would be expected in an area of retreating ice. *Mylodon* was apparently living in an open, moist, cool boggy sedge-grassland, such as that which today occurs further west in Patagonia. Then followed a rise of evergreen forest dominated by the southern beech, *Nothofagus betuloides*, which reached its maximum about 7000 years ago and then declined, being replaced by deciduous southern beech or lenga, *N. pumilio*, forest. A break in the fossil plant record, covering the human occupation of about 5600 years ago and the subsequent fossil appearance of *Mylodon* remains in the upper dung layer, is followed by evidence of mixed evergreen/deciduous forest in which lenga gradually increases in importance to give dedicuous forest, about 2500 years ago, similar to that found in the environs of the cave in historic times.

Contrary to the beliefs of Hauthal and Nordenskjiold there is no evidence of man's and *Mylodon's* presence in the cave at the same time. Man's involvement is nevertheless not necessary to explain the great accumulations of *Mylodon* dung. Free-living *Mylodon* could well have used the cave for shelter. The Epsom Salts may have been an additional attraction. The mining of minerals in caves by salt deficient mammals is still known at the present day; for example by elephants living in the forests of Mt Elgon in Kenya, previously mentioned (pp. 74–5), which go into the dark zones of caves, in search of sodium sulphate, leaving the floors strewn with their droppings.

Saxon has shown that man and *Mylodon* apparently alternated in the cave. First the cave was frequented by sloths (but man was absent); then it was occupied by man, with no evidence of sloths; then the sloths returned on their own, apparently surviving beyond 5300 years ago, more than 5000 years after the first appearance of man elsewhere in Patagonia. Man's arrival could not therefore have caused the immediate extinction of *Mylodon*.

The apparent survival of this animal until so late a date created considerable surprise among those who had previously supposed its extinction to have occurred about 10 000 years ago, and has become a topic of lively discussion among those concerned with man's potential role in the disappearance of so many large mammals at the end of the Pleistocene.

Chapter 13 **The marsupial continent**

The first European explorers to reach Australia were astonished to find that the mammals living there, most of which were later to be classified as marsupials (only distantly related to the placental mammals which predominate throughout the rest of the world), were quite unlike anything they had seen before. Although other marsupials were already known from America. Cook's kangaroo (fig. 13.1), found by the expedition of 1770, was the first to become widely known, partly from Cook's own description:

> *'In form, it is most like the Jerboa, which it also resembles in its motion, . . .; but it greatly differs in size, the Jerboa not being larger than a common rat, and this animal, when full grown, being as big as a sheep: . . . The head, neck and shoulders are very small in proportion to the other parts of the body; the tail is nearly as long as the body, thick near the rump, and tapering towards the end: the fore-legs of this individual were only eight inches long, and the hind legs two-and-twenty: its progress is by successive leaps or hops, of a great length, in an erect posture; the fore-legs are kept bent close to the breast, and seemed to be of use only for digging. . .'*

As well as kangaroos, Cook also recorded bats, a 'kind of pole-cat', 'wolves' and

> *'an animal of the Opossum tribe: it was a female, and with it he' (Banks) 'took two young ones: it was found to resemble the remarkable animal of the kind which Mons. de Buffon has described in his Natural History by the name of* Phalanger, *but it was not the same. Mons. Buffon supposes this tribe to be peculiar to America, but in this he is certainly mistaken.'*

Thus we see that, although the marsupials were not formally separated from the placental mammals by zoologists until the nineteenth century, the affinities of the American and Australian marsupials were already beginning to be noticed before the end of the eighteenth century.

The present-day mammalian fauna of Australia

Australia differs from most of the faunal areas of the world previously discussed in being an island continent of great size, surrounded today by deep water so that mammalian exchanges to and from other continents (other than by swimming and flying mammals) are highly improbable, except with the assistance of man. This isolation goes far back into geological time.

In order to understand the significance of the Australian mammalian fauna we must first consider the zoological relationships of the groups represented. Zoologists now classify living mammals into three main sub-divisions; the placentals and marsupials, already mentioned, and the monotremes.

The placentals are characterized by the young being retained in the uterus, where they are nourished by a placenta, for a considerable length of time; there is never a pouch. The young, are fed with milk. Certain structural features of the skull are also characteristic. This group is of worldwide distribution and includes

Fig. 13.1 *Pencil sketch of Captain Cook's kangaroo made by Sydney Parkinson at Endeavour River, Queensland, Australia in July 1770. (Original preserved in the British Museum (Natural History)).*

most of the familiar mammalian families including cattle, elephants, horses, cats, whales, bats, rodents and man himself.

The marsupials are basically rather similar to the placentals, though with a number of differences in the structures of the reproductive organs, skull and skeleton (including the presence of two extra bones, the epipubic bones, in the pelvic girdle). The young are born in a very immature condition and in some species are subsequently protected until well developed in a pouch, where they are milk-fed. Marsupials are most varied and numerous in Australia and adjacent islands, including New Guinea, the home of about two-thirds of the approximately 250 known species. Most of the remaining third live in South America. One species, the opossum, occurs in North America.

The monotremes feed their young with milk, but they also show a number of reptilian characters, especially in the structure of the skeleton; and they lay eggs. There are only three living genera; the spiny ant eaters of Australia and New Guinea (Irian), which have pouches; and the Australian duck-billed platypus, which has no pouch.

Even today the exact ancestral history of these three groups of mammals is only imperfectly understood, though it is generally believed that the placentals and marsupials have evolved independently since the Cretaceous (probably about 100 million years ago), when they diverged from a common ancestor. The relationship of the monotremes is more uncertain. Some workers even believe that this group evolved to the mammalian state directly from reptile stock, quite independently of the ancestral placental–marsupial mammal but achieving similar changes. Whatever the precise details of these relationships, by Pleistocene times these three mammalian groups had long been distinct from one another.

When zoologists began to classify the living Australian mammals, they found representatives of all three groups among them, although marsupials greatly outnumbered in variety both the monotremes and placentals.

In addition to the kangaroos, other important Australian marsupials include the wallabies, the cuscus, the possum, the glider (flying opossum), the koala, wombats, bandicoots, the marsupial 'mole', pouched 'mice', the banded ant-eater, native cats, the Tasmanian devil and the Tasmanian 'tiger' or 'wolf' (fig. 13.2).

Australian marsupials occupy many of the ecological niches occupied by placentals on other continents (arboreal, fruit-eating, grazing, gnawing, digging, burrowing, ant eating, insectivorous and carnivorous). In some instances they resemble their placental equivalents quite closely. The kangaroos are the dominant large grazing herbivores, occupying an ecological niche similar to the deer and antelopes, although they do not resemble these animals in appearance. The carnivore niche is occupied by the Tasmanian 'wolf', Tasmanian devil and marsupial cat. The wombat is a burrowing tail-less marsupial with rodent-like grinding teeth, resembling a marmot; and there are other marsupials resembling squirrels, moles and mice. The flying opossum can glide like a placental flying squirrel. There are no marsupial 'bats', 'seals' or 'whales'.

Although the monotremes are represented by only three genera, previously mentioned, they are of special interest in palaeogeographical studies since they are unknown outside the Australia–New Guinea (Irian) region (fig. 13.3).

Before Europeans began importing domestic mammals into Australia, relatively few placental mammals were present there; most important being bats, some rodents and seals. There was also a wild dog known as the dingo now believed to be quite a late human introduction.

Marsupial biogeography

For zoologists and geologists the question of how the marsupials and monotremes managed to establish themselves in Australia without the placentals arriving at the same time, competing for all the same ecological niches and becoming the dominant fauna, presents a striking problem. Elsewhere, when these two groups have found themselves in contact (for example after the 'Great Interchange', when North American placental carnivores came into competition with South American marsupial carnivores; and in Australia today where introduced domestic placental mammals compete with the native marsupials) the marsupials have often been the losers and extinctions have followed. The Tasmanian marsupial 'wolf' for example, though still surviving in the 1930s is now either extinct or nearly so, its decline greatly hastened by human persecution. A supposed sighting in 1982 raised hopes that a few individuals might still survive. Not all marsupials, however, are unable to survive in face of placental competition – many of the South American marsupials continue to flourish at the present day.

In our search for an answer to this problem we must look at the fossil evidence, both Pleistocene and earlier. Here problems immediately arise because the earliest Australian mammals are still totally unrepresented by fossil remains. The evidence becomes increasingly scarce as the geological record is followed back into time. The earliest Australian marsupials so far discovered are of upper Oligocene age (about 22 million years old), from Geilston Bay, Tasmania. Before this, for the time being, the fossil record peters

Fig. 13.2 *Some living Australian marsupials. a) Glider,* Petaurus, *b) Kangaroo,* Macropus, *c) Banded ant-eater,* Myrmecobius, *d) Marsupial mouse,* Sminthopsis, *e) Wombat,* Vombatus, *f) Koala,* Phascolarctos, *g) Native cat,* Dasyurus, *h) Tasmanian devil,* Sarcophilus, *i) Cuscus,* Phalanger, *j) Possum,* Trichosurus, *k) Tasmanian wolf,* Thylacinus *(living recently but probably now extinct). (After drawings by Ella Fry, in Ride, 1970).*

Fig. 13.3 *Australian monotremes. The duck-billed platypus,* Ornithorhynchus, *and spiny ant-eater,* Tachyglossus. *(After drawings by Ella Fry, in Ride, 1970).*

out. The earliest monotreme yet discovered (an early platypus) is from the Miocene. Unfortunately, it provides little information about monotreme evolution other than that monotremes were present in Australia about 15 million years ago. Only a few placentals are known from the Tertiary of Australia, notably whales and a bat from the middle Miocene (15 million years old) and murid rodents from the lower Pliocene (4·5 million years old).

Fossil marsupials are also known from some of the other continents of the world, notably North and South America and Europe. Does their geographical and chronological distribution tell us anything about the possible origin of the Australian Pleistocene and present-day fauna, which the Australian evidence does not reveal? The New World fossil record is of special interest, since the earliest marsupials found in both South and North America are of upper Cretaceous age (about 70–80 million years old), much earlier than the earliest known Australian marsupials. Could the Australian marsupials possibly have originated in America?

The earliest known European marsupials are of Eocene age (about 50 million years old) and are interpreted as immigrants of North American stock before North America and Europe had become separated by the processes of continental drift. In late Eocene or early Oligocene times one marsupial species apparently managed to cross to North Africa, where its remains were only recently discovered at one unique locality in Egypt. The European marsupials became extinct in the Miocene. Meanwhile in North America the marsupials had already become extinct by the end of the Oligocene, although they continued to thrive in South America (by then separated from North America by the sea), where the mammalian carnivore niche was wholly occupied by marsupials, including the extinct *Borhyaena.* The present-day North American opossum is not a descendant of Oligocene North American opossums, but is a Pleistocene immigrant from the south.

No indisputable marsupials are known from Asia. *Deltatheridium,* an early mammal from the upper Cretaceous of Mongolia, has some marsupial characters but these are usually interpreted on evidence of closeness of this genus to the primitive mammalian stock rather than as reasons for assigning it to the marsupials.

Although the geographical and chronological relationship between the marsupial faunas of the Americas and Europe now seems to be fairly well understood, the relationship of the Australian fauna, now isolated by deep water from the other continents was for a long time difficult to explain. Since the earliest known marsupials are American and since the only continent affording a convenient stepping off point for Australia is Asia, it was for a long time widely believed that the marsupial fauna of Australia had arrived from North America through Asia and then by island-hopping through Indonesia, probably in Cretaceous times. Somehow the placentals failed to follow; and Australia became a refuge for the marsupials, which were able to speciate in isolation from them.

Such a theory, however, was not without problems. If we exclude *Deltatheridium*, no fossil marsupials are known from the Asian mainland, where there are numerous species of placentals. In addition there is a marked faunal break (Wallace's line) running along the deep channel between Java and Kalimantan (Borneo) to the north-west (the Oriental zoogeographic region); and Sulawesi (Celebes), Irian (New Guinea) and Australia (the Australian zoogeographic region) to the south-east. Although marsupials do occur on many of the islands lying between Wallace's line and Australia there are fewer species there than on the Australian mainland, suggesting that colonization had been from Australia to the islands, rather than from the opposite direction. Placental mammals, on the other hand, were present only on the most western islands, with very little overlap of the marsupial area, suggesting spread from the Asian continent.

A further argument against an Asian source for the Australian marsupials has been introduced by recent studies of continental drift. In early discussions of marsupial origins it was assumed that the continents have always been in the same place relative to one another. As early as 1924, L. Harrison was already arguing that the marsupials of Australia had reached that continent from South America by land connections through Antarctica; but so widely separated were these land masses that his idea did not receive ready acceptance.

With ocean floor spreading and lateral movements of continents now proved to be a geological process that is still continuing at the present day, Harrison's theory gains new feasibility. Recent studies now suggest that, until lower Jurassic times, about 175 million years ago, the continents that we know today were joined as a single land mass – Pangaea; which soon afterwards divided into a northern part – Laurasia – including what are now North America, Greenland, Europe and Asia; and a southern part – Gondwana – including South America, Africa, India, Antarctica and Australia (fig. 13.4).

Subsequently these two major regions began to break into smaller parts. The Americas drifted westwards, Antarctica south, India and Australia northwards. Africa is believed to have separated from Antarctica in the early Cretaceous, about 135 million years ago; Australia from Antarctica during the Palaeocene, about 56–50 million years ago; and Antarctica from South America during the Eocene, about 40 million years ago. Before these dates direct land connections existed; and faunal exchange by island hopping would have been possible for a further period of time. Antarctica was not then the ecological barrier that it is today, since it was not a centre of glaciation. For a long time the lack of any fossil marsupial remains from Antarctica made it difficult to test Harrison's theory. The discovery, in March 1982, of remains of late Eocene age (about 40 million years old) on Seymour Island near the Antarctic Peninsula, of a marsupial of a family previously known only from South America, now adds substantial support to his hypothesis (fig. 13.5).

At the same time, in the light of the theory of continental drift, the old concept of an Asian origin for the Australian marsupials becomes even less feasible than before. In Cretaceous times the Australian continent, according to the most generally accepted reconstruction, lay far south of its present position, with a wide expanse of sea between it and the Asian continent. It was only when what is now Australia moved towards Asia that island hopping became possible and the first terrestrial placentals (rodents) arrived.

Today most palaeontologists accept an origin (i.e. differentiation from primitive mammal stock) for the

Fig. 13.4 *Probable positions of the continents during the upper Cretaceous, about 90 million years ago. Superimposed are Cretaceous mammal sites that have yielded marsupials (triangles) and those with non-marsupial mammals (circles) only. (After Owen, 1983 and Tyndale-Biscoe, 1973).*

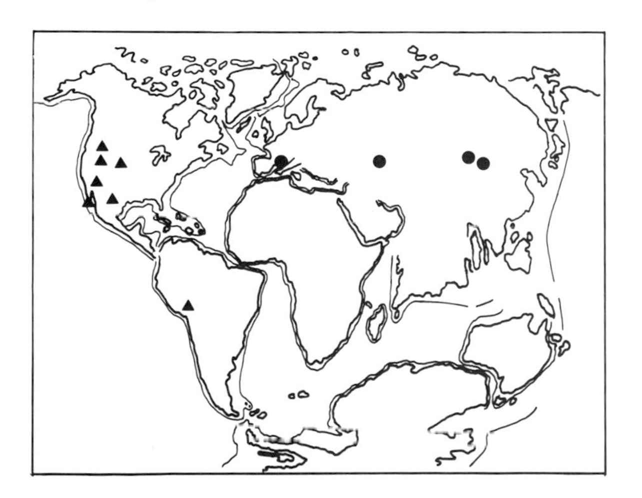

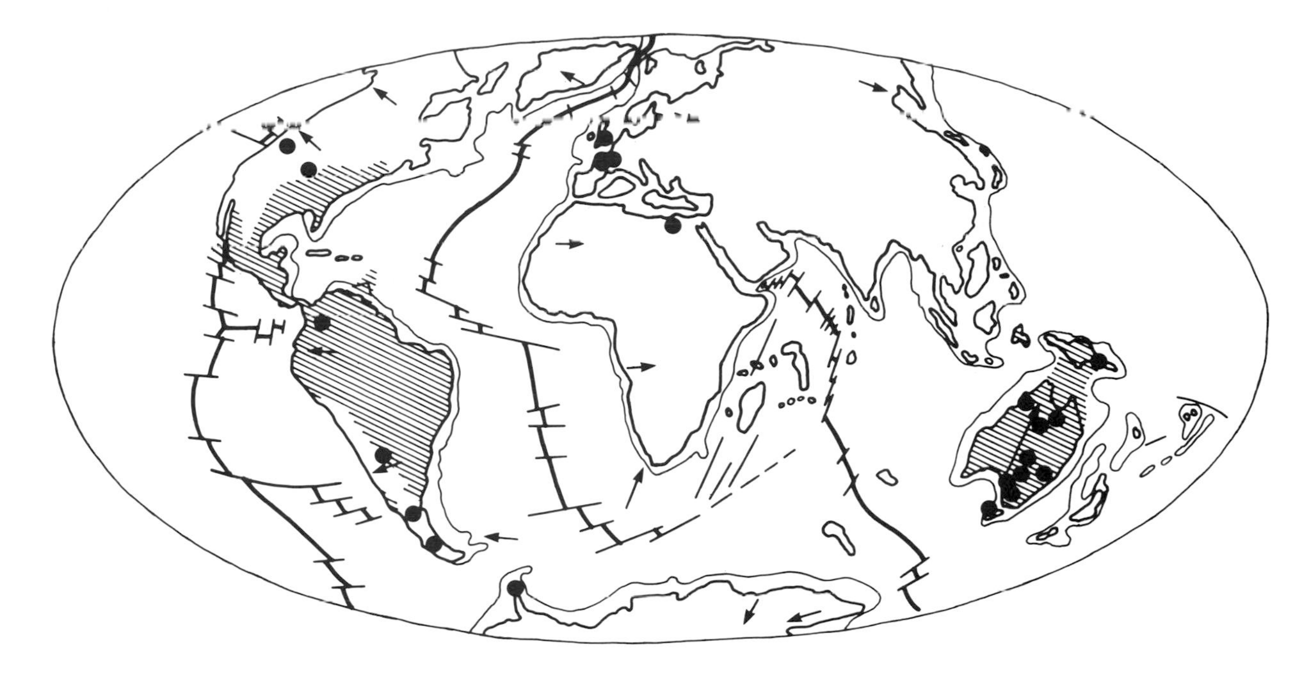

Fig. 13.5 *Living and fossil, except Cretaceous (see fig. 13.4), marsupial distribution shown on a map of the present-day world, with plate boundaries of the spreading continents indicated by solid lines in the mid-oceanic areas. Direction of spread indicated by arrows. (After Tedford, 1974; more recently discovered Tertiary localities in Egypt and Antarctica added).*

Australian marsupials during the Cretaceous, either within or through Gondwana; although all are not agreed about the details. Arguments have been presented in favour of South America, North America, Antarctica and Australia itself.

During the last decade the possibility that the marsupials had their origin in Australia has received renewed attention. If the centre of marsupial diversity is considered as the possible point of marsupial origin, then Australia could be this point; never colonized by the placentals (or where the placentals died out very early); not just a place of refuge.

The search for remains of Cretaceous mammals in Australia continues with increasing intensity, though so far without any success. In the meanwhile the discovery in Australia of a fossil flea of Lower Cretaceous age, of a type associated elsewhere in the world with furred animals and not with birds, raises the question of the identity of its host and demonstrates the imperfection of the geological record.

Despite the difficulties still associated with the interpretation of the early history of the mammalian fauna of Australia, by Quaternary times the marsupial fauna was diverse and abundant. Many fossil remains have been found that date from this period, especially in old lake and river deposits and in caves (including both pitfall and den deposits).

The Pleistocene environment

Before we look more closely at some of Australia's Pleistocene mammalian faunas we must first consider the climatic events of that region, which help provide a chronological framework for them.

As in other parts of the world, the Quaternary was a time of great fluctuations of climate. Owing to the relatively low altitude and latitude of that continent, however, glacial development (as in Africa) was possible only in the coldest stages in a few small high mountainous areas. Climatic change expressed itself principally in fluctuations of rainfall and associated changes in the level of water in lakes; in the expansion and contraction of desert areas; in changes of vegetation and, indirectly, in the distribution of the mammalian species which were dependent upon it.

At times of low sea-level, during the glacial episodes, New Guinea (Irian) and some of the other islands were connected to Australia by land, allowing exchange of mammalian species, but the sea was too deep between New Guinea and Borneo (Kalimantan)

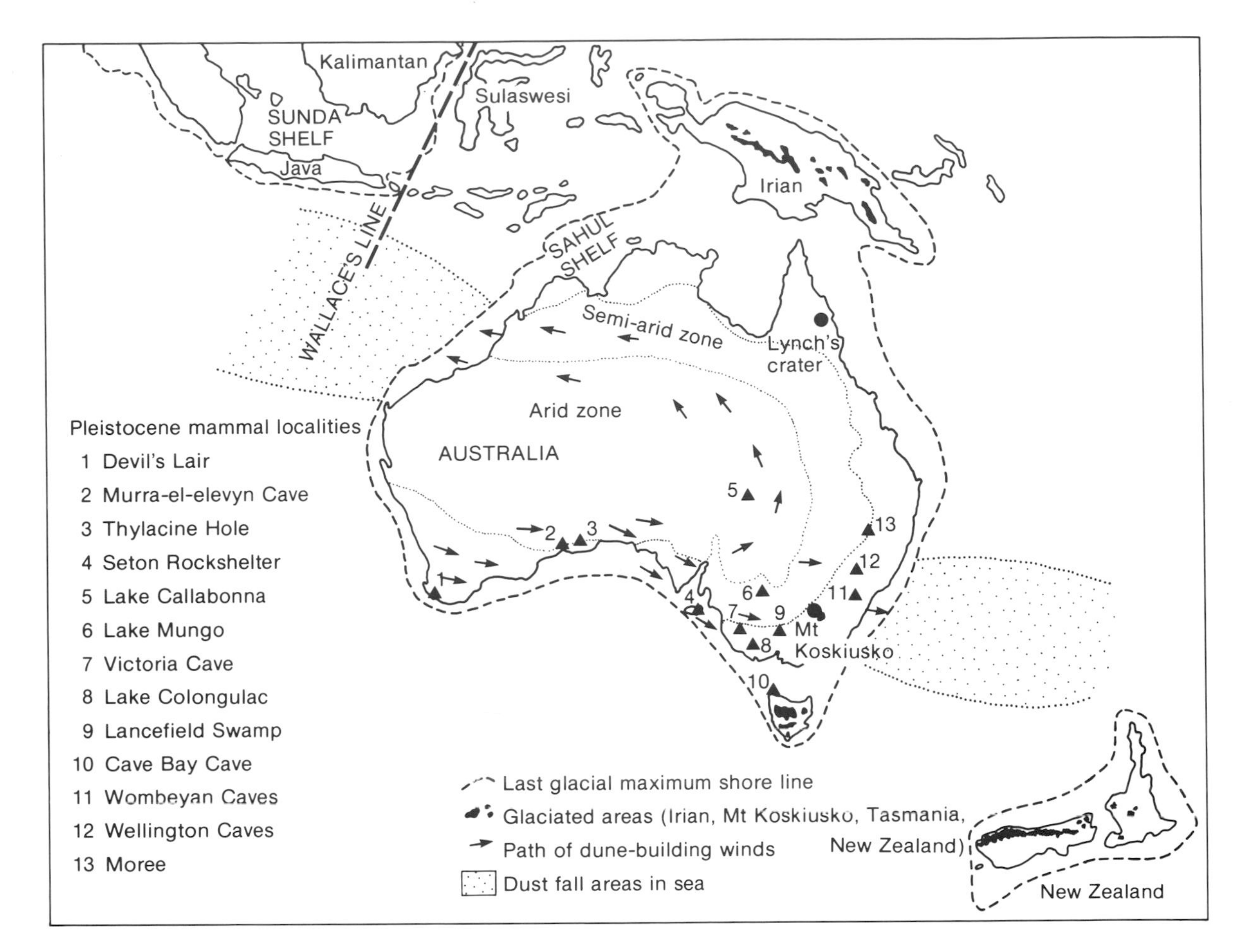

Fig. 13.6 *Map of Australia and neighbouring land areas at the time of the last glacial maximum about 20–15 000 years ago, showing localities of Pleistocene glaciation, approximate shoreline (low sea-level), areas of continental deserts and major dust-fall areas in the sea. Also shown are Lynch's Crater and fossil mammal localities (of various upper Pleistocene dates) mentioned in the text. (After Bowler, CLIMAP and others).*

for this lowering to create a land connection to the Asian mainland. The terms Sahul Shelf and Sunda Shelf are often employed to describe the shelves on the Australian and Asian sides respectively flanking this dividing zone (fig. 13.6).

One of the best records of Pleistocene climatic change in Australia comes from palaeobotanical studies of a 40-metre-long column of organic sediments extracted by boring into the lake and swamp deposits filling an old volcanic crater – Lynch's Crater – in Queensland. Comparison of the relative abundance of plants characteristic of humid and dry conditions from the various layers has provided a record of the changing pattern of rainfall in the area for over 120 000 years. The resulting diagram shows a period of relatively high rainfall during the early Holocene, decreasing again towards the present day, preceded by an arid period back to about 80 000 years ago, before which were two earlier peaks of higher rainfall at about 85 000 and 120 000 years ago. The rainfall graph shows an astonishing similarity to the temperature graph established from the deep sea sediments, the Greenland Ice Cap and Grande Pile Peat bog in France (see chapter 6); but with relatively high rainfall accompanying the Holocene and Last Interglacials; drought accompanying the Last Glaciation.

Studies of the few glacial deposits in Australia, Tasmania and New Guinea indicate a major glacial advance about 25–15 000 years ago (as in other parts of the world: the Last Glaciation maximum) although there were substantial local differences in the waning of this event. In the Snowy Mountains of Australia the ice began to retreat before 20 000 years ago, whereas in New Guinea and Tasmania this did not occur until after 15 000 years ago.

Some Australian Pleistocene mammal sites and their faunal remains

Although Australian fossil mammal localities span the whole of the Quaternary and parts of the Tertiary, those of upper Pleistocene age make up a high proportion of them. In consequence the mammalian faunas of the last 30–40 000 years are especially well known, and there is progressively less information about those of earlier age. Like those from other continents, the remains occur in deposits that have accumulated as the result of a great variety of taphonomic processes, especially in caves (where pitfall processes and occupation by carnivores, birds of prey and man are of particular importance) and in old lakes, marsh and river deposits. In many instances these mammalian sites also have geomorphological importance, for example Seton Rock Shelter on Kangaroo Island, South Australia. At the time of accumulation of the principal bone deposits, between about 16 000 and 10 000 years ago, the cave was situated on a peninsula attached to the Australian mainland. It was subsequently isolated on an island by eustatic rise of sea-level.

Some of the Pleistocene remains are of mammalian species still living. Others are of species which, although quite closely related to living forms are now extinct, the most obvious difference of some of them being their spectacular size. The extinct browsing kangaroo *Procoptodon goliah*, for example, reached a height of nearly three metres and may have weighed as much as 400 kg. A few Pleistocene forms have no close living relatives. The rhinoceros-sized *Diprotodon* (its nearest surviving relatives are the wombats) and the marsupial lion, *Thylacoleo*, which we will examine more closely shortly, are examples.

Wellington Caves, lairs for marsupial carnivores

The first important Australian locality at which bones of Pleistocene mammals were found in abundance was one of the Wellington Caves, west of Sydney in New South Wales. Lang, who first announced the discovery in 1831 and who was already familiar with Buckland's account of the hyaena den in Kirkdale Cave in England, written only eight years previously, regarded the cave as another example of a den of beasts of prey:

> *'While this very interesting discovery supplies us therefore with another convincing proof of the reality and the universality of the deluge, it supplies us also with a powerful motive of gratitude to divine providence for that long forgotten visitation. For if this territory were over run with such beasts of prey as the antediluvian inhabitants of the cave at Wellington Valley, it would not have been so eligible a place for the residence of Man as it actually is.'*

The collection of bones included kangaroo, native cat and other animals that could not at first be identified.

The idea that the Wellington bone deposit represented the left-overs of beasts of prey which had lived in the cave (nearly 100 long bones were found, not one with its ends still intact) has been amply confirmed by more recent studies in other caves. Bone remains from Pleistocene deposits in several caves near Perth have been compared with the food débris of various living carnivores, including the Tasmanian devil, *Sarcophilus*. Unlike the cave (spotted) hyaena, *Sarcophilus* cannot digest bone and fragments survive unaltered in the droppings. Splinters, believed to be of this origin, have been found in a number of caves.

There is also direct evidence that some of the marsupial carnivores used caves as places of retreat.

Fig. 13.7, pp. 194–195 *Artist's impression of a scene outside the entrance of one of the caves in the Wellington Valley, New South Wales, during late Pleistocene times. It was in this area that, early during the nineteenth century, the first Australian fossil mammalian remains were found. Ideas for the scene, which shows a fine example of karst weathering of the limestone bed rock in which the cave is situated, are based on a lithograph of 1838 in Mitchell's* Three expeditions into the interior of eastern Australia.

Seven species of mammals are shown, all represented by fossil remains in the Wellington Caves. In clockwise order, from the left foreground are: in a Eucalyptus *tree, a koala,* Phascolarctos cinereus *(length of head and body 70–80 cm)–arboreal leaf eater which still survives in eastern Australia;* Diprotodon optatum, *the largest marsupial which ever lived, 3 metres long, 2 metres high at the shoulder, extinct; the marsupial 'lion',* Thylacoleo carnifex, *1·25 metres long (note the large opposing 'thumb'), probably carnivorous, extinct; a herd of the extinct short-faced kangaroo,* Sthenurus atlas *(about 2 metres from nose to tip of tail); two Tasmanian wolves,* Thylacinus cynocephalus, *carnivorous, survived in Tasmania until at least the 1930s, but probably now extinct; a spiny ant-eater or echidna,* Zaglossus, *forages forest litter for earthworms and large insects, extinct on the Australian mainland (where a related form still occurs) but survives in the highlands of West Irian (New Guinea); centre foreground, a Tasmanian 'devil',* Sarcophilus harrisii, *carnivorous, feeds mainly on carrion, extinct on the Australian mainland but survives in Tasmania. All these mammals are marsupials except* Zaglossus, *which is a monotreme.* Thylacinus, Sarcophilus *and* Zaglossus *commonly go into caves, where their remains may be preserved as fossils (see fig. 13.8, p. 196);* Thylacoleo *may also have done so.*

Some artistic license has been allowed in this scene, since Zaglossus *and some of the other mammals are nocturnal in their habits and would seldom be seen in daylight.*

Painted by Peter Snowball, under the direction of the author, 1985, from data provided by J. Mahoney.

Fig. 13.8 *The desiccated carcass of a Tasmanian wolf,* Thylacinus, *found 150 metres inside Thylacine Hole, Western Australia. It has been shown by radiocarbon dating to be 4600 years old. Also found in the same cave chamber were remains of five other thylacines, native cats, one Tasmanian devil and various other animals. The cave had apparently acted as a natural trap from which these animals could not escape. (Photo: D. C. Lowry).*

Mummified carcasses of the Tasmanian wolf, *Thylacinus* and of *Sarcophilus* have attracted considerable attention, since their apparent freshness suggests that both these animals could have survived on the Australian mainland in relatively recent times. Radiocarbon dates obtained for some of these carcasses nevertheless gave surprisingly high figures. A well-preserved carcass of the former animal from Thylacine Hole, Western Australia, was found to be 4600 years old; and a less well-preserved carcass from Murra-el-elevyn Cave, Nullarbor Plain, Western Australia, 3300 year old – the youngest remains of *Thylacinus* yet known from the Australian mainland (fig. 13.8).

Lake Callabonna and *Diprotodon*, the largest marsupial

Among the remains from the 1831 collection from the Wellington Caves was a fragment of a jaw of an animal to which Professor Owen, in London, later gave the name *Diprotodon* (literally 'two front teeth'). By 1877 he had received sufficient additional specimens from Australia to be able to reconstruct the skeleton of a gigantic herbivorous marsupial nearly the size of a rhinoceros, although he was uncertain about the feet and overcame this problem by showing a reconstructed skeleton standing in grass tall enough to hide them.

The problem of the feet was not to be resolved for another 15 years, but when this happened it occurred in a very spectacular way. The South Australian Museum heard of large bones being found in a normally dry salt pan, now known as Lake Callabonna, in South Australia. An expedition was sent into the field in 1893 and within three months had located 360 *Diprotodon* skeletons (fig. 13.9), many of them with the feet complete, on the dry lake bed. The feet proved to be of most unusual proportions, with massive ankle and wrist bones but with very small toes. Other mammalian species, including giant kangaroos, were also represented; and skeletons of giant emus and emu-like birds, with impressions of feathers.

During recent years, investigators have also uncovered lines of footprints of *Diprotodon*, preserved between the layers of the old lake deposits. In spite of their abundance, these fossil remains are believed to represent a slow accumulation of animals trapped on the boggy flats of the lake at times of low water, rather than the product of a single catastrophe such as drought.

Associated plant remains and pollen show that, at the time when *Diprotodon* was living, the Lake Callabonna area supported a flora including *Eucalyptus* and native pines which would have required a higher rainfall than occurs at the present day. An attempt to determine the age of part of a gum tree from these deposits by radiocarbon dating showed the specimen to lie beyond the range of this method (more than 40 000 years old). Forty thousand years ago is known to have been a period of greater (not lesser) aridity in Australia, however, and it is probably necessary to go back to high rainfall period of 120–80 000 years ago to accommodate the Lake Callabonna fauna.

Lake Colongulac and *Thylacoleo*, the marsupial lion

Even more remarkable than *Diprotodon* is the so-called marsupial lion, *Thylacoleo*, Australia's most problematic large extinct marsupial. Remains of this 1·25-metre-long animal have been found throughout most of the Australian continent. First described by Owen in 1858 from a skull found about 1846 at Lake Colongulac, Victoria, *Thylacoleo* was interpreted as a marsupial beast of prey because of its secateur-like posterior cheek teeth, closely resembling the shearing teeth of a placental lion, which are known to be used for cutting flesh (fig. 13.10a).

The carnivorous role of *Thylacoleo*, however, has not been universally accepted by palaeontologists. It has been pointed out that, even though the cheek teeth were well-adapted for shearing meat, the incisors (of which there was a single closely placed forward-pointing pair in both upper and lower jaws, rather like the beak of a parrot) were inadequate for grasping live prey. They also seemed unsuitable for cropping vegetation unless perhaps the animal had been a browser,

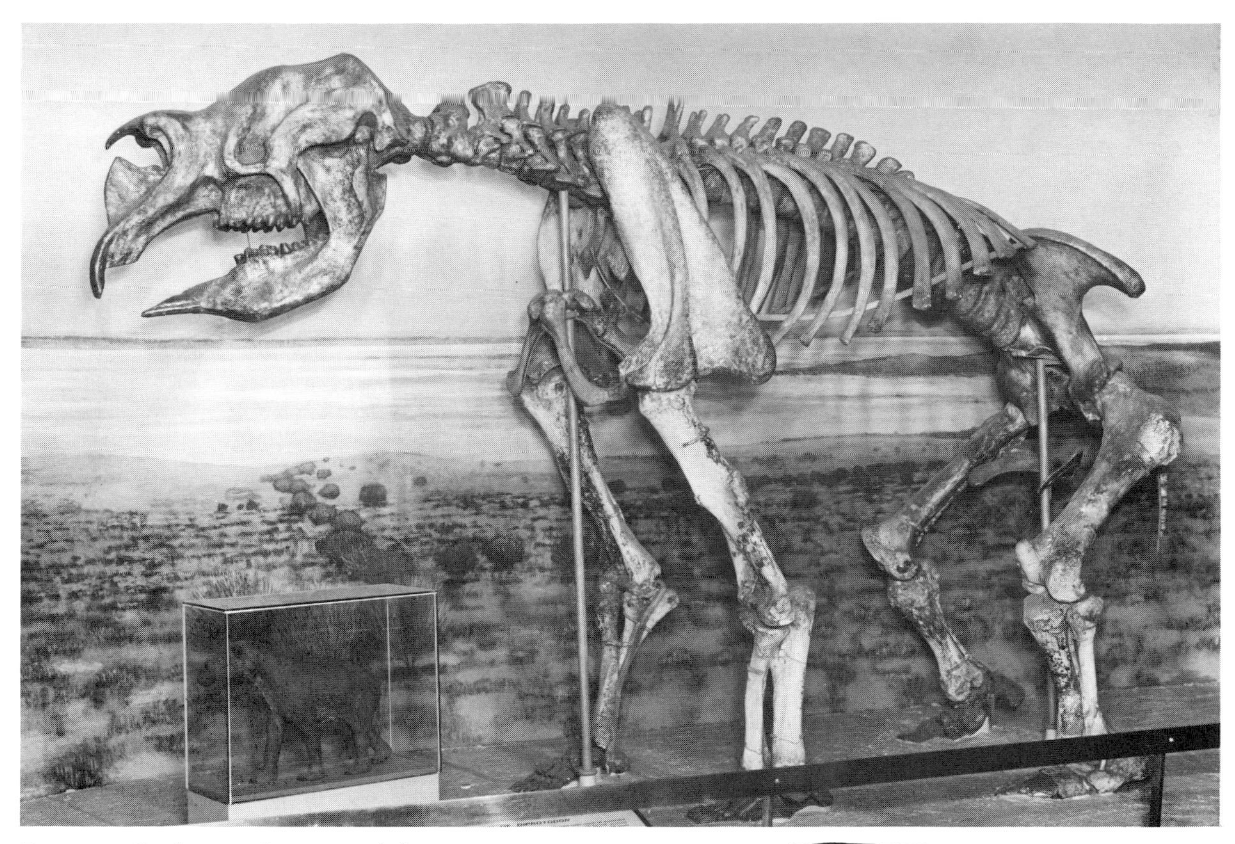

Fig. 13.9 *Partly restored composite skeleton of* Diprotodon *from Queensland, on display in the British Museum (Natural History). (Photo: BM(NH)).*

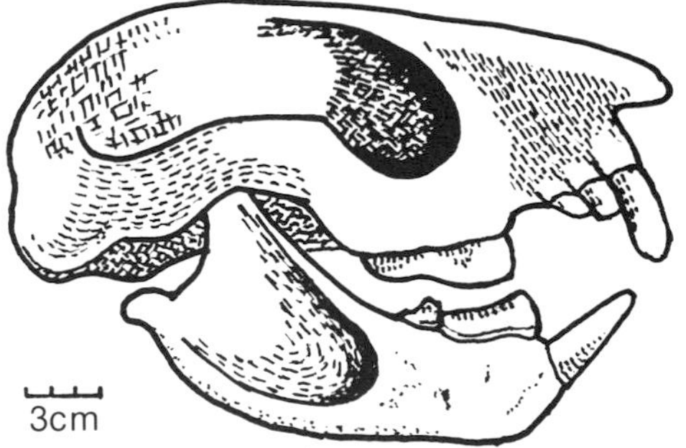

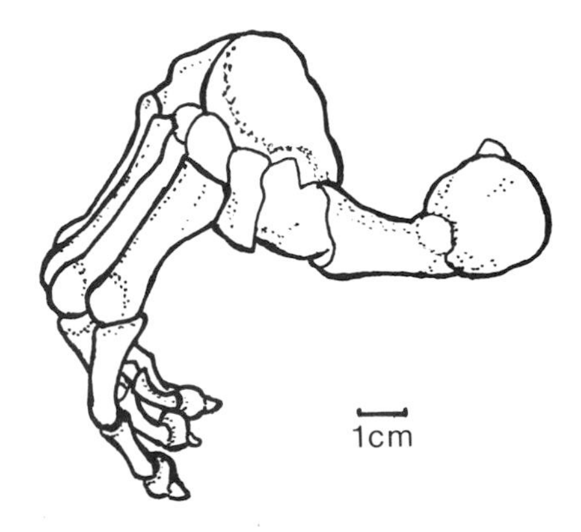

Fig. 13.10a) *Skull of* Thylacoleo, *showing the massive secateur-like cheek teeth; and the forward-pointing paired incisors which resemble the beak of a parrot.*
b) *Lateral view of the articulated bones of the right fore foot of* Thylacoleo *from Victoria Cave, Naracoorte, South Australia. Note the opposable 'thumb' with a large claw. (After Wells & Nichol, 1977).*

using its shearing cheek teeth for chopping up stems too coarse for the other vegetarian marsupials. It has also been suggested that *Thylacoleo* was insectivorous; that it was omnivorous; that it fed on melons or that it fed on eggs. Its jaw muscles were nevertheless of massive dimensions, far stronger than would be required for eating melons or eggs, so that such diets seem very improbable.

The present-day mammal with incisor teeth most closely resembling those of *Thylacoleo* is *Daubentonia*, the aye-aye, a Madagascar lemur that bites into the bark of trees in search of the grubs of insects. Its flat crowned molars are nevertheless quite unlike the shearing cheek teeth of *Thylacoleo*.

The eyes of *Thylacoleo* were placed relatively far forward on the skull, suggesting that it had binocular vision, a feature associated with active hunting of prey.

Most of these early discussions of *Thylacoleo* were based on the examination of the skulls. Very little was known about the rest of the skeleton until 1966 when part of a skeleton with fore-feet was discovered near Moree, New South Wales; and a few years later both fore and hind feet were recovered during excavations in Victoria Cave, Naracoorte, South Australia. A massive opposable thumb on the fore foot, with a large nail, shows that *Thylacoleo* could grasp strongly (fig. 13.10b), though whether this adaptation was for climbing trees or grasping prey or even for both purposes remains a subject for discussion. The closest structural affinities of the feet are found in the phalangers (an arboreal group of marsupials which includes the possums and cuscus, fig. 13.2).

On the present evidence a carnivorous role is quite widely accepted for *Thylacoleo* by palaeontologists. Its nearest relatives are probably the arboreal phalangers.

Fig. 13.11 Burramys parvus, *the mountain possum, known only from fossil remains until 1976, when a living individual was found on Mount Hotham, Victoria. (After a drawing by Ella Fry, in Ride, 1970).*

Burramys, a living fossil

Diprotodon and *Thylacoleo* apparently died out ten or more thousand years ago; and we can only speculate about their appearance from their fossil remains. There are many other extinct species of Australian marsupials, some of them so small that they have never attracted the popular interest created by their larger extinct brethren. One of these apparently extinct forms was the possum *Burramys parvus*, first recognized in 1894 by the comparative anatomist Robert Broom from remains, apparently accumulated by owls, which he found in the Wombeyan Caves, New South Wales. For over 70 years, *Burramys* was accepted as being totally extinct, to be studied only from bones in the drawers of museums; and here the story might have been expected to finish. But in August 1976 a small possum (its body measured 120 mm, its tail 150 mm) was found by visitors to a ski hut on the slopes of Mount Hotham in the Australian Alps in Eastern Victoria. As it seemed unusual it was taken for identification to the Fisheries and Wildlife Department in Melbourne, where the research officers had not seen anything like it before and sought the opinion of the Melbourne naturalist Mr Norman Wakefield who, by chance, had recently been working on the fossil remains of *Burramys* from a cave in northeast Victoria. *Burramys*, so long believed to be extinct, was still a living member of the Australian fauna! Since then several other living individuals of this animal (fig. 13.11) have been found by Australian zoologists. Will any other of the supposedly extinct Pleistocene marsupials come to light as living species, in a similar manner, in the future?

The placental mammals

Despite the predominance of marsupial remains some placental mammals are also represented by fossil or sub-fossil material, notably bats and rodents. It is known that bats had already arrived in Australia by the Miocene, the first rodents by the Pliocene. They became part of the Australian native fauna, but did not eliminate the marsupial fauna which was already present. The dingo, a dog rather smaller than a northern hemisphere wolf, is a post-Pleistocene arrival, probably taken to Australia by man only 3-4000 years ago. On the Australian mainland this animal overlapped in time with the marsupial carnivores *Sarcophilus* and *Thylacinus* and may have contributed to their extinction there. Dogs are not known to have occurred in Tasmania (where both the above mentioned animals still survive or, in the latter instance, survived until quite recently) before settlement by Europeans about 1800.

Pleistocene man

Although it was formerly believed that man reached the Australian continent only a few thousand years ago, excavations in caves and at open sites during the last two decades have now demonstrated aboriginal presence for at least 35 000 years and possibly 40 000 years. Important sites include Devil's Lair, Western Australia, with a long sequence of occupation hearths dated back to 35 000 years (also food débris of *Sarcophilus* and owls); six other caves with evidence of human activity back more than 20 000 years; Lake Mungo, New South Wales with a human burial and cremation about 30 000 years old; and Cave Bay Cave, Hunter Island, Tasmania, occupied for several millenia before 18 500 years ago.

Were these Aborigines responsible for the extinction of *Diprotodon* and *Thylacoleo,* which disappeared before the end of the Pleistocene? Of special importance in this connection is the Lancefield Swamp site in southeast Australia. Although the bone deposit (dated as 26 000 years old) extends over only a small area, a count of the remains from a part recently excavated suggests that as many as 10 000 animals may be represented there. Ninety per cent of the remains are of a single species of marsupial – the giant kangaroo *Macropus titan* – with smaller quantities of other extinct kangaroos, of *Diprotodon* and of emus. A worked quartzite blade in the same deposit shows that man was present in the area at the same time as the extinct fauna, although he does not appear to have been involved in the accumulation of the bone bed, which was a natural phenomenon associated with the swamp. The survival of these extinct marsupials, contemporaneously with man, for a period of at least 7000 years after man had first arrived in New South Wales, suggests that there were few or no sudden extinctions as the result of his coming.

Chapter 14 **Extinctions**

One of the most intriguing aspects of Pleistocene studies is that, although many of the mammals (including, in various parts of the world, such bizarre species as the woolly mammoth, the mastodon, the woolly rhinoceros, the giant sloth and the sabre-toothed cat) are now extinct, their disappearance did not occur until late in geological time. Man had already appeared on the scene; he actually saw these animals; and he even left contemporary pictorial records of some of them on cave walls and on bones and pebbles for us to study today. Sometimes, as we have already observed in the case of the frozen Siberian mammoth carcasses and dried ground sloth skin in the cave of Ultima Esperanza, the remains are so well preserved that it is difficult to believe that the animals concerned have been dead for more than a few hundred years. The disappearance of these species is even more striking than similar occurrences of remote ages, because it happened so near to our own era. The extinction of the Pleistocene mammalian fauna was nevertheless not total. Losses were significantly greater among the large mammals, but there were many other species (including some large ones) that survived and still flourish in various parts of the world at the present day.

Why did so many mammals disappear? Various possible explanations have been put forward. Principal are a great catastrophe; biological competition; epidemics; parasites; degeneration and overspecialization; climatic change, and man. The possibility that several such factors worked together, with man delivering the final *coup de grace* has also been widely considered. Whatever the cause of Pleistocene extinctions, there were many of them.

One of the earliest scientific deliberations on this subject was that of the British physician Thomas Molyneux, who discussed the cause of the extinction of the Irish giant deer, as long ago as 1697:

> *'By what means this Kind of Animal, formerly so common and numerous in this Country, should now become utterly lost and extinct, deserves our Consideration:... I know some have been apt to imagine this like all other Animals might have been destroyed from off the Face of this Country by that Flood recorded in the Holy Scripture to have happened in the time of* Noah; *which I confess is a ready and short way to solve this Difficulty, but it does not at all satisfy me: For (besides that there want not Arguments, and some of them not easily answer'd, against the Deluge being Universal)... it seems more likely to me, this kind of Animal might become extinct here from a certain ill Constitution of Air in some of the past Seasons long since the Flood, which might occasion an* Epidemick Distemper, *if we may so call it, or* Pestilential Murren, *peculiarly to affect this sort of Creature, so as to destroy at once great Numbers of 'em, if not quite ruine the Species. And this is not so groundless an Assertion as at first it may appear, if we consider this Island may very well be thought neither a Country nor Climate so truly proper and natural to this Animal, as to be perfectly agreeable to its temper... we may think some might have escaped the* Common Calamity; *but these being so few in Number, I imagine as the Country became peopled, and thickly inhabited; they were soon destroy'd, and kill'd like other Venison as well for the sake of food as Mastery and Diversion. And indeed none of these animals by reason of their* Stupendious Bulk *and* Wide Spreading horns *could possibly lye sheltered long in any Place, but must be soon discovered, and being so conspicuous and heavy were the more easily pursued and taken by their numerous Hunters, in a Country all environed by the Sea: For had they been on the wide Continent they might have fared better, and secured themselves and their* Race *till this time... But this could not be expected from those savage Ages of the World, which certainly would not have spared the rest of the* Deer Kind, *Stags and Hinds, Bucks and Does, which we still have; but that these being of much smaller Size, could shelter and conceal themselves easier under the Covert of Woods and Mountains, so as to escape utter Destruction ... and here I cannot but observe that the* Red Deer *in these our Days, is much more rare with us in Ireland, than it has been formerly ... and ... unless there be some care taken to preserve it, I believe in process of time this Kind may be lost also, like the other sort we are now speaking of'.*

Molyneux' conclusions were far in advance of the general level of scientific thought of the time. In

1697 he was already doubting the universality of the Biblical Deluge; and he did not attempt to attribute the extinction of the giant deer to a single cause but saw climate and disease as part of a chain of factors reducing the giant deer population to a size where it could be exterminated by man. He also recognized the greater risk of extinction to animals living in an island environment; the greater vulnerability of the giant deer to human hunters compared with the smaller deer; and that, in consequence of man's arrival in Ireland, the red deer was now also at risk there.

As Molyneux himself pointed out, these conclusions could only be a matter of conjecture. He had no clear chronology with which he could relate the extinction of the giant deer to changes of environment or the appearance of man and there was no information about the ecological tolerance of this animal.

Similar problems were to attend all mammalian extinction studies for a further two and a half centuries. In recent years, however, the advent of absolute dating methods; improvements in the understanding of the evidence for and effects of past climatic change; studies of the ecological requirements of living mammalian species; and the development of palaeo-ecological studies (such as the examination of fossil contents of stomachs) have made it possible to assess more effectively whether potentially related causes and effects can feasibly be connected.

Instead of resolving the earlier problems, however, these new techniques (which now add a framework of time to the study of Pleistocene extinctions) have brought with them a fresh wave of problems.

Although it was found that not all the extinctions had occurred at the same time, the end of the Pleistocene was unique in the intensity of those that occurred in many areas. Why, even though there had also been many losses earlier during the Pleistocene, did faunas which had survived several glaciations suddenly decrease at the end of the last, just when the climate might be expected to have been improving again?

Why were the large mammals more vulnerable to extinction than the small ones?

Why were the late Pleistocene extinctions often without ecological replacement, whereas the earlier ones generally did not lead to the vacation of ecological niches?

How, although there had been such a major episode of extinction of large mammals in most other parts of the world at the end of the Pleistocene, did a Pleistocene-like megafauna manage to survive in Africa until the present day?

Why did the marine mammals of the Pleistocene not experience the same accelerated extinction shown by the terrestrial mammals?

Few lines of study have created greater controversy than the problem of late Pleistocene mammalian extinctions. Though discussion has been centred principally on the effects of human interference and climatic change, agreement among scientists about the relative importance of these processes is still far away.

Some workers have pointed out that, on a world-wide basis, late Pleistocene extinction dates make no sense in terms of climatic change but that they coincide with the arrival of man too closely to be unconnected. Paul Martin, who studied the global pattern of late Pleistocene mammalian extinctions, argued that overkill by early man, though occurring at a time of climatic change, was nevertheless the overriding factor in this otherwise incomprehensible part of the geological record. In North America, for example, it was during the period about 12–10 000 years ago, characterized archaeologically by such artefact types as Clovis and Folsom, that important Pleistocene extinctions (including that of the mammoth and mastodon) were taking place (fig. 14.1).

Other workers, whilst accepting that Palaeo-Indians may have played some part in the elimination of mammals at this time, ask how, if man had been responsible for the disappearance of the mammoth, mastodon and horse from America at the end of the Pleistocene (for which there is very little direct evidence of hunting), other species which are known to have been hunted (such as the caribou, musk ox and bison) managed to survive there until the present day. Until the spread of Europeans during the last century, Plains Indians and bison existed in equilibrium, the bison supplying food, clothing and other requirements. Only with the arrival of firearms and demand for skins and meat, leading to overkill, was the bison virtually exterminated. Many writers have claimed that, in North America, increased aridity was the most likely cause of the disappearance of the large mammals.

In many parts of the world today, especially in the Americas, extensive programmes of dating relevant finds and of reviewing already available dates are in hand for the purpose of testing the feasibility of these conflicting hypotheses. Sometimes improved chronological information has shown that particular extinctions did not occur at the same time as previously considered causes. A 13000-year date for the butchered remains of a mastodon, recently excavated at Taima Taima in Venezuela (fig. 4.3, p. 37), suggests that man may already have reached South America a thousand years before the above mentioned period of apparently accelerated extinction on the North American continent. In Australia it is now believed that man had already arrived in New South Wales by 33 000 years ago, yet some of the large kangaroos which later became extinct were still flourishing many

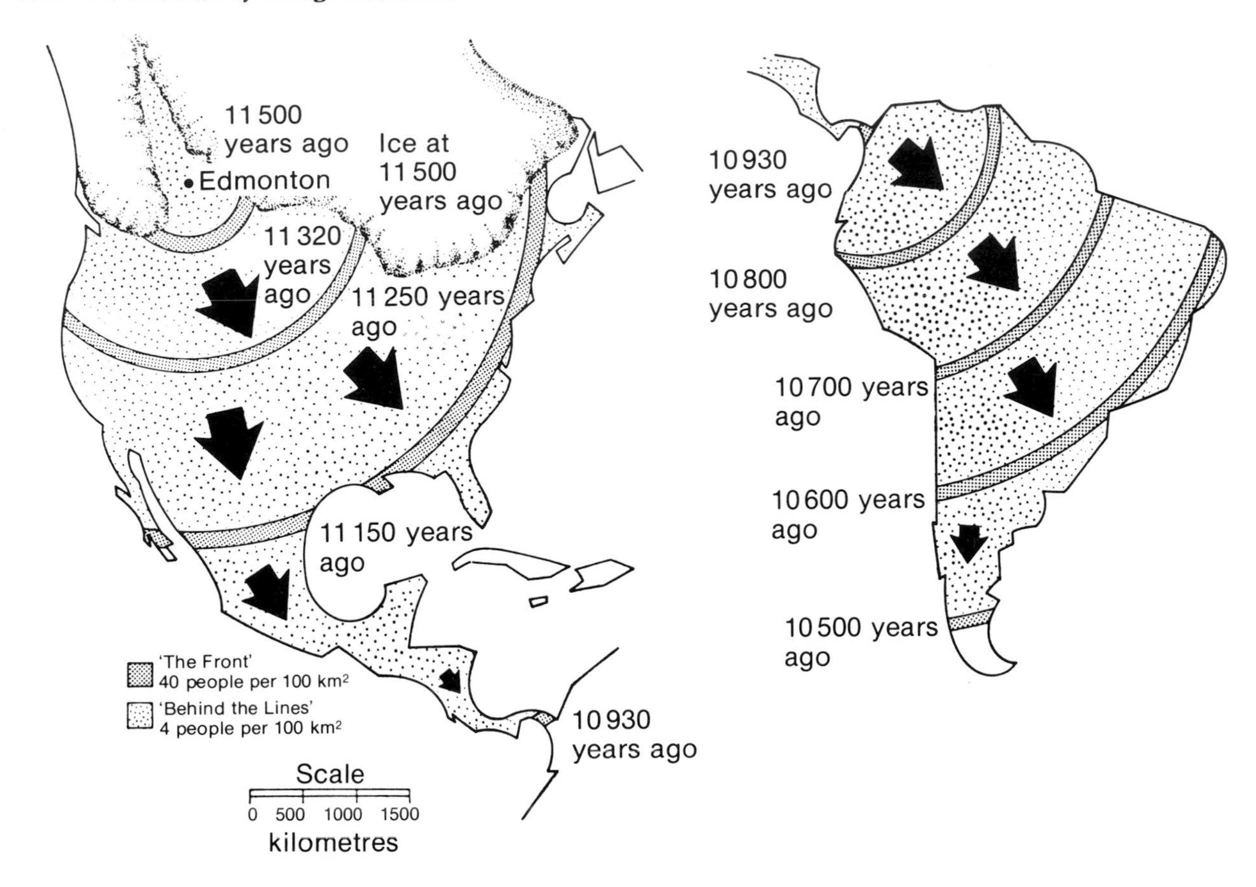

Fig. 14.1 *Suggested chronology of the advance of the human population front through North and South America. As local extinctions occurred, so the hunters moved on. The two continents are not to scale. (From Martin, 1973, The Discovery of America,* Science *179: 969).*

thousand years later. If Pleistocene man was responsible for their disappearance then this was a very lengthy process; not the sudden extermination that the hunting hypothesis would suggest.

Increasingly, the view is gaining favour that one must not look to any one general cause for extinction during the Pleistocene. There were complex ecological chains of events, the details of which are very difficult to decipher. Survival depended on a balance of circumstances. Several causes may have worked together and each extinction must be considered on its own merits. At the time of introduction of any form of stress (whether it be man-made or due to environmental events) into the accustomed life pattern of a species or assemblage of mammals, survival will depend on whether those concerned can undergo sufficiently rapid adaptation to enable them to avoid getting out of harmony with conditions to the extent that it becomes impossible to maintain an adequate breeding population. Some mammals can adapt better than others, so that extinctions can be expected to be selective.

It follows that difficulty in accounting for individual extinctions is more likely to be due to the number of choices rather than a lack of any possible single reason. 'To try to select a particular set of circumstances', the late American palaeontologist John Guilday has claimed, 'is fun but futile'. In any attempt to establish the reasons for a particular extinction we must, therefore, be prepared to consider together all the factors that could have led to a numerical reduction in the size of the population of that particular species. Let us examine, in more detail, the mechanisms of the most important processes.

Faunal turnover versus true extinction

The gradual evolution of one species into another will result in the species name of the ancestral animal disappearing from faunal lists. For example, *Mimomys*, an extinct rodent genus evolved in the mid-Pleistocene into *Arvicola*, which still survives today as *Arvicola terrestris*, the water vole (see p. 49). This 'faunal turnover' must not be confused with a true extinction in which no descendants are left. We are only concerned here with the latter.

Stress factors that lead to decreasing populations

Factors that may affect mammalian survival are of two sorts – internal and external. The former are inherent within the animals themselves and the latter are related to the environment, including interference by man.

Internal stress factors

At a time of stress some mammals are likely to be more adversely affected than others. Deteriorating ecological conditions, for example, may throw co-existing species into greater competition as the result of decrease of habitat. At such a time, large mammals are likely to be more adversely affected than small ones because of their greater demand on space, food and cover. Differential extinction is therefore to be expected and is indeed demonstrated by the geological evidence. An additional factor that is weighted against the large mammals is that many, though not all, have long gestation periods, thus making it more difficult to sustain their numbers by breeding; and mutations favourable to adaptation to changed environmental conditions are less likely to occur. In the living elephants, for example, gestation (leading to a single birth) lasts for nearly two years, whereas that of some rodents and shrews is only three weeks, allowing the birth of several generations of many individuals each in a single year. It has been estimated that a single pair of brown rats, with enough food and without any casualties, could theoretically increase in ten years to 48 million, million, million individuals.

Ability to produce young at the most favourable time of year is another important survival factor, for which living mammals show a variety of adaptations, ranging from all the year round breeding in some tropical areas to hibernation and delayed fertilization in such mammals as the bats and badgers of the higher latitudes.

Mammals that take a great time to reach sexual maturity; or which have fewer generations in a given time; or which are unable to adapt the time of arrival of their young to the most favourable season are the least able to survive changes of environment.

External stress factors

Catastrophes

Before Darwin, most geologists would have attributed extinctions to the Biblical Deluge or to some other great catastrophe. Molyneux was an unusual exception to this attitude. Although the Deluge may provide a convenient explanation for the great accumulation of fossil mammalian bones sometimes found, it nevertheless cannot be invoked on documentary grounds as a cause of extinction, since one pair of every species of animal is recorded as having survived on the Ark to repopulate the world when the flood subsided. Furthermore, the Deluge recorded in the Biblical account is now known not to have been universal; it has been shown that extinctions in different parts of the world did not all occur at the same time; and scientific studies have shown that it is no longer necessary to invoke death by sudden freezing to explain the occurrence of frozen carcasses of mammoths and other animals in Siberia and Alaska. Darwin considered that extinctions were probably slow, animals becoming rarer and rarer until finally lost. The cataclysm hypothesis can receive no support, other than to explain local disappearances of fauna.

Epidemics and parasites

Epidemics have sometimes been invoked to explain the extinction of mammalian species, for example by Molyneux in his account of the Irish giant deer, written in 1697, previously mentioned.

Without doubt, epidemics sometimes have very sudden and dramatic effects, especially where previously separated mammalian faunas come into contact for the first time, such as may occur after the disappearances of a geographical barrier (for example, water or ice) or as the result of new introductions by man. Two recent examples of disease, respectively accidentally and intentionally introduced by man, are *rinderpest* which spread from domestic cattle to the wild ungulates in Africa; and *myxomatosis*, a flea-carried disease of South American origin that has decimated the rabbit populations of Australia and Britain since the 1950s.

Infectious disease is nevertheless unlikely to cause extinction, since different individuals have different susceptibilities to disease and scattered populations of surviving resistant individuals can breed to restore the previous numbers (as has happened in the case of myxomatosis). Parasites and disease could be preconditioning agents for extinction, but are unlikely to cause it, except locally.

Climatic change

One of the most widely invoked explanations for Pleistocene mammalian extinctions is climatic change. Let us consider, in isolation from other factors, the potential effect of climatic change on mammalian populations. There can be no doubt from the geological evidence that changes have repeatedly occurred during Pleistocene and post-Pleistocene times that have been so far reaching that they have transformed once lushly vegetated areas able to support rich mammalian faunas into areas where the same mammals could no longer find a living. The reverse changes also occurred. The most extreme instances of ecological deterioration are the regions overridden by advancing ice sheets, those submerged by rising sea-level following the melting of continental ice sheets and those transformed into hot or cold unvegetated deserts. No mammal can find sustenance on the top of a continental ice sheet, only those most specially adapted for drought in a moderately dry desert. Yet we know from rock paintings and other evidence in the Sahara that only about 6000 years ago (during the Neolithic humid period) there was sufficient rainfall for elephants, giraffes and other mammals to range over most of the now desert region which at times was reduced to only relict areas isolated by forest and swamps. Such fluctuations of climate during the Pleistocene also had far reaching effects on the mammals of the areas concerned. Sometimes these faced extinction unless they could adapt or migrate.

Opportunities for successful migration in response to climatic change vary regionally and the ability of the affected mammalian species to survive depends on the nature of the environment encountered in their new territory. The simplest situation is that where climatic change leads to a simple northwards or southwards displacement of a vegetational zone (biome) and, if the change is not too sudden, the fauna can follow without any great inconvenience. In Europe, for example, such tundra species as the reindeer (once common as far south as France) have been able to survive subsequent climatic amelioration by moving towards the arctic regions; their former terrain being taken over by more southerly mammals, also migrating northwards. The survival, until the present day, in face of changes of rainfall, of so many large African mammals may have been achieved in a similar manner; and this has sometimes been used as an argument that man (who had long coexisted with these animals) was not a major cause of extinctions.

Whilst it was generally assumed until quite recently that the changes of climate that occurred during the Pleistocene took place gradually, recent studies have shown that, in fact, they may sometimes have been very rapid. The recent southward spread of the dune fields of West Sahara, principally because of lower rainfall (though poor agricultural practices by man have probably also contributed to this situation) is a striking example. Three or four hundred years ago Portuguese explorers still observed summer rains at 24°N. Today rain hardly reaches 19°. As we have observed previously the droughts that accompany such desertification can sometimes cause mammalian mortality of appalling extent. During the Sahelian drought from 1972–4, and now tragically renewed, more than 100 000 people and 6 000 000 cattle and many camels, sheep and other animals died of starvation. In the Tsavo National Parks (East) in Kenya alone 5900 elephants died from the same cause. Whilst the extent of this catastrophe was extensively influenced by man's control over the ability of the animals concerned to move out of the drought area (much of the human population also failed to move) the event vividly illustrates the effect which increased aridity could have had during Pleistocene times.

Similar population crashes, during exceptionally severe winters, can sometimes be observed among mammals living at high latitudes at the present day. On Bathurst and Melville Islands, Northwest Territories, Canada (75–77°N), the severe winter of 1973–4 reduced the caribou and musk oxen populations to less than half their previous numbers. Exceptionally deep snowfall which froze on its surface to form an ice crust, made access to the underlying plants difficult; and many of the caribou and musk oxen died of starvation. Fertility of the animals is apparently linked with physical condition, which in turn is largely dependent on the availability of winter forage, and there were consequently very few births in 1974. There was nevertheless a significant recovery of the population in 1977.

The decline of mammalian populations as the result of climatic change need not necessarily be accompanied by spectacular crashes such as those just described. Decreased equability (change from a humid climate with relatively uniform temperatures to dryer conditions with a greater summer heat and winter cold), even if it does not affect the adult mammalian population, can affect breeding patterns and lead to a long-term decline of numbers.

Neither the drought nor the severe winter described above caused the complete elimination of the mammalian species concerned, but it would only be a small step further for a series of such events to cause local extinctions.

Sometimes mammals that were fully capable of migrating at a time of climatic change found that their accustomed habitat was not available to them at any new locality and, when this occurred, their liability to extinction was very great. Mountain barriers sometimes stood in their way, some were forced on to peninsulas and sometimes small fragments of a population were separated and isolated on the upper parts of mountains. Mammals living on islands were especially vulnerable to climatic change since they were often very specifically adapted to relatively few ecological niches and their island situation provided them with no avenue of escape.

Although, on first consideration, it might be expected that the amelioration of climate which accompanied the final melting of the Pleistocene ice sheets about 10 000 years ago would generally have created more favourable living conditions for the surviving mammalian species, this was not the case everywhere. After the melting of each successive Pleistocene ice sheet vast areas that had been land (for example the southern part of the North Sea, the area on either side of the Bering Straits and the region along the north coast of Asia) were submerged by the rising sea and the resident mammals were displaced from them.

There are instances where biomes, instead of shifting in response to climatic amelioration were pinched out and disappeared, leaving the mammals that had been dependent on them without any suitable habitat. The replacement of 'Arctic steppe' with boggy tundra has already been discussed as a possible cause of mammoth extinction in chapter 9.

Even increased rainfall in a desert area does not invariably lead to increased mammalian food supply. Recent instances have occurred, for example in Australia, where irrigation and abnormally heavy rainfalls have led to a rise of the saline water table and destruction of existing vegetation. Cycles of similar soil salinization are believed to have occurred during the Pleistocene.

It follows from the considerations outlined above that climatic changes can sometimes have devastating effects on mammalian populations. Even if they are not so far-reaching that they cause extinction, then at least they may reduce numbers to an extent where the survivors might succumb to other causes. It must not be assumed, however, that times of general deterioration or improvement of world climate during the Pleistocene were synonymous with times of stress and expansion respectively of all the mammalian populations. Sometimes amelioration of climate resulted in decrease of forage. Consideration must be given to local conditions when trying to assess the likely influence of climatic change in individual extinctions.

Predator–prey equilibrium

Although predators cannot exterminate their prey, since their own numbers are governed by the quantity available to them, they can themselves become extinct because of the disappearance of the prey species, through other causes. The American sabre-toothed cat, *Smilodon*, may have disappeared in this way when the large mammalian species which had formed its prey died out.

In general, carnivore predation, by eliminating sick animals, maintains a vigorous stock, so that it is more likely to prevent extinction than to cause it. There have been many opportunities for present-day zoologists to study the effect of disturbance on predator–prey balance among mammalian populations, especially in areas where man has conducted programmes of predator extermination. Often this has led to overpopulation of the prey species and winter starvation.

The zoologists, Klein and Olson, who compared the pattern of mortality of deer in areas frequented by wolves with that on islands from which wolves were absent in southeast Alaska, found that in the wolf-free areas the deer population exceeded the winter forage capacity. The ranges were in poor condition and there was very high winter mortality. In contrast, in the areas where wolves were also present, the winter ranges were usually in good condition and winter mortality of deer was low. Indeed the wolf-populated areas, during the period of their study, were able to support a greater annual hunter harvest of deer per unit of area under comparable hunting pressure than the wolf-free areas.

As another example, when moose arrived on Isle Royale in northern Lake Superior, Canada, during the early part of the present century, the population quickly multiplied and caused damage to the island vegetation which resulted in mass starvation. After wolves appeared on the island in 1947–8 (probably by walking across the ice from Ontario) the population was believed to have stabilized at one fifth of its previous peak number, although recent reports about the island tell of further complicated changes in the flora and in the fauna (including a decrease in the number of beavers, another important item of wolf diet, and of food shortages among the wolves themselves).

Overpopulation and mass starvation may also have occurred in Pleistocene times in areas without carnivores. It has been suggested by Sondaar, for instance, that the occurrence of remains of deer showing a disease known as osteoporosis in Pleistocene deposits on the island of Crete may be an instance of such starvation on a carnivore-free island.

Interspecific competition (other than from man)

In stable mammalian populations, competition between comparable species (for example, among antelopes) is substantially reduced by their ability to exploit the greatest variety of ecological niches, so that no food source remains unused. Competition can nevertheless be very severe and only a slight change of circumstances (for example climatic change or mixing of previously separated populations following the establishment of land bridges) could be sufficient to tip the balance against the least well-adapted species.

Vegetational changes resulting from climatic change can bring competition from herbivorous species previously unable to colonize an area. The mammals already present may then have to yield territory, not because the changed vegetation is initially unsuitable for their survival but because the new mammals are dominant.

The instance of mixing of mammalian faunas that has attracted greatest attention among palaeontologists is the re-establishment of the land bridge between North and South America at the end of the Pliocene, previously mentioned in chapter 12. Until this event the only carnivorous mammals in South America had been marsupials, including *Borhyaena*. Before the end of the Pliocene this genus was extinct, apparently in direct consequence of competition with the newly arrived placental carnivores from North America. Many of the other mammals of South American stock, which are now extinct, nevertheless continued to flourish long after the re-establishment of the Panamanian land bridge and their ultimate disappearance cannot be attributed to competition since often there was no ecological replacement for them. The ground sloth *Megalonyx*, which survived until the end of the Pleistocene, even spread to Alaska. Other South American mammals which became extinct at this time were other ground sloths and also *Glyptodon, Toxodon* and *Macrauchenia*.

Like the marsupial species of South America, those of Australia were and still are exceptionally vulnerable to competition from placental mammals arriving from elsewhere; and some extinctions resulting from this cause probably also occurred there.

Today, the ecological niche left vacant by the extinction of the Pleistocene megafauna is filled in many parts of the world (though not efficiently) by man's domestic animals.

Competition from man

The relative importance of man's interference as a cause of mammalian extinction remains a cause of much controversy, since it is often difficult to distinguish between his agency and natural causes. Would the same pattern of extinction have occurred in the absence of man; or did he give the final blow to species already in decline because of climatic stress?

Man's influence could have been either direct or indirect. From Palaeolithic times until the present day he has hunted mammals with varying degrees of intensity. We can distinguish a series of stages in the development of his hunting activities. Earliest man had no agriculture and, except where he coexisted with a wide range of mammalian species, he could not hunt to extinction since he would have cut off his own food supply by so doing. He may nevertheless sometimes have caused a general reduction in the sizes of the hunted species by killing the largest individuals, and causing a genetic selection of those that matured most quickly, which tended to be smaller.

With the advent of agriculture in post-Pleistocene times man could afford to be more reckless with his hunting, leading to a situation where he no longer killed only for his own use but where he began to trade mammalian products in quantity for items supplied by communities living outside the range of the animals concerned; and finally to the use of firearms and the massive decrease of the mammalian population of the world which has occurred during the present century. There can be no doubt about man's role in causing the extinction of some mammals during recent times. Steller's sea cow, for example, formerly restricted to a few islands in the Bering Sea, was exterminated in 1769, only 27 years after it had first been discovered by Europeans, as the direct result of slaughter by sailors for food and other purposes.

Even if Palaeolithic hunters were limited in the number of individual mammals that they could kill without depleting their own food supplies, some Old World sites have nevertheless revealed astonishing quantities of remains of slaughtered animals. In the Ukraine skeletal parts of mammoths, apparently obtained at least partly as a by-product of hunting, were sometimes used to provide supports for tents. One of these dwellings, at Mezhirich (fig. 14.2) was found to contain 385 bones, tusks and jaws representing 95 individual mammoths. Three similar dwellings at Krakow, Poland (dated as 21 000–23 000 years old) contained remains of 60 mammoths. Although it is sometimes claimed that hunting on such a scale by Palaeolithic man could have been a cause of extinction such was not the case in the last mentioned Polish instance, since the mammoth survived in Siberia for a further ten thousand years.

Although, in the Old World, upper Palaeolithic man hunted some mammals intensively without causing their extinction, such a relationship was not necessarily universal. Mammalian faunas that have evolved independently of human contact are especially vulnerable at the time of man's first arrival, since they have not yet learned to protect themselves from this new predator. Charles Darwin recorded that, on the Island of San Pedro, Chile (which he visited from the *Beagle*), the foxes had so little fear of man that he was able to kill one of them for scientific study by hitting it on the head with his geological hammer. Could upper Palaeolithic hunters, spreading through North and South America into previously unpopulated areas, have exterminated some of the indigenous mammals in a similar manner, before these had acquired (through natural selection) the ability to flee from man; whereas those of the Old World had not been so caught out, since they had co-existed with man throughout the development of his hunting technology?

For mammalian populations living in areas invaded by man or already inhabited by him, hunting was not the only threat. Man can also change the environment and this can be even more damaging. Probably Palaeolithic hunters caused little environmental damage but, with the development of agriculture in post-Pleistocene times, man started on a massive course of habitat destruction which is proceeding at its most alarming rate ever at the present day. Agricultural practices lead to the cutting down or burning of forests and the introduction of domestic grazing animals. Man may also introduce predator species such as cats, dogs and rats; embark upon programmes of pest extermination; and introduce diseases to the wild mammalian population.

Such indirect interference may be very far-reaching in its effects. The herding of a mammalian population into a small area, for example, can cause a rapid deterioration in the physical condition of the individuals concerned. Nowhere has this been more dramatically demonstrated than in the present-day plight of the East African lowland elephant population. By preference elephants are animals of the forest. They

Fig. 14.2 *Above, a mammoth bone shelter from Mezhirich, Ukraine, displayed after restoration in the Palaeontological Museum of the Ukrainian Academy of Sciences, probably about 20–25 000 years old. (Photo: Novosti Press Agency).*
Below, an excavation in 1971 of the remains of one of the houses at Mezhirich. On the left is Academician I. G. Pidoplichko. (Photo: N. K. Vereshchagin).

are shade-adapted, vulnerable to dehydration and sunstroke, and thrive on a diet which includes not only herbs but also arboreal bark, wood fibres and roots. Salt is also important.

In recent years many of the lowland herds have been deprived of parts of their previous territory by human settlement, the displaced animals moving especially into the National Parks where (their previous migration routes closed) they have become isolated in far greater numbers than can be supported by the natural forage of these areas. This has been followed by the mass uprooting of trees by the elephants, which have then found themselves dependent on a diet of grass and without shade. This has been followed, apparently in direct consequence, by widespread heart and arterial disease, fewer births and increased calf mortality – the first stages of a population crash – recently intensified by drought conditions and poaching. The elephants of the mountain forest areas nevertheless still remain in good health and do not suffer from the arterial complaints of their lowland relatives.

The introduction of domestic mammals to an area usually reduces the wild fauna, although this does not follow invariably. When sheep and cattle were introduced to the Alice Springs region of Australia there was a decline in numbers of wallabies and bandicoots as a result of the grazing down of tall grass tussocks which they previously used for shelter. The larger previously relatively uncommon euro and red kangaroos, however, increased greatly in abundance as the result of vegetational changes caused by the grazing. This situation is not necessarily stable and it has been suggested from zoological studies that in the long run the euro kangaroo (which can survive on a nitrogen-poor diet) has better prospects in competition with domestic grazing animals than the red kangaroo, which requires green nitrogenous herbage and is more susceptible to the effects of drought.

Conclusions

We have seen that, although there have been many attempts in the past to attribute the extinction of mammalian species to specific causes, such as climatic change or man, such extinctions are more likely to be the result of complex chains of ecological events, which differed from one instance to another. The details can only be understood after a vast amount of data has been collected concerning the ecological requirements of the extinct mammals in question; concerning the exact timing of past fluctuations of climate and their effect on the vegetation of the areas being studied; and concerning the movements of Palaeolithic man. Whilst modern methods of scientific study are constantly providing an ever greater wealth of relevant information, it is still too early to try to be sure about the causes of extinction in most instances. It is sometimes easy to find a convincing explanation for an extinction in a particular area, only to discover after further consideration that the same stress did not occur throughout the whole of the range of the mammal concerned and that a totally different reason must be found to account for its extinction beyond the primary area of study. The problem of the extinction of the woolly mammoth is such an instance. In the Russian plains man's considerable influence on the mortality of the mammoth has been strikingly demonstrated on archaeological grounds, but there is little evidence of hunting activity in the frozen north where the mammoth apparently died out without the intervention of man.

Initially, the extinctions that took place on islands are likely to be the simplest to explain; and it is from islands that the first undeniable deductions will probably come in the future. It is tempting to attribute the extinction of the pigmy hippopotamus and the large lemurs of Madagascar to man. Man appears to have already existed on this island by the end of the first millenium AD and many of the extinctions (for example the giant lemur, *Megaladapis*, remains of which have been found in association with human occupation débris) post-date his arrival. There is in the British Museum (Natural History) a skull of *Archaeolemur majori* with a depressed fracture of the left frontal bone which could only have been made by an axe-like implement, while flesh was still intact (fig. 14.3).

Extinction of the lemurs (less than 1000 years ago) is found to have been selective. Surviving lemurs are mainly small, active, mainly arboreal and often nocturnal, whilst all large lemurs (which were also mainly diurnal and slow-moving or ground-living and would thus have been easier to hunt and kill) have disappeared. Increased aridity in parts of Madagascar may have contributed to their extinction but would not account for their disappearance from the central highlands, which still has a high rainfall at the present day. Deforestation and hunting by man stand foremost among the possible causes of their extinction.

A striking instance of island extinction during the Pleistocene was the disappearance of the giant deer, *Megaceros*, from Ireland at the end of a minor warm stage, about 11 000 years ago, already discussed in chapter 10. Geological evidence has shown that a brief cold stage followed when a small ice cap formed in the west of Scotland; the archaeological evidence shows that man did not arrive in Ireland for another two thousand years; and there is an almost overwhelming argument that, trapped on an island, the giant deer (which had flourished on a diet of lush herbaceous vegetation and birches) succumbed to periglacial conditions. But *Megaceros* also disappeared at about the same time from England, Scandinavia and France, where man was also present and from where escape by migration might have been possible.

Did more than one process cause the final extinction of the giant deer? In his study of extinctions, as in his other studies, the mammalian palaeontologist is just approaching the beginning of his task.

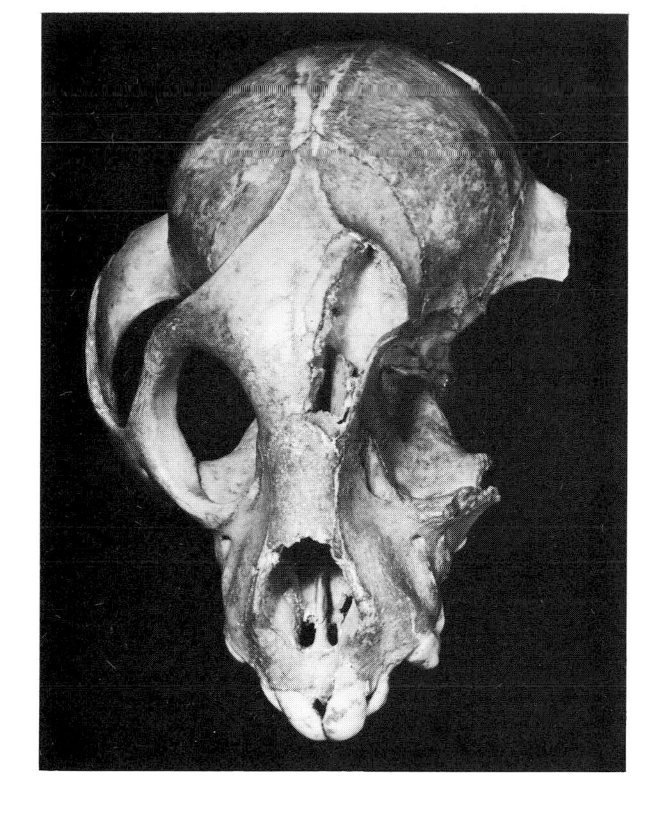

Fig. 14.3 *Skull of the extinct lemur* Archaeolemur majori *from Androhomana, Madagascar, showing a fracture apparently made by an axe-like instrument. (Photo: BM(NH)).*

Glossary

activity horizon horizon within a stratified sequence of deposits at an archaeological or palaeontological site, with activity débris or footprints of man or animals

Ailuropoda giant panda; its fossil remains common in Chinese caves, teeth from which are sold as 'dragons' teeth

ante-diluvian dating from before the Biblical Deluge

Aurignacian an upper Palaeolithic industry

australopithecines Plio-Pleistocene early hominids, the so called 'man-apes'

badlands erosion Channelling mainly by water, often seasonal, to create intricate gulley systems in vegetation-poor regions

benthos forms of marine life that are bottom dwelling or deep water

Bergmann's Rule In zoology, the statement that warm-blooded animals tend to be larger in colder parts of their environment than in warmer

biomass total weight of all organisms in a particular habitat area; the term is also used to designate total weight of a particular species or group of species

biome a climax community that characterizes a particular natural region, especially a particular type of vegetation, climatically bounded, which dominates a large geographic area

Carboniferous geological time unit (see figure 2.1)

chalicotheres clawed Old World ungulates which spread to North America during the Miocene. One genus, *Ancylotherium*, survived in Africa until the late Pliocene

Clovis type of North American projectile point, from the period about 12–11 000 years ago, associated with mammoth fauna (see also Folsom)

conspecific belonging to the same species

continental drift lateral movement of large plates of continental crust relative to one another

coprolite fossil vertebrate excrement

crag shelly sandy deposits of Pliocene and Pleistocene age found in eastern England

Cretaceous geological time unit (see figure 2.1)

Cro-Magnon man the general name given to the European peoples who produced upper Palaeolithic artifacts and the cave art

Deinotherium a proboscidean characterized by absence of tusks in the upper jaw and downturned tusks in the lower jaw. From the Tertiary of Europe, Asia and Africa, surviving until the end of the Pliocene in Africa

diachronous said of a sedimentary formation which becomes laterally younger in the direction in which deposition was being displaced; for instance at a changing shore line or ice front

echo-location the process by which an animal such as a bat or a dolphin orients itself through the emission of high frequency sounds

ecological niche the functional role of an organism in a community. If two species occupy the same niche competition occurs. A similar niche may be occupied by different species in different areas. Conversely one type of organism may evolve by adaptive radiation to fill several niches

ecology the study of the relationships between organisms and their environment

elk a large European deer, *Alces*, conspecific with the North American moose. The term is also used in North America for a large form of red deer, *Cervus elaphus*

feral having escaped from domestication and bred successfully in the wild

floodplain a surface of relatively level land adjacent to a river channel, periodically covered with water when the river overflows its banks

fluviatile related to rivers

Folsom type of North American projectile point, from the period about 11–10 000 years ago, associated with bison fauna (see also Clovis)

Foraminifera small single-celled marine organisms with calcareous tests, of the phylum Protozoa, much used by geologists in deep sea oxygen isotope studies

Gastropoda class of molluscs including snails and winkles

glaci-fluvial related to meltwater streams flowing from wasting glacier ice

glyptodonts mammals belonging to the order Edentata, which had an armoured carapace composed of a mosaic of bony polygons covered by a horny scale-like epithelium, as in the closely related armadillos. The tail and head were similarly armoured. Originating in South America during the Tertiary the group survived there until the end of the Pleistocene, briefly invading southern North America during the late Pliocene or early Pleistocene

Hipparion three-toed horse. Widely distributed in Eurasia and Africa from the late Miocene, surviving into the Pleistocene in Africa *hominid* common name for the zoological family which contains man and his fossil relatives

hygroscopic tending to absorb moisture

insolation radiation reaching the Earth's surface

Inuit North Canadian race of people; also known as Eskimos (a name not accepted by the Inuit themselves)

lacustrine related to lakes

leaching the dissolving out of soluble constituents of a rock by the natural action of percolating water

litopterns an extinct order of South American ungulates, with habits probably similar to those of horses. They flourished during the Tertiary and became extinct during the Pleistocene, possibly as the result of migration of carnivores from North America

Magdalenian European upper Palaeolithic industry

mastodons a general term for members of the families Gomphotheriidae and Mammutidae of the suborder Elephantoidea of the Proboscidea (the elephants comprise a third family). Widely distributed in the Tertiary of Eurasia, Africa and North America, reaching South America in the late Pliocene or early Pleistocene. The North American genus *Mammut* survived until at least the end of the Pleistocene

megafauna In discussion of mammalian extinctions the large mammals, such as the elephants and sloths – in contrast to, for example, the rodents and insectivores

Mesolithic Middle Stone Age (see table 1)

Miocene geological time unit (see figure 2.1)

moraine A mound or ridge of unsorted débris laid down chiefly by the direct action of glacier ice

Neanderthal man a form of fossil human known from Europe and western Asia, usually associated with Mousterian artifacts (see table 3)

Neolithic New Stone Age (see table 1)

nuée ardente a swiftly flowing turbulent gaseous cloud, sometimes incandescent, erupted from a volcano and containing ash and other pyroclastics in its lower part

Oldowan industry lower Palaeolithic industry first identified at Olduvai Gorge, Tanzania (see figure 11.9)

Ordovician geological time unit (see figure 2.1)

overkill in discussion of extinction, the killing by man of a greater proportion of an animal population than will allow it to maintain its numbers

Palaeolithic Old Stone Age (see table 1)

Permian geological time unit (see figure 2.1)

petrology branch of geology dealing with origin, occurrence, structure, and history of rocks

photosynthesis process by which plants use the energy of sunlight in order to build up sugars from carbon dioxide and water

plankton aquatic organisms that drift or swim weakly

Pliocene geological time unit (see figure 2.1)

primate belonging to the zoological group containing man, apes, monkeys and 'lower primates'

Proboscidea order of mammals including the elephants, mastodons and deinotheres

puparium the thickened, pigmented barrel-like last larval skin inside which the true pupa of most carrion-feeding flies is formed

pyroclastic pertaining to the fragmentary material ejected during a volcanic emption

Sahel Semi-arid zone along the southern edge of the Sahara Desert

savanna an open grassy, at most sparsely wooded, plain, especially as developed in tropical and sub-tropical regions

Sivatherium a giraffid characterized by large branched horns. Widely distributed in the Tertiary of Eurasia and Africa. It survived in Africa until late in the Pleistocene

Solutrean European upper Palaeolithic industry

sensu stricto 'In a narrow sense'

speleotravertine calcium carbonate (travertine) deposited in a cave as flowstone, stalactites, stalagmites etc. (speleothems). General term for the substance itself

Sphagnum 'Bog-moss'

Stegodon extinct Proboscidean of the same family as the living elephants, but with lower crowned teeth. From the Pliocene and Pleistocene of Asia. Teeth from Chinese caves are sold as 'dragons' teeth'

stratigraphy the science of rock strata

talus rock fragments of any size or shape derived from or lying at the base of a cliff or steep rocky slope

Tertiary period of time between the Cretaceous and Quaternary, comprising the Palaeocene, Eocene, Oligocene, Miocene and Pliocene (see figure 2.1)

till, glacial mainly unsorted and unstratified deposit, usually unconsolidated, laid down directly by and underneath a glacier without reworking by melt water

tuff consolidated volcanic ash

tundra sparsely vegetated arctic biome, underlain by permafrost. It merges northwards and locally into arctic desert

Further reading

Chapter 1

BRYSON, R.A. & MURRAY, T. J. 1977. *Climates of hunger.* 171pp. University of Wisconsin Press.

GRIBBIN, J. (ed.) 1978. *Climatic change.* 280pp. Cambridge University Press, Cambridge.

LADURIE, E. le R. 1971. *Times of feast, times of famine.* 413pp. Doubleday, New York.

LAMB, H.H. 1977. *Climate, present, past and future.* **2.** 835pp. Methuen, London.

MATTHEWS, S.W. 1976. *What's happening to our climate? Nat. geogr. Mag.* **150:** 576–615.

NATIONAL ACADEMY OF SCIENCES 1977a. *Climate, climatic change and water supply* (collection of 8 papers). 132pp. Washington D.C.

NATIONAL ACADEMY OF SCIENCES 1977b. *Energy and climate* (collection of 10 papers). 158pp. Washington D.C.

SMITH, C.D. & PARRY, M. (eds) 1981. *Consequences of climatic change.* 143pp. Nottingham University.

SYMPOSIUM ON CLIMATIC CHANGE AND FOOD PROBLEMS, TOKYO, 1980. 1981. *GeoJourn.* **5:** 98–203.

Chapter 2

BOWEN, D.Q. 1978. *Quaternary geology: a stratigraphic framework for multidisciplinary work.* 221pp. Pergamon Press, London.

CHARLESWORTH, J.K. 1957. *The Quaternary era.* 1700pp. Edward Arnold, London.

CHORLTON, W. & EDITORS OF TIME-LIFE BOOKS, 1983. *Planet earth: ice ages.* 176pp. Time-Life Books, Amsterdam.

CORNWALL, I. 1970. *Ice ages, their nature and effects.* 180pp. John Baker, Humanites Press, London.

DENTON, G.H. & HUGHES, T.J. (eds) 1981. *The last great ice sheets.* 484pp. John Wiley & Sons, New York and Chichester.

FLINT, R.F. 1971. *Glacial and Quaternary geology.* 892pp. John Wiley & Sons, New York and Chichester.

FRENCH, H.M. 1976. *The periglacial environment.* 309pp. Longmans, London.

GEIKIE, J. 1874. *The great ice age.* 573pp. W. Isbiter, London. 3rd ed. 1894, 850pp. Edward Stamford, London.

GEORGI, C.E. (ed.) 1978. The ice age – when did it begin and has it ended? *Trans. Nebr. Acad. Sci.* **6:** 147pp.

JOHN, B.S. 1977. *The ice age: past and present.* 245pp. Collins, London.

JOHN, B.S. (ed.) 1979. *The winters of the world; earth under the ice ages.* 256pp. David & Charles, Newton Abbott.

KAHLKE, H.D. 1981. *Das Eiszeitalter.* 192pp. Urania-Verlag, Leipzig.

KURTEN, B. 1972. *The ice age.* 179pp. Rupert Hart-Davis, London.

LOWE, J.J. & WALKER, M.J.C. 1984. *Reconstructing Quaternary environments.* 389pp. Longman, London.

NILSSON, T. 1983. *The Pleistocene: geology and life in the Quaternary ice age.* 651pp. Reidel Publishing Co. Dordrecht.

RABASSA, J. (ed.) 1983. *Quaternary of South America and Antarctic Peninsula.* 156pp. Balkema, Rotterdam.

SARNTHEIN, M. 1978. Sand deserts during glacial maximum and climatic optimum. *Nature, Lond.* **272:** 43–46.

VELICHKO, A.A. (ed.) 1985. *Late Quaternary environments of the Soviet Union.* 327pp. University of Minnesota Press/Longman.

WALKER, D. & GUPPY, J.C. (eds) 1978. *Biology and Quaternary environments: based on the symposium on biological problems . . .* 264pp. Australian Academy of Science, Canberra.

WRIGHT, A.E. & MOSELEY 1974. *Ice ages: ancient and modern.* 320pp. Seel House Press, Liverpool.

WRIGHT, W.B. 1936. *The Quaternary ice age* (2nd ed.) 478pp. Macmillan, London.

ZEUNER, F.E. 1958. *Dating the past: an introduction to geochronology* (4th ed.) 516pp. Methuen, London.

ZEUNER, F.E. 1959. *The Pleistocene period* (2nd ed.) 447pp. Hutchinson, London.

Chapter 3

ABEL, O. 1939. Vorzeitliche Tierreste im Deutschen Mythus, Brauchtum und Volksglauben. 304pp. Fischer in Jena.

COLBERT, E.D. & HOOIJER, D.A. 1953. Pleistocene mammals from the limestone fissures of Szechwan, China. *Bull. Am. Mus. nat. Hist.* **102:** 1–134.

KAHLKE, H.D. 1961. On the complex of the *Stegodon–Ailuropoda* fauna of South China and the chronological position of *Gigantopithecus blacki* V. Koenigswald. *Vertebr. palasiat.* **2:** 104–5.

KIRCHER, A. 1678. *Mundus Subterraneus* 3rd ed. Amsterdam.

KOCH, A. 1841. *Description of the Missourium or Missouri Leviathan.* 16pp. St Louis.

KOENIGSWALD, G.H.R. VON 1956. *Meeting Pleistocene Man.* 216pp. Thames & Hudson, London.

LEIBNITZ, G.W. 1749. *Protogaea.* Goettingae.

OAKLEY, K.P. 1975. Decorative and symbolic uses of vertebrate fossils. *Occ. Pap. Technol. Pitt Rivers Mus.* **12:** 60pp.

VALENTINI, M.B. 1704–14. *Museum Museorum.* Frankfurt am Mäyn.

WEN-CHUNG, P. 1965. Excavation of the Luching *Gigantopithecus* cave and exploration of other caves in Kwangsi. *Mem. Inst. Vertebr. Palaeont. Paleoanthrop, Peking* **7:** 39–54.

Chapter 4

BEHRENSMEYER, A.K. 1975. The taphonomy and palaeoecology of Plio-Pleistocene vertebrate assemblages east of Lake Rudolf, Kenya. *Bull. Mus. Comp. Zool.* **146:** 473–578.

BEHRENSMEYER, A.K. & HILL, A.P. 1980. *Fossils in the making. Vertebrate taphonomy and paleoecology.* 338pp. Chicago University Press, Chicago & London.

BINFORD, L.R. 1981. *Bones. Ancient men and modern myths.* 320pp. Academic Press, New York and London.

BONNICHSEN, R. 1979. Pleistocene bone technology in the Beringian Refugium. *Pap. archaeol. Surv. Can.* **89:** 280pp.

BRAIN, C.K. 1981. *The hunters or the hunted? An introduction to African cave taphonomy.* 365pp. Chicago University Press. Chicago and London.

CONYBEARE, A. & HAYNES, G. 1984. Observations on elephant mortality and bones in water holes. *Quaternary Res.* **22:** 189–200.

GLOB, P.V. 1971. *The bog people.* 142pp. Paladin, Frogmore, St Albans.

MORLAN, R.E. 1980. Taphonomy and archaeology in the upper Pleistocene of the northern Yukon Territory: a glimpse of the peopling of the New World. *Pap. archaeol. Surv. Can.* **94:** 380pp.

MUNTHE, K. & McLEOD, S.A. 1975. Collection of taphonomic information from fossil and recent vertebrate specimens, with a selected bibliography. *Paleobios* **19:** 1–12.

SHIPMAN, P. 1981. *Life history of a fossil: an introduction to taphonomy and paleoecology.* 222pp. Harvard University Press, Cambridge, Mass. and London.

Chapter 5

BRADLEY, R.S. 1985. *Quaternary paleoclimatology.* 472pp. Allen & Unwin, Boston and London.

BURLEIGH, R. (ed.) 1980. Progress in scientific dating methods. *British Museum Occasional Paper* **21:** 96pp.

IVANOVICH, M. & HARMON, R.S. (eds) 1982. *Uranium series disequilibrium: applications to environmental problems.* 591pp. Clarendon Press, Oxford.

MASTERS, P.M. & FLEMING, N.C. (eds) 1983. *Quaternary coastlines and marine archaeology: towards the prehistory of land bridges and continental shelves.* 641pp. Academic Press, London.

MOORE, P.D. & WEBB, J.A. 1978. *An illustrated guide to pollen analysis.* 133pp. Hodder & Stoughton, London.

OAKLEY, K.P. 1969. *Frameworks for dating fossil man.* (3rd ed.) 366pp. Weidenfeld & Nicholson, London.

Chapter 6

BOULTON, G.S. 1979. A model of Weichselian glacier fluctuations in the North Atlantic region. *Boreas* **8:** 373–395.

CLIMAP PROJECT MEMBERS, 1984. The last interglacial ocean. *Quat. Res.* **21:** 123–224.

DANSGAARD, W. *et al.* 1971. Climatic record revealed by the Camp Century ice core. pp. 37–56 in Turekian, K.K. (ed.) *The late Caenozoic glacial ages.* 606pp. Yale University Press, New Haven and London.

HAYS, J.D., IMBRIE, J. & SHACKLETON, N.J. 1976. Variations in the earth's orbit: pacemaker of the ice ages. *Science* **194:** 1121–1132.

IMBRIE, J. & IMBRIE, K.P. 1979. *Ice ages: solving the mystery.* 224pp. Macmillan, London.

KUKLA, C.J. 1977. Pleistocene land–sea correlations. I. Europe. *Earth. Sci. Rev.* **13:** 307–374.

MISC. 1978. Climatology. Special supplement. *Nature Lond.* **276:** 327–359.

RICHMOND, G.M. 1976. Pleistocene stratigraphy and chronology in the mountains of western Wyoming. pp. 353–379 in Mahaney, W.C. (ed.) *Quaternary stratigraphy of North America.* 512pp. Dowden, Hutchinson & Ross, Stroudsburg.

RUDDIMAN, W.F. & McINTYRE, A. 1976. North-east Atlantic paleoclimatic changes over the past 600 000 years. *Mem. geol. Soc. Am.* **145:** 111–146.

SHACKLETON, N.J. 1977. The oxygen isotope stratigraphic record of the late Pleistocene. *Phil. Trans. R. Soc.* B **280:** 169–182.

TURON, J. 1984. Direct land/sea correlations in the last interglacial complex. *Nature, Lond.* **309:** 673–676.

WIJMSTRA, T.A. & HAMMEN, T. VAN DER 1974. The last interglacial–glacial cycle: state of affairs of correlation between data obtained from the land and from the sea. *Geol. en Mijn.* **53:** 386–392.

WOILLARD, G.M. 1978. Grande Pile peat bog: A continuous pollen record for the last 140 000 years. *Quat. Res.* **9:** 1–21.

Chapter 7

ABEL, O. & KYRLE, G. 1931. Speläolog. Monogr. **7–8.** *Die Drachenhöhle bei Mixnitz.* 953pp. Ossterr. Staatsdruckerei, Wien.

BUCKLAND, W. 1823. *Reliquiae Diluvianae.* 303pp. John Murray, London.

CASTERET, N. 1939. *Ten years under the earth.* 240pp. J. M. Dent, London.

CULLINGFORD, C.H.D. (ed.) 1953. *British caving.* 468pp. Routledge & Kegan Paul, London.

DAWKINS, W.B. 1874. *Cave hunting.* 455pp. McMillan & Co., London.

FORD, T.D. & CULLINGFORD, C.H.D. (eds) 1976. *The science of speleology.* 593pp. Academic Press, London.

GREEN, S.H. 1984. *Pontnewydd Cave. A lower Palaeolithic hominid site in Wales.* 227pp. National Museum of Wales, Cardiff.

GUILDAY, J.E., MARTIN, P.S. & McRADY, A.D. 1964. New Paris No. 4: A Pleistocene cave deposit in Bedford County, Pennsylvania. *Bull. natn. speleol. Soc.* **26:** 121–194.

KURTEN, B. 1976. *The cave bear story.* 163pp. Columbia University Press, New York.

LUMLEY, H. de 1972. La grotte de l'Hortus. *Études Quaternaires, Geol. Pal. Prehist.* **1:** 668pp. Université de Provence, Marseille.

MOHR, C.E. & POULSON, T.L. 1966. *The life of the cave.* Our living world of nature series. 232pp. McGraw Hill, New York.

Chapter 8

ADAM, K.D. 1980. *Eiszeitkunst im süddeutschen Raum.* 161pp. Konrad Theiss Verlag, Stuttgart. Stuttgart.

BEGOUEN, H. & BREUIL, H. 1958. *Les Cavernes du Volp.* 124pp. Arts et Métiers Graphiques, Paris.

BREUIL, H. 1952. *Four hundred centuries of cave art.* 414pp. Montignac.

BREUIL, H. & OBERMAIER, H. 1935. *The Cave of Altamira.* 223pp. Tipographia de Archivos, Madrid.

GRAZIOSI, P. 1960. *Palaeolithic art.* 278pp. Faber, London.

LEROI-GOURHAN, A. 1968. *The art of prehistoric man.* 543pp. Thames & Hudson, London.

POWERS, R. & STRINGER, C. B. 1975. Palaeolithic cave art fauna. *Stud. Speleol.* **2:** 266–298.

SIEVEKING, A. 1979. *The cave artists.* 221pp. Thames & Hudson, London.

UCKO, P.J. & ROSENFELD, A. 1967. *Palaeolithic cave art.* 256pp. Weidenfeld & Nicholson, London.

ZERVOS, C. 1959. *L'Art de l'Epoque du Renne.* 459pp. Ed. 'Cahiers d'Art', Paris.

Chapter 9

AUGUSTA, J. 1962. *A book of mammoths.* Illustrated by Z. Burian. 50pp. Hamlyn, London.

DIGBY, B. 1926. *The mammoth and mammoth-hunting in north-east Siberia.* 224pp. Witherby, London.

FARRAND, W.R. 1961. Frozen mammoths and modern geology. *Science* **133:** 729–735.

GARUTT, W.E. 1964. *Das Mammut.* 141pp. Die neue Brehm-Bucherei. Kosmos Verlag, Stuttgart.

PFIZENMAYER, E.W. 1939. *Siberian man and mammoth.* (First published in German in 1926). 256pp. Blackie & Son, London.

PIDOPLICHKO, I.G. 1969. *Late palaeolithic dwellings of mammoth bone houses.* (In Russian with English summaries). 164pp. 'Naukova Dumka', Kiev.

SLOANE, J.M. 1728. An account of elephants teeth and bones from underground. *Phil. Trans. R. Soc.* **35:** 457–471.

STEWART, J.M. 1977. Frozen mammoths from Siberia. *Smithsonian* **10** (6): 125–126.

TOLMACHOFF, I.P. 1929. The carcasses of the mammoth and rhinoceros found in the frozen ground of Siberia. *Trans. Am. phil. Soc. N.S.* **23:** 1–74.

VERESHCHAGIN, N.K. 1974. The mammoth 'cemeteries' of north-east Siberia. *Polar Record* **17** (106): 3–12. Reviewed in the *Times* 2 March 1974.

VERESHCHAGIN, N.K. 1979. *Why the mammoths died out.* (In Russian). 194pp. Nauka, Leningrad.

VERESHCHAGIN, N.K. & MIKHELSON, V.M. (eds) 1981. *The Magadan baby mammoth.* (In Russian). 296pp. 'Nauka', Leningrad.

Chapter 10

BISHOP, M.J. 1982. The mammalian fauna of the early middle Pleistocene cavern infill site of Westbury-sub-Mendip, Somerset. *Special Pap. Palaeont.* **28:** 1–108.

COOK, J. *et al.* 1982. A review of the chronology of the European middle Pleistocene record. *Yb. Phys. Anthrop.* **25:** 19–65.

DAWKINS, W.B. & REYNOLDS, S.H. 1872–1939. British Pleistocene Mammalia, 3 (Artiodactyla). *Palaeontogr. Soc. (Monogr.),* London.

DAWKINS, W.B. & SANFORD, W.A. 1866–72. British Pleistocene Mammalia, 1 (Felidae). 194pp. *Palaeontogr. Soc. (Monogr.),* London.

FRASER, F.C. & KING, J.E. 1954. Faunal remains. pp. 70–95 in Clark, J.G.D. *Excavations at Star Carr*, Cambridge University Press, Cambridge.

GRAY, M. 1983. Update: the Quaternary ice age (with particular reference to Britain). *Queen Mary College Papers in geography, Spec. Pub.*, **5.** Queen Mary College, London.

GREEN, C.P. *et al.* 1984. Evidence of two temperate episodes in late Pleistocene deposits at Marsworth, UK. *Nature, Lond.* **309:** 778–781.

HINTON, M.A.C. 1926. *Monograph of the voles and lemmings (Microtinae), living and extinct.* 488pp. British Museum, London.

KURTÉN, B. 1968. *Pleistocene mammals of Europe.* 316pp. Weidenfeld & Nicholson, London.

MITCHELL, G.F. 1976. *The Irish landscape.* 240pp. Collins, London.

MITCHELL, G.F., PENNY, L.F., SHOTTON, F.W. & WEST, R.G. 1973. A correlation of Quaternary deposits in the British Isles. *Spec. Rep. geol. Soc. Lond.* **4:** 99pp.

OWEN, R. 1846. *A history of British fossil mammals and birds.* 560pp. John van Voorst, London.

REYNOLDS, S.H. 1902–1912. The British Pleistocene Mammalia, 2 (Hyaenidae, Ursidae, Canidae and Mustelidae). *Palaeontogr. Soc. (Monogr.)*, London.

SHOTTON, F.W. 1977. *British Quaternary studies: recent advances.* 298pp. Clarendon Press, Oxford.

SPARKS, B.W. & WEST, R.G. 1972. *The ice age in Britain.* 302pp. Methuen, London.

SPENCER, H.E.P. 1966–71. A contribution to the geological history of Suffolk. *Trans. Suffolk Nat. Soc.* **13:** 197–209, 290–313, 366–389, **15:** 148–196, 279–363, 517–519.

STUART, A. 1982. *Pleistocene vertebrates of the British Isles.* 212pp. Longman, London.

STUART, A. J. 1983. Pleistocene bone caves in Britain and Ireland. *Stud. Speleol.* **4:** 9–36.

SUTCLIFFE, A.J. & KOWALSKI, K. 1976. Pleistocene rodents of the British Isles. *Bull. Br. Mus. nat. Hist.* (Geol) **27** (2): 33–147.

WEST, R.G. 1977. *Pleistocene geology and biology.* 440pp. Longman, London.

WEST, R.G. with contributions from Norton, P.E.P., Sparks, B.N. and Wilson, D.G. 1980. *The pre-glacial Pleistocene of the Norfolk and Suffolk coasts.* 218pp. Cambridge University Press, Cambridge.

WYMER, J. J. 1968. *Lower Palaeolithic archaeology in Britain.* 429pp. John Baker, London.

Chapter 11

BISHOP, W.W. (ed.) 1978. *Geological background to fossil man.* 564pp. Scottish Academic Press, Edinburgh.

COLE, S. 1975. *Leakey's luck.* 448pp. Collins, London.

COPPENS, Y. *et al.* 1976. *Earliest man and environments in the Lake Rudolf basin.* 615pp. University of Chicago Press, Chicago and London.

HAY, R.L. 1976. *Geology of the Olduvai Gorge.* 203pp. University of California Press, Berkley.

HOPWOOD, A.T. & HOLLYFIELD, J.P. 1954. An annotated bibliography of the fossil mammals of Africa (1742–1950). *Foss. Mammals. Afr.* London **8:** 1–194.

KLEIN, R.G. 1984. *Southern African prehistory and palaeoenvironments.* 416pp. Balkema, Rotterdam.

LEAKEY, L.S.B. 1951. *Olduvai Gorge* 164pp. Cambridge University Press, Cambridge.

LEAKEY, L.S.B. 1965. *Olduvai Gorge, 1951–61. Volume 1, A preliminary report on the geology and fauna.* 118pp. Cambridge University Press, Cambridge.

LEAKEY, M.D. 1971. *Excavations in Beds I and II, Olduvai Gorge,* **3:** 306pp, Cambridge University Press.

LEAKEY, M.D. 1979. *Olduvai Gorge: my search for early man.* 187pp. Collins, London.

LEAKEY, M.G. & LEAKEY, R.E. (eds) 1978. *Koobi Fora Research Project, 1, Fossil hominids and an introduction to their context.* 191pp. Clarendon Press, Oxford.

LEAKEY, R.E. & LEWIN, R. 1977. *Origins.* 264pp. Macdonald & James, London.

MAGLIO, J. & COOKE, H.B.S. (eds) 1978. *Evolution of African mammals.* 641pp. Harvard University Press, Cambridge and London.

Chapter 12

BRYAN, A.L. (ed.) 1978. Early man in America. *Occ. Pap.* **1:** 327pp. Department of Anthropology, University of Alberta, Edmonton.

GUTHRIE, R.D. 1972. Recreating a vanished world. *Nat. geogr. Mag.* **141:** 294–301.

HAUTHAL, R., ROTH, S. & LEHMANN-NITSCHE, R. 1899. El mamifero misterioso de la Patagonia, *Grypotherium domesticum. Revista del Museo la Plata,* **9:** 409–474.

JEFFERSON, T. 1799. A memoir on the discovery of certain bones of a quadruped of the clawed kind in the western parts of Virginia. *Trans. Am. phil. Soc.* **4:** 246–260.

KURTEN, B. & ANDERSON, E. 1980. *Pleistocene mammals of North America.* 443pp. Columbia University Press, New York.

MOORE, D.M. 1978. Post-glacial vegetation in the South Patagonia territory of the giant sloth, *Mylodon. Bot. J. Linn. Soc.* **77:** 177–202.

SAXON, E.C. 1979. Natural prehistory: the geology of Fuego–Patagonian ecology. *Quaternaria* **21:** 329–356.

SHUTLER, R. (ed.) *Early man in the New World.* 223pp. Sage Publications, Beverly Hills.

SIMPSON, G.G. 1980. *Splendid isolation.* 266pp. Yale University Press, Newhaven and London.

STOCK, C. 1956. Rancho La Brea. A record of Pleistocene life in California. *Sci. Ser. Los Ang. Mus.* **20,** Palaeo. **11:** 1–181.

Chapter 13

ARCHER, M. 1981. A review of the origins and radiation of Australian mammals. pp. 1437–1488 in Keast, A. (ed.) Ecological biogeography of Australia. *Monogr. biol.* **41.** Junk, The Hague.

ARCHER, M. (ed.) 1982. *Carnivorous marsupials.* 804pp. Royal Zoo. Soc. N.S.W., Mosman.

ARCHER, M. & CLAYTON, G. 1984. *Vertebrate zoology and evolution in Australasia.* 1206pp. Hesperian Press, Carlisle, Australia.

BALME, J., MERRILEES, D. & PORTER, J.K. 1978. Late Quaternary mammal remains, spanning about 30 000 years, from excavations in Devil's Lair, Western Australia. *J. Roy. Soc. W. Aust.* **61:** 33–65.

KEAST, A. 1972. Australian mammals: zoogeography and evolution. pp. 195–246 in Keast, A., Erk, F.C. & Glass, B. (eds) *Evolution, mammals and southern continents.* 543pp. University of New York Press, Albany.

KIRSCH, J.A.W. 1977. The comparative serology of Marsupialia, and classification of marsupials. *Aust. J. Zool. Suppl. Ser.* **52:** 1–152.

LANE, E.A. & RICHARDS, A.M. 1963. The discovery, exploration and scientific investigation of the Wellington Caves, New South Wales. *Helictite* **2:** 53pp.

LOWRY, J.W. & MERRILEES, D. 1969. Age of the desiccated carcass of a thylacine (Marsupialia, Dasyuroidea) from Thylacine Hole, Nullarbor Region, Western Australia. *Helictite* **7:** 15–16.

MAHONEY, J.A. & RIDE, W.D.L. 1975. Index to the genera and species of fossil mammalia described from Australia and New Guinea between 1838 and 1968. *Spec. Publs. West. Aust. Mus.* **6:** 1–250.

QUIRK, S. & ARCHER, M. (eds) 1983. *Prehistoric animals of Australia.* 80pp. Australian Museum, Sydney.

RIDE, W.D.L. 1963. A review of Australian fossil marsupials. *J.R. Soc. W. Aust.* **47:** 97–131.

SIMPSON, G.G. 1977. Too many lines; the limits of the oriental and Australian geographic regions. *Proc. Am. phil. Soc.* **121:** 107–120.

STONEHOUSE, B. & GILMORE, D. (eds) 1977. *The biology of the marsupials.* 486pp. Unwin, Old Woking.

TEDFORD, R.H. 1974. Marsupials and the new palaeogeography. pp. 109–126 in Ross, C.A. (ed.) *Palaeogeographic provinces and provinciality. Spec. Publ. Soc. Econ. Pal. & Min.* **21.**

TYNDALE-BISCOE, H. 1973. *Life of marsupials.* 254pp. Edward Arnold, London.

Chapter 14

MARTIN, P.S. & KLEIN, R.G. (eds) 1984. *Quaternary extinctions, a prehistoric revolution.* 892pp. University of Arizona Press, Tuscon.

MOLYNEUX, T. 1697. A discourse concerning the large horns frequently found under ground in Ireland . . . *Phil. Trans. R. Soc.* **9:** 485–512.

Journals

Some periodicals specializing in the Quaternary.

Boreas Universitets for laget, Oslo.

Bulletin de l'Association Française pour l'Etude du Quaternaire Universite Pierre et Marie Curie, Paris.

Eiszeitalter und Gegenwart Verlag Hohenlohe, Öhringen.

Geographie Physique et Quaternaire Université de Montreal.

Journal of Quaternary Science (Provisional title; a new journal to be issued by the Quaternary Research Association, Cambridge, England; first number not yet published).

Quartär Ludwig Röhrscheid, Bonn.

Quartärpaläontologie Institute of Quaternary Palaeontology, Weimar.

Quaternaria Roma.

Quaternary Newsletter Quaternary Research Association, Cambridge.

Quaternary Research Academic Press, San Diego & London.

Quaternary Research Tokyo (In Japanese with English contents lists and some English summaries).

Quaternary Science Reviews Pergamon Press, Oxford.

Quaternary Studies in Poland Institute of Geology, Warsaw.

Voprosy Chetvertichnoi Geologii (Problems of Quaternary geology; in Russian) Ministry of Geology of the USSR. Moscow.

Index